Birkhäuser

# Systems & Control: Foundations & Applications

*Series Editor*
Tamer Başar, University of Illinois at Urbana-Champaign, Urbana, IL,
USA

*Editorial Board*
Karl Johan Åström, Lund University of Technology, Lund, Sweden
Han-Fu Chen, Academia Sinica, Beijing, China
Bill Helton, University of California, San Diego, CA, USA
Alberto Isidori, University of Rome, Rome, Italy;
    Washington University, St. Louis, MO, USA
Miroslav Krstic, University of California, San Diego, CA, USA
Alexander Kurzhanski, University of California, Berkeley, CA, USA;
    Russian Academy of Sciences, Moscow, Russia
H. Vincent Poor, Princeton University, Princeton, NJ, USA
Mete Soner, ETH Zürich, Zürich, Switzerland;
    Swiss Finance Institute, Zürich, Switzerland

For further volumes:
http://www.springer.com/series/4895

Daniel Hernández-Hernández
J. Adolfo Minjárez-Sosa

Editors

# Optimization, Control, and Applications of Stochastic Systems

In Honor of Onésimo Hernández-Lerma

 Birkhäuser

*Editors*
Daniel Hernández-Hernández
Department of Probability and Statistics
Center for Research in Mathematics
Mineral de Valenciana, GJ, México

J. Adolfo Minjárez-Sosa
Department of Mathematics
University of Sonora
Hermosillo, SO, México

ISBN 978-0-8176-8336-8    ISBN 978-0-8176-8337-5 (eBook)
DOI 10.1007/978-0-8176-8337-5
Springer New York Heidelberg Dordrecht London

Library of Congress Control Number: 2012942475

Mathematics Subject Classification (2010): 62C05, 60J05, 91A15, 91B70

Printed on acid-free paper

Springer is part of Springer Science+Business Media
(www.birkhauser-science.com)

Onesimo Hernández-Lerma, Mexico City, 2011

# Foreword

Professor Onésimo Hernández-Lerma has made a great number of fundamental contributions in the field of stochastic control systems during the last 30 years. He has contributed to the development of this theory in many aspects, which include adaptive control, parametric estimation, recursive algorithms, infinite dimensional linear programming, ergodic properties of Markov processes, measure theory, discrete- and continuous-time controlled Markov processes, and game theory. Also, his research work has been concerned with important applications to the areas of queuing systems, control of population, and management sciences.

He has authored more than 140 papers, some of which still have a great influence in the study of controlled stochastic systems, since he was the first to mathematically formalize many fundamental results. He recently received the Scopus Prize (2008) and the Thomson Reuters Award (2009), recognizing the influence and importance of his work.

Professor Hernández-Lerma has written or coauthored 11 books or monographs on different topics. In the field of stochastic control, his book *Adaptive Markov Control Processes* (1989) soon became a reference for researchers and graduate students, and today it is considered a classic. In 1996 and 1999 he jointly wrote with J. B. Lasserre the books *Discrete-Time Markov Control Processes: Basic Optimality Criteria* and *Further Topics on Discrete-Time Markov Control Processes*, giving a systematic and deep study to controlled Markov processes on Borel spaces with several optimality criteria. The techniques required to deal with discrete- and continuous-time models are different, but in many aspects the general intuition from one can be applied to the other. This intuition was firmly grounded in his most recent books *Continuous-Time Markov Decision Processes: Theory and Applications* (with X. P. Guo) and *Selected Topics on Continuous-Time Controlled Markov Chains and Markov Games* (with T. Prieto-Rumeau).

Among the diverse optimality criteria analyzed in the work of Professor Hernández-Lerma, one of the most important is the *average* or *ergodic* index, where the asymptotic behavior of the controlled stochastic process needs to be well understood to verify the existence of solutions of the optimal control problem and to characterize the value function in terms of the optimality equation. The book

*Markov Chains and Invariant Probabilities*, written jointly with J. B. Lasserre and published in 2003, is precisely a well-established work on the ergodic behavior of Markov chains in metric spaces. Average optimality represents an interesting combination of ergodic theory and stochastic optimal control and requires effective techniques and clever ideas to deal with problems from adaptive control, partially observed processes, linear programming and approximating procedures, among others, and on this topic we can find a good number of papers by Professor Hernández-Lerma. A complete list of his publications is given below.

Professor Hernández-Lerma received his Ph.D. from Brown University in 1978, and since then he has been a regular faculty member of the Mathematics Department at the Centro de Investigación y de Estudios Avanzados (Cinvestav) of the Instituto Politécnico Nacional, where he has carried out most of his research and teaching activities. His work had a pioneering character in Mexico, where he was the first expert in stochastic control. Groups of mathematicians in Mexico were quite small in those years and only few people could see the potential of having a strong group in the field of stochastic optimal control.

At Cinvestav he has always been recognized by his students for his excellent ability to lecture on many topics such as real analysis, probability theory, Markov processes, and stochastic calculus. Up to now he has graduated 17 Ph.D. students, a record among Mexican mathematicians, and his former students recognize his generosity, sensitivity, and his ability to propose cutting-edge research projects. The results of most of those Ph.D. theses have been published in well-established international journals. Since 1986 he has regularly organized the Workshop on Stochastic Control, which is an important forum for the group of stochastic control in Mexico to present new research and to interact with experts from other countries. This group has an intense research activity thanks to the inspiration of Professor Hernández-Lerma, who has also influenced a significant number of other young mathematicians and visiting postdoctoral fellows who have visited Cinvestav in these years.

He has participated in numerous editorial boards, for journals like the SIAM Journal on Control and Optimization, International Journal on Stochastic Analysis, and Journal of Dynamics and Games. He has also served as a guest editor of special volumes in top journals. Additionally, Professor Hernández-Lerma has always showed a genuine interest in probability and statistics education in Mexico and has written two monographs in on these topics.

The contributions of Professor Hernández-Lerma to the development of applied mathematics in his country were recognized by the Government of Mexico in 2001, when he received the Sciences and Arts National Award, being the third mathematician to obtain such a high distinction. Also, in 2003 he received a *doctor honoris causa* from the Universidad de Sonora, and in 2008 he was honored with the Medalla Lázaro Cárdenas by the Instituto Politécnico Nacional.

Guanajuato, México                                                    Daniel Hernández-Hernández
Guanajuato, México                                                        J. Adolfo Minjárez-Sosa

# Preface

This volume presents a collection of papers by friends and colleagues of Professor Onésimo Hernández-Lerma in his honor. The first idea to compile the book arose during Onésimo's Fest Symposium held in San Luis Potosí, Mexico, during March 16–18, 2011, to honor the 65th birthday of Professor Onésimo Hernández-Lerma, who has been an important contributor to stochastic optimal control.

Thereafter, a group of colleagues whose research interests are in the areas of stochastic optimal control, optimization theory, and probability were invited to collaborate on this project. As a result, 33 authors from all over the world have contributed 18 chapters to the book. All papers have been peer-reviewed and give a general view on the state-of-the-art of the art of several topics on the covered fields. In particular, the book presents recent developments on discrete-time Markov control or decision processes under different contexts: discounted and average criterion as well as sample-path and constrained optimality; continuous-time controlled problems for diffusion processes, jump Markov processes, and semi-Markov processes; optimal stopping problems; global optimization; and the existence of solutions of stochastic partial differential equations. Additionally the book contains important applications to inventory control problems and financial systems. Chapters are presented in alphabetical order by first author.

We express our deep gratitude to all the people who have collaborated to make this special volume a success. Mainly, we thank all the authors for their excellent and important contributions, as well as all the reviewers for their time and effort. We would also like to thank the Centro de Investigación en Matemáticas (CIMAT) and the Departamento de Matemáticas, Universidad de Sonora.

Guanajuato, México                                    Daniel Hernández-Hernández
Guanajuato, México                                      J. Adolfo Minjárez-Sosa

# Contents

# Contributors

**Ari Arapostathis** Department of Electrical and Computer Engineering, The University of Texas at Austin, Austin, TX, USA

**Lankere Benkherouf** Faculty of Science, Department of Statistics and Operations Research, Kuwait University, Safat, Kuwait

**Alain Bensoussan** International Center for Decision and Risk Analysis, School of Management, University of Texas at Dallas, Richardson, TX, USA

WCU Distinguished Professor, The Hong Kong Polytechnic University, Hung Hom, Kowloon, Hong Kong Ajou University, San S. Woncheon-Dong, Yeontong-Gu, Suwon, Korea

**Rolando Cavazos-Cadena** Departamento de Estadística y Cálculo, Universidad Autónoma Agraria Antonio Narro, Buenavista, Saltillo, Coahuila, México

**François Dufour** Institut de Mathématiques de Bordeaux, Université Bordeaux I, Talence, France

INRIA Bordeaux Sud Ouest, Team CQFD, Bordeaux, France

**Eugene A. Feinberg** Department of Applied Mathematics and Statistics, Stony Brook University, Stony Brook, NY, USA

**Michael C. Fu** The Robert H. Smith School of Business & Institute for Systems Research, University of Maryland, College Park, MD, USA

**Mrinal K. Ghosh** Department of Mathematics, Indian Institute of Science, Bangalore, India

**T. E. Govindan** ESFM, Instituto Politécnico Nacional, México D.F., México

**Xianping Guo** Sun Yat-Sen University, Guangzhou, China

**Gerardo Hernandez-del-Valle** Statistics Department, Columbia University, New York, NY, USA

**Jiaqiao Hu** Department of Applied Mathematics and Statistics, State University at Stony Brook, Stony Brook, NY, USA

**Yonghui Huang**  Sun Yat-Sen University, Guangzhou, China

**Héctor Jasso-Fuentes**  Departamento de Matemáticas, CINVESTAV-IPN, México D.F., México

**Steven I. Marcus**  Department of Electrical and Computer Engineering & Institute for Systems Research, University of Maryland, College Park, MD, USA

**Raúl Montes-de-Oca**  Departamento de Matemáticas, Universidad Autónoma Metropolitana–Iztapalapa, México D.F., México

**Alexey Piunovskiy** Department of Mathematical Sciences, University of Liverpool, Liverpool, UK

**Tomás Prieto-Rumeau**  Department of Statistics, UNED, Madrid, Spain

**Rosario Romera**  University Carlos III de Madrid, Madrid, Spain

**Luz del Carmen Rosas-Rosas**  Departamento de Matemáticas, Universidad de Sonora, Hermosillo, Sonora, México

**Wolfgang Runggaldier**  Dipartimento di Matematica Pura ed Applicata, University of Padova, Padova, Italy

**Subhamay Saha** Department of Mathematics, Indian Institute of Science, Bangalore, India

**Lukasz Stettner**  Institute of Mathematics Polish Academy of Sciences, Warsaw, Poland

**Richard H. Stockbridge**  Department of Mathematical Sciences, University of Wisconsin at Milwaukee, Milwaukee, WI, USA

**Yuemeng Sun**  Cornell University, Ithaca, NY, USA

**Óscar Vega-Amaya** Departamento de Matemáticas, Universidad de Sonora, Hermosillo, Sonora, México

**Yongqiang Wang**  Department of Electrical and Computer Engineering & Institute for Systems Research, University of Maryland, College Park, MD, USA

**Qingda Wei**  Sun Yat-Sen University, Guangzhou, China

**George Yin** Department of Mathematics, Wayne State University, Detroit, MI, USA

**Junyu Zhang**  Sun Yat-Sen University, Guangzhou, China

**Yi Zhang** Deparment of Mathematical Sciences, University of Liverpool, Liverpool, UK

**Enlu Zhou**  Department of Industrial and Enterprise Systems Engineering, University of Illinois at Urbana-Champaign, IL, USA

**Chao Zhu** Department of Mathematical Sciences, University of Wisconsin at Milwaukee, Milwaukee, WI, USA

# Resume of Professor Onésimo Hernández-Lerma

**Present Position:**  Full Professor "3F" (top level), Department of Mathematics, CINVESTAV-IPN

- Member of the Mexican National Research System (SNI): Area I (Basic Sciences), Level 3 (top level).

## Special Awards

- *Sciences and Arts National Award* by the Government of México (2001).
- Doctor *Honoris Causa* by the Universidad de Sonora (2003).
- *Lázaro Cárdenas Medal* (IPN) 2008.
- *Scopus Prize* (Elsevier) 2008.
- *Thomson Reuters Award* 2009.

**Research interests:**  Optimal control of stochastic systems, Multiobjective control problems, Stochastic games, Infinite–dimensional linear programming, Markov processes.

## Academic Background

- Mexican Air Force College, 1964.
- Escuela Superior de Física y Matemáticas (ESFM) of the Instituto Politécnico Nacional (IPN), Licenciado en Física y Matemáticas, 1970.
- Division of Applied Mathematics, Brown University, M.Sc., 1976.
- Division of Applied Mathematics, Brown University, Ph.D., 1978.

## Employment

- Mexican Air Force, 1964–1968.
- Escuela Superior de Ingeniería Mecánica y Eléctrica (ESIME) of the IPN, 1968–1988.
- Universidad Autónoma Metropolitana, Azcapotzalco, 1974–1975.
- Centro de Investigación y de Estudios Avanzados, Mathematics Department, 1978—present. Chairman, 1992–1997, 2011–2015.

## Visiting Appointments

- Universidad de los Andes (Mérida, Venezuela), 1971–1972.
- University of Texas at Austin, 1982–1983. One-month visits: 1984, 1985, 1986, 1987.
- LAAS-CNRS (Toulouse, France). Several one-month visits per year, during 1986–2001.
- Texas Tech University, 1987–1988.
- University of Padova (Padova, Italy). Two months, 1991.
- Universidad Carlos III de Madrid. Two months, 2000, 2002.

## Visiting Postdoctoral Fellows

- Nadine Hilgert, INRA–ENSA.M, Montpellier, France, January–August 1999
- Xianping Guo, Zhongshan University, Guangzhou, P.R. China, August 2000–2002
- Tomás Prieto Rumeau, Universidad Complutense de Madrid, Spain, August 2003–January 2004

## Former Ph.D. Students

| | | |
|---|---|---|
| Rolando Cavazos Cadena | CINVESTAV | 1985 |
| Roberto S. Acosta Abreu | CINVESTAV | 1987 |
| Sergio O. Esparza Núñez | Texas Tech Univ. | 1989 |
| Daniel Hernández Hernández | CINVESTAV | 1993 |
| Raúl Montes de Oca Machorro | UAM–Iztapalapa | 1994 |
| Fernando Luque Vásquez | FC–UNAM | 1997 |
| Oscar Vega Amaya | UAM–Iztapalapa | 1998 |
| Juan González Hernández | FC–UNAM | 1998 |
| César E. Villarreal Rodríguez | CINVESTAV | 1998 |
| J. Rigoberto Gabriel Argüelles | CINVESTAV | 2000 |

| | | |
|---|---|---|
| Raquiel R. López Martínez | CINVESTAV | 2001 |
| Jorge Alvarez Mena | CINVESTAV | 2002 |
| Guadalupe Carrasco Licea | FC–UNAM | 2003 |
| Mario A. Villalobos Arias | CINVESTAV | 2005 |
| Héctor Jasso Fuentes | CINVESTAV | 2007 |
| Armando F. Mendoza Pérez | CINVESTAV | 2008 |
| Beatris A. Escobedo Trujillo | CINVESTAV | 2011 |

## Associate Editor Positions

- SIAM Journal on Control and Optimization (1991–1996)
- Boletín de la Sociedad Matemática Mexicana, 3a. serie (1995–2005)
- Journal of Mathematical Systems, Estimation, and Control (1992–1998)
- Applicationes Mathematicae (Warsaw) (1999–present)
- Top, journal of the Spanish Society of Statistics and Operations Research, SEIO (2001–2006)
- Information Technology for Economics and Management (ITEM) (2001–2005)
- Morfismos, journal of the CINVESTAV Mathematics Department (1997–present)
- Revista Mexicana de Economía y Finanzas (ReMEF) (2002-2007)
- International Journal on Stochastic Analysis (2010–present)
- Journal of Dynamics and Games (2011-)

# List of Publications by Onésimo Hernández-Lerma

## Books and Monographs

1. *Métodos de Fourier en la Física y la Ingeniería*, Editorial Trillas, México, 1974.
2. *Elementos de Probabilidad y Estadística*, Fondo de Cultura Económica, México, 1979; 2nd. printing, 1982.
3. *Adaptive Markov Control Processes*. Springer–Verlag, New York, 1989.
4. *Annals of Operations Research* Special Volumes **28** and **29** (1991) on Markov Decision Processes, with J.B. Lasserre, Guest Editors.
5. *Lectures on Continuous–Time Markov Control Processes*. Aportaciones Matemáticas, Advanced Texts series Vol. 3, Sociedad Matemática Mexicana, 1994.
6. *Discrete–Time Markov Control Processes: Basic Optimality Criteria*. Springer–Verlag, New York, 1996, with J.B. Lasserre.
7. *Further Topics on Discrete–Time Markov Control Processes*. Springer–Verlag, New York, 1999, with J.B. Lasserre.
8. *Markov Chains and Invariant Probabilities*. Birkhäuser, Basel, 2003, with J.B. Lasserre.
9. *Elementos de Probabilidad y Estadística*, Sociedad Matemática Mexicana, 2003, with A. Hernández del Valle.
10. *Continuous–Time Markov Decision Processes: Theory and Applications*. Springer–Verlag, 2009, with X.P. Guo.
11. *Selected Topics on Continuous–Time Controlled Markov Chains and Markov Games*. Imperial College Press, 2012, with T. Prieto–Rumeau.

## Papers

1. *Series de Fourier: Breve introducción histórica*, Miscelánea Matemática No. 7, 1974, pp. 13–24.
2. *Probabilidad y estadística en el nivel primario*, Matemáticas y Enseñanza No. 2, 1975, pp. 23–42, with L.G. Gorostiza.
3. *Lyapunov criteria for stability of differential equations with Markov parameters*, Bol. Soc. Mat. Mexicana **24** (1979) pp. 27–48.

4. *Exit probabilities for a class of perturbed degenerate systems*, SIAM J. Control Optim. **19** (1981), pp. 39–51.
5. *Control óptimo de procesos de difusión markovianos*, Ciencia (journal of the Mexican Academy of Sciences) **32** (1981), pp. 39–55, with W.H. Fleming.
6. *Modelos matemáticos en amibiasis*, Sigma **7** (1981), pp. 131–138, with R. Cano Mancera and R. López–Revilla.
7. *Modelos matemáticos de adhesión celular con aplicaciones a la adhesión de trofozoítos de Entamoeba histolytica*, Ciencia (journal of the Mexican Academy of Sciences) **33** (1982), pp. 107–117, with R. Cano Mancera and R. López Revilla.
8. *Adaptive control of service in queueing systems*, Syst. Control Lett. **3** (1983), pp. 283–289, with S.I. Marcus.
9. *Identification and approximation of queueing systems*, IEEE Trans. Automatic Control **AC–29** (1984), pp. 472–474, with S.I. Marcus.
10. *Modelos matemáticos de la adhesión de trofozoítos de Entamoeba histolytica a eritrocitos humanos*, Acta Mexicana de Ciencia y Tecnología **II** (1984), No. 5, pp. 65–75, with R. Cano Mancera and R. López–Revilla.
11. *Control adaptable iterativo de sistemas markovianos con costo promedio*, Acta Mexicana de Ciencia y Tecnología **II**, No. 6 (1984), pp. 63–68, with R.S. Acota–Abreu.
12. *Modelado, estimación y control de recursos pesqueros, I. Modelos de poblaciones*, Acta Mexicana de Ciencia y Tecnología, **II** No. 7 (1984), pp. 51–61.
13. *Optimal adaptive control of priority assignment in queueing systems*, Syst. Control Lett. **4** (1984) 65–72, with S.I. Marcus.
14. *Adaptive control of discounted Markov decision chains*, J. Optim. Theory Appl. **46** (1985), pp. 227–235, with S.I. Marcus.
15. *Nonstationary value–iteration and adaptive control of discounted semi–Markov processes*, J. Math. Anal. Appl. **112** (1985), pp. 435–445.
16. *Approximation and adaptive policies in discounted dynamic programming*, Bol. Soc. Mat. Mexicana **30** (1985), pp. 25–35.
17. *Iterative adaptive control of denumerable state average–cost Markov systems*, Control & Cybernetics **14**, No. 4 (1985), pp. 313–322, with R.S. Acosta Abreu.
18. *Finite–state approximations for denumerable multidimensional state discounted Markov decision processes*, J. Math. Anal. Appl. **113** (1986), pp. 382–389.
19. *Filtrado estadístico*, Acta Mexicana de Ciencia y Tecnología **4**, Nos. 13–14 (1986), pp. 49–55.
20. *Adaptive control of Markov processes with incomplete state information and unknown parameters*, J. Optim. Theory Appl. **52** (1987), pp. 227–241, with S.I. Marcus.
21. *Control adaptable de sistemas inciertos a tiempo discreto*, Ciencia (journal of the Mexican Academy of Sciences) **38** (1987), Part I, pp. 69–80; Part II, pp. 147–154; Part III, pp. 217–228, with M. España and D.B. Hernández.

22. *Adaptive policies for discrete–time stochastic control systems with unknown disturbance distribution*, Syst. Control Lett. **9** (1987), pp. 307–315, with S.I. Marcus.

23. *Estimación de parámetros en sistemas compartamentales parcialmente observables*, Acta Mexicana de Ciencia y Tecnología **5**, No. 17 (1987), pp. 33–43, with F. Maldonado Etchegaray.

24. *Approximation and adaptive control of Markov processes: average reward criterion*, Kybernetika (Prague) **23** (1987), pp. 265–288.

25. *Continuous dependence of stochastic control models on the noise distribution*, Appl. Math. Optim. **17** (1988), pp. 79–89, with R. Cavazos–Cadena.

26. *Controlled Markov processes: recent results on approximation and adaptive control*, Texas Tech University Mathematics Series, Visiting Scholars Lectures 1986–1987, **15** (1988), 91–117.

27. *A forecast horizon and a stopping rule for general Markov decision processes*, J. Math. Anal. Appl. **132** (1988), 388–400, with J.B. Lasserre.

28. *Recursive nonparametric estimation of nonstationary Markov processes*, Bol. Soc. Mat. Mexicana **33** (1988), 57–69, with S.O. Esparza and B.S. Duran.

29. *Nonparametric adaptive control of discrete–time partially observable stochastic systems*, J. Math. Anal. Appl. **137** (1989), pp. 312–334, with S.I. Marcus.

30. *On rolling horizon procedures for Markov control processes*, Aportaciones Matemáticas (Soc. Mat. Mexicana), Serie Notas de Investigación 4 (1989), 73–88.

31. *Discretization procedures for adaptive Markov control processes*, J. Math. Anal. Appl. **137** (1989), pp. 485–514, with S.I. Marcus.

32. *Adaptive policies for priority assignment in discrete time queues — discounted cost criterion*, Control and Cybernetics **19** (1990), 149–177, with R. Cavazos Cadena.

33. *Error bounds for rolling horizon policies in discrete–time Markov control processes*, IEEE Trans. Automatic Control **35** (1990), 1118–1124, with J.B. Lasserre.

34. *Average cost optimal policies for Markov control processes with Borel state space and unbounded costs*, Syst. Control Lett. **15** (1990), 349–356, with J.B. Lasserre.

35. *Density estimation and adaptive control of Markov processes: average and discounted criteria*, Acta Appl. Math. **20** (1990), 285–307, with R. Cavazos–Cadena.

36. *Average cost Markov decision processes: optimality conditions*, J. Math. Anal. Appl. **158** (1991), 396–406, with J.C. Hennet and J.B. Lasserre.

37. *Recursive adaptive control of Markov decision processes with the average reward criterion*, Appl. Math. Optim. **23** (1991), 193–207, with R. Cavazos–Cadena.

38. *Recurrence conditions for Markov decision processes with Borel state space: A survey*, Ann. Oper. Res. **28** (1991), 29–46, with R. Montes de Oca and R. Cavazos–Cadena.

39. *Average optimality in dynamic programming on Borel spaces — unbounded costs and controls*, Syst. Control Lett. **17** (1991), 237–242.

40. *On integrated square errors of recursive nonparametric estimates of nonstationary Markov processes*, Prob. Math. Stat. **12** (1991), 25–33.

41. *Discrete–time Markov control processes with discounted unbounded costs: Optimality criteria*, Kybernetika (Prague) **28** (1992), 191–213, with M. Muñoz de Ozak.

42. *Equivalence of Lyapunov stability criteria in a class of Markov decision processes*, Appl. Math. Optim. **26** (1992), 113–137, with R. Cavazos–Cadena.

43. *Control óptimo estocástico y programación lineal infinita*, Aportaciones Matemáticas (Soc. Mat. Mexicana). *Serie Notas de Investigación* **7** (1992), 109–120.

44. *Value iteration and rolling plans for Markov control processes with unbounded rewards*, J. Math. Anal. Appl. **177** (1993), 38–55, with J.B. Lasserre.

45. *Existence of average optimal policies in Markov control processes with strictly unbounded cost*, Kybernetika (Prague). **29** (1993), 1–17.

46. *Monotone approximations for convex stochastic control problems*, J. Math. Syst., Estimation, and Control **4** (1994), 99–140, with W. Runggaldier.

47. *Linear programming and average optimality of Markov control processes on Borel spaces–unbounded costs*, SIAM J. Control Optim. **32** (1994), 480–500, with J.B. Lasserre.

48. *Weak conditions for average optimality in Markov control processes*, Syst. Control Lett. **22** (1994), 287–291, with J.B. Lasserre.

49. *Discounted cost Markov decision processes on Borel spaces: The linear programming formulation*, J. Math. Anal. Appl. **183** (1994), 335–351, with D. Hernández–Hernández.

50. *Conditions for average optimality in Markov control processes on Borel spaces*, Bol. Soc. Mat. Mexicana **39** (1994), 39–50, with R. Montes–de–Oca and J.A. Minjárez–Sosa.

51. *Average cost Markov control processes with weighted norms: value iteration.* Appl. Math. (Warsaw) **23** (1995), 219–237, with E. Gordienko.

52. *Average cost Markov control processes with weighted norms: existence of canonical policies.* Appl. Math. (Warsaw) **23** (1995), 199–218, with E. Gordienko.

53. *Linear programming and infinite horizon problems of deterministic control theory.* Bol. Soc. Mat. Mex. (3rd series) **1** (1995), 59–72, with D. Hernández–Hernández.

54. *Numerical aspects of monotone approximations in convex stochastic control problems.* Ann. Oper. Res. **56** (1995), 135–156, with C. Piovesan and W.J. Runggaldier.

55. *Conditions for average optimality in Markov control processes with unbounded costs and controls*, J. Math. Syst., Estimation, and Control **5** (1995), 459–477, with R. Montes de Oca.

56. *A counterexample on the semicontinuity of minima.* Proc. Amer. Math. Soc. **123** (1995), 3175–3176, with F. Luque Vásquez.

57. *Invariant probabilities for Feller–Markov chains.* J. Appl. Math. and Stoch. Anal. **8** (1995), 341–345, with J.B. Lasserre.

58. *Value iteration in average cost Markov control process on Borel spaces.* Acta Appl. Math. **42** (1996), 203–222, with R. Montes–de–Oca.

59. *Average optimality in Markov control processes via discounted cost problems and linear programming.* SIAM J. Control Optim. **34** (1996), 295–310, with J.B. Lasserre.

60. *Existence of bounded invariant probability densities for Markov chains.* Statist. Prob. Lett. **28** (1996), 359–366, with J.B. Lasserre.

61. *The linear programming approach to deterministic optimal control problems.* Appl. Math. (Warsaw) **24** (1996), 17–33, with D. Hernández–Hernández and M. Taksar.

62. *An extension of the Vitali–Hahn–Saks Theorem.* Proc. Amer. Math. Soc. **124** (1996), 3673–3676; correction, ibid. **126** (1998), p. 949, with J.B. Lasserre.

63. *The linear programming formulation of optimal control problems,* Texas Tech University, Math. Series, Visiting Scholars Lectures, 1993–1996, **19** (1997), 1–17.

64. *Policy iteration for average cost Markov control processes on Borel spaces.* Acta Appl. Math. **47** (1997), 125–154, with J.B. Lasserre.

65. *Cone–constrained linear equations in Banach spaces.* J. Convex Anal. **4** (1997), 149–164, with J.B. Lasserre.

66. *Infinite linear programming and multichain Markov control processes in uncountable spaces.* SIAM J. Control Optim. **36** (1998), 313–335, with J. González–Hernández.

67. *Existence and uniqueness of fixed points for Markov operators and Markov processes.* Proc. London Math. Soc. **76** (1998), 711–736, with J.B. Lasserre.

68. *Infinite–horizon Markov control processes with undiscounted cost criteria: from average to overtaking optimality.* Appl. Math. (Warsaw) **25** (1998), 153–178, with O. Vega–Amaya.

69. *Linear programming approximations for Markov control processes in metric spaces.* Acta Appl. Math. **51** (1998), 123–139, with J.B. Lasserre.

70. *Approximation schemes for infinite linear programs.* SIAM J. Optim. **8** (1998), 973–988, with J.B. Lasserre.

71. *Ergodic theorems and ergodic decomposition of Markov chains.* Acta Appl. Math. **54** (1998), 99–119, with J.B. Lasserre.

72. *Envelopes of sets of measures, tightness, and Markov control processes.* Appl. Math. Optim. **40** (1999), 377–392, with J. González–Hernández.

73. *Semi–Markov control models with average costs.* Appl. Math. (Warsaw) **26** (1999), 315–331, with F. Luque–Vásquez.

74. *Markov control processes with the expected total–cost criterion: optimality, stability, and transient models.* Acta Appl. Math. **59** (1999), 229–269, with G. Carrasco and R. Pérez–Hernández.

75. *Sample–path optimality and variance–minimization of average cost Markov control processes.* SIAM J. Control Optim. **38** (2000), 79–93, with O. Vega–Amaya and G. Carrasco.

76. *Fatou and Lebesgue convergence theorems for measures.* J. Appl. Math. Stoch. Anal. **13** (2000), 137–146, with J.B. Lasserre.
77. *Limiting optimal discounted–cost control of a class of time–varying stochastic systems.* Syst. Control Lett. **40** (2000), 37–42, with N. Hilgert.
78. *Constrained Markov control processes in Borel spaces: the discounted case.* Math. Meth. Oper. Res. **52** (2000), 271–285, with J. González–Hernández.
79. *On the classification of Markov chains via occupation measures.* Appl. Math. (Warsaw) **27** (2000), 489–498, with J.B. Lasserre.
80. *Zero–sum stochastic games in Borel spaces: average payoff criteria.* SIAM J. Control Optim. **39** (2001), 1520–1539, with J.B. Lasserre.
81. *Further criteria for positive Harris recurrence of Markov chains.* Proc. Amer. Math. Soc. **129** (2001), 1521–1524, with J.B. Lasserre.
82. *Strong duality of the general capacity problem in metric spaces.* Math. Meth. Oper. Res. **53** (2001), 25–34, with J.R. Gabriel.
83. *Limiting average cost problems in a class of discrete–time stochastic systems.* Appl. Math. (Warsaw) **28** (2001), 111–123, with N. Hilgert.
84. *Limiting discounted–cost control of partially observable stochastic systems.* SIAM J. Control. Optim. **40** (2001), 348–369, with R. Romera.
85. *On the probabilistic multichain Poisson equation.* Appl. Math. (Warsaw) **28** (2001), 225–243, with J.B. Lasserre.
86. *A multiobjective control approach to priority queues.* Math. Meth. Oper. Res. **53** (2001), 265–277, with L.F. Hoyos–Reyes.
87. *Nonstationary continuous–time Markov control processes with discounted costs on infinite horizon.* Acta Appl. Math. **67** (2001), 277–293, with T.E. Govindan.
88. *The Lagrange approach to infinite linear programs.* Top **9** (2001), 293–314, with J.R. Gabriel and R.R. López–Martínez.
89. *The linear programming approach,* Chapter 12 in *Handbook of Markov Decision Processes,* edited by E. Feinberg and A. Shwartz, Kluwer, 2002, with J.B. Lasserre.
90. *Strong duality of the Monge–Kantorovich mass transfer problem in metric spaces.* Math. Zeit. **239** (2002), 579–591, with J.R. Gabriel.
91. *Convergence of the optimal values of constrained Markov control processes.* Math. Meth. Oper. Res. **55** (2002), 461–484, with J. Alvarez–Mena.
92. *Continuous–time controlled Markov chains.* Ann. Appl. Prob. **13** (2003), 363–388, with X.P. Guo.
93. *Performance analysis and optimal control of the Geo/Geo/c queue.* Performance Evaluation **52** (2003), 15–39, with J.R. Artalejo.
94. *Minimax control of discrete–time stochastic systems.* SIAM J. Control Optim. **41** (2003), 1626–1659, with J.I. González–Trejo and L.F. Hoyos–Reyes.
95. *Drift and monotonicity conditions for continuous–time controlled Markov chains with an average criterion.* IEEE Trans. Automatic Control **48** (2003), 236–245, with X.P. Guo.
96. *Constrained continuous–time Markov control processes with discounted criteria.* J. Stoch. Anal. Appl. **21** (2003), 379–399, with X.P. Guo.

97. *Zero–sum games for continuous–time Markov chains with unbounded transition and average payoff rates.* J. Appl. Prob. **40** (2003), 327–345, with X.P. Guo.

98. *Bias optimality versus strong O–discount optimality in Markov control processes with unbounded costs.* Acta Appl. Math. **77** (2003), 215–235, with N. Hilgert.

99. *Constrained average cost Markov control processes in Borel spaces.* SIAM J. Control Optim. **42** (2003), 442–468, with J. González–Hernández and R.R. López–Martínez.

100. *Continuous–time controlled Markov chains with discounted rewards.* Acta Appl. Math. **79** (2003), 195–216, with X.P. Guo.

101. *The Lagrange approach to constrained Markov control processes.* Invited paper, Morfismos **7** (2003), 1–26, with R.R. López–Martínez.

102. *Zero–sum games for nonhomogeneous Markov chains with expected average payoff criteria.* Applied and Computational Math., **3** (2004), 10–22, with X.P. Guo.

103. *The scalarization approach to multiobjective Markov control problems: why does it work?*, Appl. Math. Optim. **50** (2004), 279–293, with R. Romera.

104. *Multiobjective Markov control processes: a linear programming approach.* Invited paper. Morfismos, **8** (2004), 1–33, with R. Romera.

105. *The Laurent series, sensitive discount and Blackwell optimality for continuous–time controlled Markov chains*, Math. Meth. Oper. Res. **61** (2005), 123–145, with T. Prieto–Rumeau.

106. *Convergence and approximation of optimization problems*, SIAM J. Optim. **15** (2005), 527–539, with J. Alvarez–Mena.

107. *Bias and overtaking equilibria for zero–sum continuous–time Markov games*, Math. Meth. Oper. Res. **61** (2005), 437–454, with T. Prieto–Rumeau.

108. *Nonzero–sum games for continuous–time Markov chains with unbounded discounted payoffs*, J. Appl. Prob. **42** (2005), 303–320, with X.P. Guo.

109. *Extreme points of sets of randomized strategies in constrained optimization and control problems*, SIAM J. Optim. **15** (2005), 1085–1104, with J. González–Hernández.

110. *Approximation of general optimization problems.* Invited paper. Morfismos, **9** (2005), 1–20, with J. Alvarez–Mena.

111. *Zero–sum continuous–time Markov games with unbounded transition and discounted payoff rates*, Bernoulli, **11** (2005), 1009–1029, with X.P. Guo.

112. *Bias optimality for continuous–time controlled Markov chains*, SIAM J. Control Optim., **45** (2006), 51–73, with T. Prieto–Rumeau.

113. *On solutions to the mass transfer problem*, SIAM J. Optim., **17** (2006), 485–499, with J. González–Hernández and J.R. Gabriel.

114. *Existence of Nash equilibria for constrained stochastic games*, Math. Meth. Oper. Res. **63** (2006), 261–285, with J. Alvarez–Mena.

115. *Asymptotic convergence of a simulated annealing algorithm for multiobjective optimization problems*, Math. Meth. Oper Res., **64** (2006), 353–362, with M. Villalobos–Arias and C.A. Coello Coello.

116. *Asymptotic convergence of metaheuristics for multiobjective optimization problems*, Soft Computing **10** (2006), 1001–1005, with M. Villalobos–Arias and C.A. Coello Coello.

117. *A survey of recent results on continuous–time Markov decision processes*, Top **14** (2006), 177–261, with X.P. Guo and T. Prieto–Rumeau.

118. *A unified approach to continuous–time discounted Markov control processes*. Invited paper. Morfismos **10** (2006), 1–40, with T. Prieto–Rumeau.

119. *Zero–sum games for continuous–time jump Markov processes in Polish spaces: discounted payoffs*, Adv. Appl. Probab. **39** (2007), 645–668, with X.P. Guo.

120. Comments on: "Dynamic priority allocation via restless bandit marginal productivity indices" [TOP 15 (2007), 161–198] by J. Niño–Mora. TOP **15** (2007), 208–210.

121. *Ergodic control of continuous–time Markov chains with pathwise constraints*, SIAM J. Control Optim. **47** (2008), 1888–1908, with T. Prieto–Rumeau.

122. *Characterizations of overtaking optimality for controlled diffusion processes*, Appl. Math. Optim. **57** (2008), 349–369, with H. Jasso–Fuentes.

123. *Existence and regularity of nonhomogeneous $Q(t)$–processes under measurability conditions*, J. Theoret. Probab. **21** (2008), 604–627, with L. Ye and X.P. Guo.

124. *The vanishing discount approach to average reward optimality: the strongly and the weakly continuous cases*, Morfismos **12** (2008), 1–15, with T. Prieto–Rumeau

125. *Ergodic control, bias, and sensitive discount optimality for Markov diffusion processes*, J. Stoch. Anal. Appl. **27** (2009), 363–385, with H. Jasso–Fuentes.

126. *Blackwell optimality for controlled diffusion processes*, J. Appl. Probab. **46** (2009), 372–391, with H. Jasso–Fuentes.

127. *Variance minimization and the overtaking optimality approach to continuous–time controlled Markov chains*, Math. Meth. Oper. Res. **70** (2009), 527–540, with T. Prieto–Rumeau.

128. *Markov control processes with pathwise constraints*, Math. Meth. Oper. Res. **71** (2010), 477–502, with A.F. Mendoza–Pérez.

129. *The vanishing discount approach to constrained continuous–time controlled Markov chains*, Syst. Control Lett. **59** (2010), 504–509, with T. Prieto–Rumeau.

130. *Policy iteration and finite approximations to discounted continuous–time controlled* Markov chains, in: Modern Trends in Controlled Stochastic Processes, edited by A.B. Piunovskiy, Luniver Press, Frome, U.K., 2010, pp. 84–101, with T. Prieto–Rumeau.

131. *Overtaking optimal strategies and sensitive discount optimality for controlled diffusions*, in: Modern Trends in Controlled Stochastic Processes, edited by A.B. Piunovskiy, Luniver Press, Frome, U.K., 2010, pp. 102–122, with H. Jasso–Fuentes.

132. *Nonstationary discrete–time deterministic and stochastic control systems with infinite horizon*, International Journal of Control **83** (2010), 1751–1757, with X.P. Guo and A. Hernández–del–Valle.

133. *Asymptotic normality of discrete–time Markov control processes.* J. Applied Probab. **47** (2010), 778–795, with A.F. Mendoza–Pérez.

134. *New optimality conditions for average–payoff continuous–time Markov games in Polish spaces.* Science China Math. **54** (2011), pp. 793–816, with X.P. Guo.

135. *Variance–minimization of Markov control processes with pathwise constraints.* Optimization. Published electronically: 30 March 2011. DOI:10.1080/02331934.2011.565762, with A.F. Mendoza–Pérez.

136. *Nonstationary discrete–time deterministic and stochastic control systems: bounded and unbounded cases.* Syst. Control Lett. **60** (2011), pp. 503–509, with X.P. Guo and A. Hernández–del–Valle.

137. *Overtaking optimality for controlled Markov–modulated diffusions.* Optimization, with B.A. Escobedo–Trujillo. Published electronically: May 3, 2011. DOI:10.1080/02331934.2011.565417.

138. *An inverse optimal problem in discrete–time stochastic control.* Journal of Difference Equations and Applications, with D. González–Sánchez. Electronic version published: 29 September 2011. DOI:10.1080/10236198.2011.613596.

139. *Bias and overtaking equilibria for zero–sum stochastic differential games.* J. Optim. Theory Appl., with B.A. Escobedo–Trujillo and J.D. López–Barrientos. Published online: December 2011. DOI: 10.1007/s10957-011-9974-4.

## To appear

140. *Deterministic optimal policies for Markov control processes with pathwise constraints.* Applicationes Mathematicae (Warsaw), with A.F. Mendoza–Pérez.

141. *First passage problems for nonstationary discrete–time stochastic control systems.* Euro. J. Control, with X.P. Guo and A. Hernández–del–Valle.

# Chapter 1
# On the Policy Iteration Algorithm for Nondegenerate Controlled Diffusions Under the Ergodic Criterion

**Ari Arapostathis**

*Dedicated to Onésimo Hernández-Lerma on the occasion of his 65th birthday.*

## 1.1 Introduction

The policy iteration algorithm (PIA) for controlled Markov chains has been known since the fundamental work of Howard [2]. For controlled Markov chains on Borel state spaces, most studies of the PIA rely on blanket Lyapunov conditions [9]. A study of the PIA that treats the model of near-monotone costs can be found in [11], some ideas of which we follow closely. An analysis of the PIA for piecewise deterministic Markov processes has appeared in [6].

In this chapter we study the PIA for controlled diffusion processes $X = \{X_t, \, t \geq 0\}$, taking values in the $d$-dimensional Euclidean space $\mathbb{R}^d$, and governed by the Itô stochastic differential equation

$$dX_t = b(X_t, U_t)\, dt + \sigma(X_t)\, dW_t. \tag{1.1}$$

All random processes in (1.1) live in a complete probability space $(\Omega, \mathfrak{F}, \mathbb{P})$. The process $W$ is a $d$-dimensional standard Wiener process independent of the initial condition $X_0$. The control process $U$ takes values in a compact, metrizable set $\mathbb{U}$, and $U_t(\omega)$ is jointly measurable in $(t, \omega) \in [0, \infty) \times \Omega$. Moreover, it is *nonanticipative*: for $s < t$, $W_t - W_s$ is independent of

$$\mathfrak{F}_s := \text{the completion of } \sigma\{X_0, U_r, W_r, \, r \leq s\} \text{ relative to } (\mathfrak{F}, \mathbb{P}).$$

A. Arapostathis (✉)
Department of Electrical and Computer Engineering,
The University of Texas at Austin, Austin, TX 78712, USA
e-mail: ari@mail.utexas.edu

D. Hernández-Hernández and A. Minjárez-Sosa (eds.), *Optimization, Control, and Applications of Stochastic Systems*, Systems & Control: Foundations & Applications,
DOI 10.1007/978-0-8176-8337-5_1, © Springer Science+Business Media, LLC 2012

Such a process $U$ is called an *admissible control*, and we let $\mathfrak{U}$ denote the set of all admissible controls.

We impose the following standard assumptions on the drift $b$ and the diffusion matrix $\sigma$ to guarantee existence and uniqueness of solutions to (1.1).

(A1) *Local Lipschitz continuity:* The functions

$$b = \left[b^1,\ldots,b^d\right]^{\mathsf{T}} : \mathbb{R}^d \times \mathbb{U} \mapsto \mathbb{R}^d \quad \text{and} \quad \sigma = \left[\sigma^{ij}\right] : \mathbb{R}^d \mapsto \mathbb{R}^{d \times d}$$

are locally Lipschitz in $x$ with a Lipschitz constant $K_R$ depending on $R > 0$. In other words, if $B_R$ denotes the open ball of radius $R$ centered at the origin in $\mathbb{R}^d$, then for all $x, y \in B_R$ and $u \in \mathbb{U}$,

$$|b(x,u) - b(y,u)| + \|\sigma(x) - \sigma(y)\| \le K_R |x - y|,$$

where $\|\sigma\|^2 := \text{trace}\left(\sigma\sigma^{\mathsf{T}}\right)$.

(A2) *Affine growth condition:* $b$ and $\sigma$ satisfy a global growth condition of the form

$$|b(x,u)|^2 + \|\sigma(x)\|^2 \le K_1\left(1 + |x|^2\right), \quad \forall(x,u) \in \mathbb{R}^d \times \mathbb{U}.$$

(A3) *Local nondegeneracy:* For each $R > 0$, there exists a positive constant $\kappa_R$ such that

$$\sum_{i,j=1}^{d} a^{ij}(x)\xi_i\xi_j \ge \kappa_R|\xi|^2, \quad \forall x \in B_R,$$

for all $\xi = (\xi_1,\ldots,\xi_d) \in \mathbb{R}^d$, where $a := \frac{1}{2}\sigma\sigma^{\mathsf{T}}$.

We also assume that $b$ is continuous in $(x,u)$.

In integral form, (1.1) is written as

$$X_t = X_0 + \int_0^t b(X_s, U_s)\,\mathrm{d}s + \int_0^t \sigma(X_s)\,\mathrm{d}W_s. \tag{1.2}$$

The second term on the right-hand side of (1.2) is an Itô stochastic integral. We say that a process $X = \{X_t(\omega)\}$ is a solution of (1.1) if it is $\mathfrak{F}_t$-adapted, continuous in $t$, defined for all $\omega \in \Omega$ and $t \in [0,\infty)$, and satisfies (1.2) for all $t \in [0,\infty)$ at once a.s.

With $u \in \mathbb{U}$ treated as a parameter, we define the family of operators $L^u : \mathscr{C}^2(\mathbb{R}^d) \mapsto \mathscr{C}(\mathbb{R}^d)$ by

$$L^u f(x) = \sum_{i,j} a^{ij}(x)\frac{\partial^2 f}{\partial x_i \partial x_j}(x) + \sum_i b^i(x,u)\frac{\partial f}{\partial x_i}(x), \quad u \in \mathbb{U}. \tag{1.3}$$

We refer to $L^u$ as the *controlled extended generator* of the diffusion.

Of fundamental importance in the study of functionals of $X$ is Itô's formula. For $f \in \mathscr{C}^2(\mathbb{R}^d)$ and with $L^u$ as defined in (1.3),

$$f(X_t) = f(X_0) + \int_0^t L^{U_s} f(X_s)\,\mathrm{d}s + M_t, \quad \text{a.s.,} \tag{1.4}$$

where

$$M_t := \int_0^t \langle \nabla f(X_s), \sigma(X_s) \, dW_s \rangle$$

is a local martingale. Krylov's extension of the Itô formula [10, p. 122] extends (1.4) to functions $f$ in the Sobolev space $\mathscr{W}_{loc}^{2,p}(\mathbb{R}^d)$.

Recall that a control is called *stationary Markov* if $U_t = v(X_t)$ for a measurable map $v : \mathbb{R}^d \mapsto \mathbb{U}$. Correspondingly, the equation

$$X_t = x_0 + \int_0^t b(X_s, v(X_s)) \, ds + \int_0^t \sigma(X_s) \, dW_s \tag{1.5}$$

is said to have a *strong solution* if given a Wiener process $(W_t, \mathfrak{F}_t)$ on a complete probability space $(\Omega, \mathfrak{F}, \mathbb{P})$, there exists a process $X$ on $(\Omega, \mathfrak{F}, \mathbb{P})$, with $X_0 = x_0 \in \mathbb{R}^d$, which is continuous, $\mathfrak{F}_t$-adapted, and satisfies (1.5) for all $t$ at once, a.s. A strong solution is called *unique*, if any two such solutions $X$ and $X'$ agree $\mathbb{P}$-a.s. when viewed as elements of $\mathscr{C}([0, \infty), \mathbb{R}^d)$. It is well known that under Assumptions (A1)–(A3), for any stationary Markov control $v$, (1.5) has a unique strong solution [8].

Let $\mathfrak{U}_{SM}$ denote the set of stationary Markov controls. Under $v \in \mathfrak{U}_{SM}$, the process $X$ is strong Markov, and we denote its transition function by $P^v(t, x, \cdot)$. It also follows from the work of [5, 12] that under $v \in \mathfrak{U}_{SM}$, the transition probabilities of $X$ have densities which are locally Hölder continuous. Thus $L^v$ defined by

$$L^v f(x) = \sum_{i,j} a^{ij}(x) \frac{\partial^2 f}{\partial x_i \partial x_j}(x) + \sum_i b^i(x, v(x)) \frac{\partial f}{\partial x_i}(x), \quad v \in \mathfrak{U}_{SM},$$

for $f \in \mathscr{C}^2(\mathbb{R}^d)$ is the generator of a strongly continuous semigroup on $\mathscr{C}_b(\mathbb{R}^d)$, which is strong Feller. We let $\mathbb{P}_x^v$ denote the probability measure and $\mathbb{E}_x^v$ the expectation operator on the canonical space of the process under the control $v \in \mathfrak{U}_{SM}$, conditioned on the process $X$ starting from $x \in \mathbb{R}^d$ at $t = 0$.

In Sect. 1.2 we define our notation. Sect. 1.3 reviews the ergodic control problem for near-monotone costs and the basic properties of the PIA. Sect. 1.4 is dedicated to the convergence of the algorithm.

## 1.2 Notation

The standard Euclidean norm in $\mathbb{R}^d$ is denoted by $|\cdot|$, and $\langle \cdot, \cdot \rangle$ stands for the inner product. The set of nonnegative real numbers is denoted by $\mathbb{R}_+$, $\mathbb{N}$ stands for the set of natural numbers, and $\mathbb{I}$ denotes the indicator function. We denote by $\tau(A)$ the *first exit time* of the process $\{X_t\}$ from the set $A \subset \mathbb{R}^d$, defined by

$$\tau(A) := \inf \{t > 0 : X_t \notin A\}.$$

The open ball of radius $R$ in $\mathbb{R}^d$, centered at the origin, is denoted by $B_R$, and we let $\tau_R := \tau(B_R)$, and $\breve{\tau}_R := \tau(B_R^c)$.

The term *domain* in $\mathbb{R}^d$ refers to a nonempty, connected open subset of the Euclidean space $\mathbb{R}^d$. We introduce the following notation for spaces of real-valued functions on a domain $D \subset \mathbb{R}^d$. The space $\mathscr{L}^p(D)$, $p \in [1, \infty)$, stands for the Banach space of (equivalence classes) of measurable functions $f$ satisfying $\int_D |f(x)|^p \, dx < \infty$, and $\mathscr{L}^\infty(D)$ is the Banach space of functions that are essentially bounded in $D$. The space $\mathscr{C}^k(D)$ ($\mathscr{C}^\infty(D)$) refers to the class of all functions whose partial derivatives up to order $k$ (of any order) exist and are continuous, and $\mathscr{C}_c^k(D)$ is the space of functions in $\mathscr{C}^k(D)$ with compact support. The standard Sobolev space of functions on $D$ whose generalized derivatives up to order $k$ are in $\mathscr{L}^p(D)$, equipped with its natural norm, is denoted by $\mathscr{W}^{k,p}(D)$, $k \geq 0$, $p \geq 1$.

In general if $\mathscr{X}$ is a space of real-valued functions on $D$, $\mathscr{X}_{\text{loc}}$ consists of all functions $f$ such that $f\varphi \in \mathscr{X}$ for every $\varphi \in \mathscr{C}_c^\infty(D)$. In this manner we obtain the spaces $\mathscr{L}_{\text{loc}}^p(D)$ and $\mathscr{W}_{\text{loc}}^{2,p}(D)$.

Let $h \in \mathscr{C}(\mathbb{R}^d)$ be a positive function. We denote by $\mathscr{O}(h)$ the set of functions $f \in \mathscr{C}(\mathbb{R}^d)$ having the property

$$\limsup_{|x| \to \infty} \frac{|f(x)|}{h(x)} < \infty, \tag{1.6}$$

and by $\mathfrak{o}(h)$ the subset of $\mathscr{O}(h)$ over which the limit in (1.6) is zero.

We adopt the notation $\partial_i := \frac{\partial}{\partial x_i}$ and $\partial_{ij} := \frac{\partial^2}{\partial x_i \partial x_j}$. We often use the standard summation rule that repeated subscripts and superscripts are summed from 1 through $d$. For example,

$$a^{ij} \partial_{ij} \varphi + b^i \partial_i \varphi := \sum_{i,j=1}^d a^{ij} \frac{\partial^2 \varphi}{\partial x_i \partial x_j} + \sum_{i=1}^d b^i \frac{\partial \varphi}{\partial x_i}.$$

## 1.3   Ergodic Control and the PIA

Let $c \colon \mathbb{R}^d \times \mathbb{U} \to \mathbb{R}$ be a continuous function bounded from below. As well known, the ergodic control problem, in its *almost sure* (or *pathwise*) formulation, seeks to a.s. minimize over all admissible $U \in \mathfrak{U}$

$$\limsup_{t \to \infty} \frac{1}{t} \int_0^t c(X_s, U_s) \, ds. \tag{1.7}$$

A weaker, *average* formulation seeks to minimize

$$\limsup_{t \to \infty} \frac{1}{t} \int_0^t \mathbb{E}^U \left[ c(X_s, U_s) \right] ds. \tag{1.8}$$

We let $\rho^*$ denote the infimum of (1.8) over all admissible controls. We assume that $\rho^* < \infty$.

We assume that the cost function $c \colon \mathbb{R}^d \times \mathbb{U} \to \mathbb{R}_+$ is continuous and locally Lipschitz in its first argument uniformly in $u \in \mathbb{U}$. More specifically, for some function $K_c \colon \mathbb{R}_+ \to \mathbb{R}_+$,

$$\big|c(x,u) - c(y,u)\big| \le K_c(R)|x - y| \qquad \forall x, y \in B_R, \ \forall u \in \mathbb{U},$$

and all $R > 0$.

An important class of running cost functions arising in practice for which the ergodic control problem is well behaved is the *near-monotone* cost functions. Let $M^* \in \mathbb{R}_+ \cup \{\infty\}$ be defined by

$$M^* := \liminf_{|x| \to \infty} \min_{u \in \mathbb{U}} c(x,u).$$

The running cost function $c$ is called near-monotone if $\rho^* < M^*$. Note that inf-compact functions $c$ are always near-monotone.

We adopt the following abbreviated notation. For a function $g \colon \mathbb{R}^d \times \mathbb{U} \to \mathbb{R}$ and $v \in \mathfrak{U}_{\mathrm{SSM}}$ we let

$$g_v(x) := g\big(x, v(x)\big), \quad x \in \mathbb{R}^d.$$

The ergodic control problem for near-monotone cost functions is characterized as follows

**Theorem 1.3.1.** *There exists a unique function $V \in \mathscr{C}^2(\mathbb{R}^d)$ which is bounded below in $\mathbb{R}^d$ and satisfies $V(0) = 0$ and the Hamilton–Jacobi–Bellman (HJB) equation*

$$\min_{u \in \mathbb{U}} \big[L^u V(x) + c(x,u)\big] = \rho^*, \quad x \in \mathbb{R}^d.$$

*The control $v^* \in \mathfrak{U}_{\mathrm{SM}}$ is optimal with respect to the criteria (1.7) and (1.8) if and only if it satisfies*

$$\min_{u \in \mathbb{U}} \left[\sum_{i=1}^d b^i(x,u)\frac{\partial V}{\partial x_i}(x) + c(x,u)\right] = \sum_{i=1}^d b^i_{v^*}(x)\frac{\partial V}{\partial x_i}(x) + c_{v^*}(x)$$

*a.e. in $\mathbb{R}^d$. Moreover, with $\breve{\tau}_r = \tau(B_r^c)$, $r > 0$, we have*

$$V(x) = \limsup_{r \downarrow 0} \ \inf_{v \in \mathfrak{U}_{\mathrm{SSM}}} \mathbb{E}_x^v\left[\int_0^{\breve{\tau}_r} \big(c_v(X_t) - \rho^*\big)\,\mathrm{d}t\right], \quad x \in \mathbb{R}^d.$$

A control $v \in \mathfrak{U}_{\mathrm{SM}}$ is called *stable* if the associated diffusion is positive recurrent. We denote the set of such controls by $\mathfrak{U}_{\mathrm{SSM}}$. Also we let $\mu_v$ denote the unique invariant probability measure on $\mathbb{R}^d$ for the diffusion under the control $v \in \mathfrak{U}_{\mathrm{SSM}}$.

Recall that $v \in \mathfrak{U}_{\mathrm{SSM}}$ if and only if there exists an inf-compact function $\mathscr{V} \in \mathscr{C}^2(\mathbb{R}^d)$, a bounded domain $D \subset \mathbb{R}^d$, and a constant $\varepsilon > 0$ satisfying

$$L^v \mathscr{V}(x) \leq -\varepsilon \qquad \forall x \in D^c . \tag{1.9}$$

It follows that the optimal control $v$ in Theorem 1.3.1 is stable. For $v \in \mathfrak{U}_{\mathrm{SSM}}$ we define

$$\rho_v := \limsup_{t \to \infty} \frac{1}{t} \int_0^t \mathbb{E}^v \big[ c_v(X_s) \big] \, ds .$$

A difficulty in synthesizing an optimal control $v \in \mathfrak{U}_{\mathrm{SM}}$ via the HJB equation lies in the fact that the optimal cost $\rho^*$ is not known. The PIA provides an iterative procedure for obtaining the HJB equation at the limit. In order to describe the algorithm we, first need to review some properties of the Poisson equation

$$L^v V(x) + c_v(x) = \rho , \quad x \in \mathbb{R}^d . \tag{1.10}$$

We need the following definition.

**Definition 1.3.1.** For $v \in \mathfrak{U}_{\mathrm{SSM}}$, and provided $\rho_v < \infty$, define

$$\Psi^v(x) := \lim_{r \downarrow 0} \mathbb{E}_x^v \left[ \int_0^{\breve{\tau}_r} \big( c_v(X_t) - \rho_v \big) \, dt \right] , \quad x \neq 0 .$$

For $v \in \mathfrak{U}_{\mathrm{SM}}$ and $\alpha > 0$, let $J_\alpha^v$ denote the $\alpha$-discounted cost

$$J_\alpha^v(x) := \mathbb{E}_x^v \left[ \int_0^\infty \mathrm{e}^{-\alpha t} c_v(X_t) \, dt \right] , \quad x \in \mathbb{R}^d .$$

We borrow the following result from [1, Lemma 7.4]. If $v \in \mathfrak{U}_{\mathrm{SSM}}$ and $\rho_v < \infty$, then there exists, a function $V \in \mathscr{W}_{\mathrm{loc}}^{2,p}(\mathbb{R}^d)$, for any $p > 1$, and a constant $\rho \in \mathbb{R}$ which satisfies (1.10) a.e. in $\mathbb{R}^d$ and such that, as $\alpha \downarrow 0$, $\alpha J_\alpha^v(0) \to \rho$ and $J_\alpha^v - J_\alpha^v(0) \to V$ uniformly on compact subsets of $\mathbb{R}^d$. Moreover,

$$\rho = \rho_v \quad \text{and} \quad V(x) = \Psi^v(x).$$

We refer to the function $V(x) = \Psi^v(x) \in \mathscr{W}_{\mathrm{loc}}^{2,p}(\mathbb{R}^d)$ as the *canonical solution* of the Poisson equation $L^v V + c_v = \rho_v$ in $\mathbb{R}^d$.

It can be shown that the canonical solution $V$ to the Poisson equation is the unique solution in $\mathscr{W}_{\mathrm{loc}}^{2,p}(\mathbb{R}^d)$ which is bounded below and satisfies $V(0) = 0$. Note also that (1.9) implies that any control $v$ satisfying $\rho_v < M^*$ is stable.

The PIA takes the following familiar form:

*Algorithm (PIA).*

1. *Initialization. Set $k = 0$ and select any $v_0 \in \mathfrak{U}_{\mathrm{SM}}$ such that $\rho_{v_0} < M^*$.*
2. *Value determination. Obtain the canonical solution $V_k = \Psi^{v_k} \in \mathscr{W}_{\mathrm{loc}}^{2,p}(\mathbb{R}^d)$, $p > 1$, to the Poisson equation*

$$L^{v_k}V_k + c_{v_k} = \rho_{v_k}$$

in $\mathbb{R}^d$.

3. *If $v_k(x) \in \text{Arg min}_{u\in\mathbb{U}} \left[b^i(x,u)\partial_i V_k(x) + c(x,u)\right]$ x-a.e., return $v_k$.*
4. *Policy improvement. Select an arbitrary $v_{k+1} \in \mathfrak{U}_{SM}$ which satisfies*

$$v_{k+1}(x) \in \text{Arg min}_{u\in\mathbb{U}} \left[\sum_{i=1}^{d} b^i(x,u)\frac{\partial V_k}{\partial x_i}(x) + c(x,u)\right], \qquad x \in \mathbb{R}^d.$$

Since $\rho_{v_0} < M^*$ it follows that $v_0 \in \mathfrak{U}_{SSM}$. The algorithm is well defined, provided $v_k \in \mathfrak{U}_{SSM}$ for all $k \in \mathbb{N}$. This follows from the next lemma which shows that $\rho_{v_{k+1}} \leq \rho_{v_k}$, and in particular that $\rho_{v_k} < M^*$, for all $k \in \mathbb{N}$.

**Lemma 1.3.1.** *Suppose $v \in \mathfrak{U}_{SSM}$ satisfies $\rho_v < M^*$. Let $V \in \mathscr{W}_{loc}^{2,p}(\mathbb{R}^d)$, $p > 1$, be the canonical solution to the Poisson equation*

$$L^v V + c_v = \rho_v, \quad in\ \mathbb{R}^d.$$

*Then any measurable selector $\hat{v}$ from the minimizer*

$$\text{Arg min}_{u\in\mathbb{U}} \left[b^i(x,u)\partial_i V(x) + c(x,u)\right]$$

*satisfies $\rho_{\hat{v}} \leq \rho_v$. Moreover, the inequality is strict unless $v$ satisfies*

$$L^v V(x) + c_v(x) = \min_{u\in\mathbb{U}} \left[L^u V(x) + c(x,u)\right] = \rho_v, \quad for\ almost\ all\ x. \qquad (1.11)$$

*Proof.* Let $\mathscr{V}$ be a Lyapunov function satisfying $L^v \mathscr{V}(x) \leq k_0 - g(x)$, for some inf-compact $g$ such that $c_v \in o(g)$ (see [1, Lemma 7.1]). For $n \in \mathbb{N}$, define

$$\hat{v}_n(x) = \begin{cases} \hat{v}(x) & \text{if } x \in B_n \\ v(x) & \text{if } x \in B_n^c. \end{cases}$$

Clearly, $\hat{v}_n \to \hat{v}$ as $n \to \infty$ in the topology of Markov controls (see [1, Section 3.3]). It is evident that $\mathscr{V}$ is a stochastic Lyapunov function relative to $\hat{v}_n$, i.e., there exist constants $k_n$ such that $L^{\hat{v}_n} \mathscr{V}(x) \leq k_n - g(x)$, for all $n \in \mathbb{N}$. Since $V \in o(\mathscr{V})$, it follows that (see [1, Lemma 7.1])

$$\frac{1}{t}\mathbb{E}_x^{\hat{v}_n}[V(X_t)] \xrightarrow[t\to\infty]{} 0 \qquad (1.12)$$

Let

$$h(x) := \rho_v - \min_{u\in\mathbb{U}} \left[L^u V(x) + c(x,u)\right], \quad x \in \mathbb{R}^d.$$

Also, by definition of $\hat{v}_n$, for all $m \leq n$, we have

$$L^{\hat{v}_n} V(x) + c_{\hat{v}_n}(x) \leq \rho_v - h(x) \mathbb{I}_{B_m}(x). \tag{1.13}$$

By Itô's formula we obtain from (1.13) that

$$
\frac{1}{t}\left(\mathbb{E}_x^{\hat{v}_n}[V(X_t)] - V(x)\right) + \frac{1}{t}\mathbb{E}_x^{\hat{v}_n}\left[\int_0^t c_{\hat{v}_n}(X_s)\,\mathrm{d}s\right]
$$

$$
\leq \rho_v - \frac{1}{t}\mathbb{E}_x^{\hat{v}_n}\left[\int_0^t h(X_s)\mathbb{I}_{B_m}(X_s)\,\mathrm{d}s\right], \tag{1.14}
$$

for all $m \leq n$. Taking limits in (1.14) as $t \to \infty$ and using (1.12), we obtain

$$\rho_{\hat{v}_n} \leq \rho_v - \int_{\mathbb{R}^d} h(x)\mathbb{I}_{B_m}(x)\,\mu_{\hat{v}_n}(\mathrm{d}x). \tag{1.15}$$

Note that $v \mapsto \rho_v$ is lower semicontinuous. Therefore, taking limits in (1.15) as $n \to \infty$, we have

$$\rho_{\hat{v}} \leq \rho_v - \limsup_{n \to \infty} \int_{\mathbb{R}^d} h(x)\mathbb{I}_{B_m}(x)\,\mu_{\hat{v}_n}(\mathrm{d}x). \tag{1.16}$$

Since $c$ is near-monotone and $\rho_{\hat{v}_n} \leq \rho_v < M^*$, there exists $\hat{R} > 0$ and $\delta > 0$, such that $\mu_{\hat{v}_n}(B_{\hat{R}}) \geq \delta$ for all $n \in \mathbb{N}$. Then with $\psi_{\hat{v}_n}$ denoting the density of $\mu_{\hat{v}_n}$ Harnack's inequality [7, Theorem 8.20, p. 199] implies that there exists a constant $C_H = C_H(R)$ such that for every $R > \hat{R}$, with $|B_R|$ denoting the volume of $B_R \subset \mathbb{R}^d$, it holds that

$$\inf_{B_R} \psi_{\hat{v}_n} \geq \frac{\delta}{C_H |B_R|}, \qquad \forall n \in \mathbb{N}.$$

By (1.16) this implies that $\rho_{\hat{v}} < \rho_v$ unless $h = 0$ a.e.      $\square$

## 1.4   Convergence of the PIA

We start with the following lemma.

**Lemma 1.4.2.** *The sequence $\{V_k\}$ of the PIA has the following properties:*

 (i) *For some constant $C_0 = C_0(\rho_{v_0})$ we have $\inf_{\mathbb{R}^d} V_k > C_0$ for all $k \geq 0$.*
 (ii) *Each $V_k$ attains its minimum on the compact set*

$$\mathscr{K}(\rho_{v_0}) := \left\{x \in \mathbb{R}^d : \min_{u \in \mathbb{U}} c(x,u) \leq \rho_{v_0}\right\}.$$

 (iii) *For any $p > 1$, there exists a constant $\tilde{C}_0 = \tilde{C}_0(R, \rho_{v_0}, p)$ such that*

$$\left\|V_k\right\|_{\mathscr{W}^{2,p}(B_R)} \leq \tilde{C}_0 \qquad \forall R > 0.$$

(iv) *There exist positive numbers $\alpha_k$ and $\beta_k$, $k \geq 0$, such that $\alpha_k \downarrow 1$ and $\beta_k \downarrow 0$ as $k \to \infty$ and*

$$\alpha_{k+1} V_{k+1}(x) + \beta_{k+1} \leq \alpha_k V_k + \beta_k \quad \forall k \geq 0.$$

*Proof.* Parts (i) and (ii) follow directly from [3, Lemmas 3.6.1 and 3.6.4].

For part (iii) note first that the near-monotone assumption implies that

$$\mu_{v_k}\left(\mathscr{K}\left(\tfrac{M^*+\rho_{v_k}}{2}\right)\right) \geq \frac{M^* - \rho_{v_k}}{M^* + \rho_{v_k}} \quad \forall k \geq 0.$$

Consequently,

$$\mu_{v_k}\left(\mathscr{K}\left(\tfrac{M^*+\rho_{v_0}}{2}\right)\right) \geq \frac{M^* - \rho_{v_0}}{M^* + \rho_{v_0}} \quad \forall k \geq 0.$$

uniformly on compact subsets of $\mathbb{R}^d$. Hence, since $J_\alpha^{v_k} - J_\alpha^{v_k}(0) \to V_k$ weakly in $\mathscr{W}^{2,p}(B_R)$ for any $R > 0$, (iii) follows from [3, Theorem 3.7.4].

Part (iv) follows as in [11, Theorem 4.4]. [1]                                           □

As the corollary below shows, the PIA always converges.

**Corollary 1.4.1.** *There exist a constant $\hat{\rho}$ and a function $\hat{V} \in \mathscr{C}^2(\mathbb{R}^d)$ with $\hat{V}(0) = 0$, such that, as $k \to \infty$, $\rho_{v_k} \downarrow \hat{\rho}$ and $V_k \to \hat{V}$ weakly in $\mathscr{W}^{2,p}(B_R)$, $p > 1$, for any $R > 0$. Moreover, $(\hat{V}, \hat{\rho})$ satisfies the HJB equation*

$$\min_{u \in \mathbb{U}} \left[ L^u \hat{V}(x) + c(x, u) \right] = \hat{\rho}, \quad x \in \mathbb{R}^d. \tag{1.17}$$

*Proof.* By Lemma 1.3.1, $\rho_{v_k}$ is decreasing monotonically in $k$ and hence converges to some $\hat{\rho} \geq \rho^*$. By Lemma 1.4.2 (iii), the sequence $V_k$ is weakly compact in $\mathscr{W}^{2,p}(B_R)$, $p > 1$, for any $R > 0$, while by Lemma 1.4.2 (iv), any weakly convergent subsequence has the same limit $\hat{V}$. Also repeating the argument in the proof of Lemma 1.3.1, with

$$h_k(x) := \rho_{v_{k-1}} - \min_{u \in \mathbb{U}} \left[ L^u V_{k-1}(x) + c(x, u) \right], \quad x \in \mathbb{R}^d,$$

we deduce that for any $R > 0$ there exists some constant $K(R)$ such that

$$\int_{B_R} h_k(x)\, dx \leq K(R)\left(\rho_{v_{k-1}} - \rho_{v_k}\right) \quad \forall k \in \mathbb{N}.$$

---

[1] Theorem 4.4 in [11] applies to Markov chains on Borel state spaces. Also the model in [11] involves only inf-compact running costs. Nevertheless, the essential arguments can be followed to adapt the proof to controlled diffusions. We skip the details.

Therefore, $h_k \to 0$ weakly in $\mathcal{L}^1(D)$ as $k \to \infty$ for any bounded domain $D$. Taking limits in the equation

$$\min_{u \in \mathbb{U}} \left[ L^u V_{k-1}(x) + c(x,u) \right] = \rho_{v_{k-1}} - h_k(x)$$

and using [3, Lemma 3.5.4] yields (1.17).                                                          □

It is evident that $v \in \mathfrak{U}_{SM}$ is an equilibrium of the PIA if it satisfies $\rho_v < M^*$ and

$$\min_{u \in \mathbb{U}} \left[ L^u \Psi^v(x) + c(x,u) \right] = \rho_v, \quad x \in \mathbb{R}^d. \tag{1.18}$$

For one-dimensional diffusions, one can show that (1.18) has a unique solution, and hence, this is the optimal solution with $\rho_v = \rho^*$. For higher dimensions, to the best of our knowledge there is no such result. There is also the possibility that the PIA converges to $\hat{v} \in \mathfrak{U}_{SSM}$ which is not an equilibrium. This happens if (1.17) satisfies

$$L^{\hat{v}} \hat{V}(x) + c_{\hat{v}}(x) = \min_{u \in \mathbb{U}} \left[ L^u \hat{V}(x) + c(x,u) \right] = \hat{\rho} > \rho_{\hat{v}}, \quad x \in \mathbb{R}^d. \tag{1.19}$$

This is in fact the case with the example in [4]. In this example the controlled diffusion takes the form $dX_t = U_t \, dt + dW_t$, with $\mathbb{U} = [-1,1]$ and running cost $c(x) = 1 - e^{-|x|}$. If we define

$$\xi_\rho := \log \frac{3}{2} + \log(1-\rho), \quad \rho \in [1/3, 1)$$

and

$$V_\rho(x) := 2 \int_{-\infty}^{x} e^{2|y-\xi_\rho|} dy \int_{-\infty}^{y} e^{-2|z-\xi_\rho|} (\rho - c(z)) \, dz, \quad x \in \mathbb{R},$$

then direct computation shows that

$$\tfrac{1}{2} V_\rho''(x) - |V_\rho'(x)| + c(x) = \rho \qquad \forall \rho \in [1/3, 1) \, ,$$

and so the pair $(V_\rho, \rho)$ satisfies the HJB. The stationary Markov control corresponding to this solution of the HJB is $w_\rho(x) = -\operatorname{sign}(x - \xi_\rho)$. The controlled process under $w_\rho$ has invariant probability density $\varphi_\rho(x) = e^{-2|x-\xi_\rho|}$. A simple computation shows that

$$\int_{-\infty}^{\infty} c(x) \varphi_\rho(x) \, dx = \rho - \tfrac{9}{8}(1-\rho)(3\rho - 1) < \rho, \quad \forall \rho \in (1/3, 1) \, .$$

Thus, if $\rho > 1/3$, then $V_\rho$ is not a canonical solution of the Poisson equation corresponding to the stable control $w_\rho$. Therefore, this example satisfies (1.19) and shows that in general we cannot preclude the possibility that the limiting value of the PIA is not an equilibrium of the algorithm.

In [11, Theorem 5.2], a blanket Lyapunov condition is imposed to guarantee convergence of the PIA to an optimal control. Instead, we use Lyapunov analysis to characterize the domain of attraction of the optimal value.

We need the following definition.

**Definition 1.4.2.** Let $v^*$ be an optimal control as characterized in Theorem 1.3.1. Let $\mathfrak{V}$ denote the class of all nonnegative functions $\mathscr{V} \in \mathscr{C}^2(\mathbb{R}^d)$ satisfying $L^{v^*}\mathscr{V} \le k_0 - h(x)$ for some nonnegative, inf-compact $h \in \mathscr{C}(\mathbb{R}^d)$ and a constant $k_0$. We denote by $\mathfrak{o}(\mathfrak{V})$ the class of inf-compact functions $g$ satisfying $g \in \mathfrak{o}(\mathscr{V})$ for some $\mathscr{V} \in \mathfrak{V}$.

The theorem below asserts that if the PIA is initialized at a $v_0 \in \mathfrak{U}_{\mathrm{SSM}}$ whose associated canonical solution to the Poisson equation lies in $\mathfrak{o}(\mathfrak{V})$ then it converges to an optimal $v^* \in \mathfrak{U}_{\mathrm{SSM}}$.

**Theorem 1.4.2.** *If $v_0 \in \mathfrak{U}_{\mathrm{SSM}}$ satisfies $\Psi^{v_0} \in \mathfrak{o}(\mathfrak{V})$, then $\rho_{v_k} \to \rho^*$ as $k \to \infty$.*

*Proof.* The proof is straightforward. By Lemma 1.4.2 (iv), $\hat{V} \in \mathfrak{o}(\mathfrak{V})$. Also by (1.17), we have

$$L^{v^*}\hat{V}(x) + c_{v^*}(x) \ge \hat{\rho}, \quad x \in \mathbb{R}^d,$$

and applying Dynkin's formula, we obtain

$$\frac{1}{t}\left(\mathbb{E}_x^{v^*}\left[\hat{V}(X_t)\right] - V(x)\right) + \frac{1}{t}\mathbb{E}_x^{v^*}\left[\int_0^t c_{v^*}(X_s)\,\mathrm{d}s\right] \ge \hat{\rho}, \tag{1.20}$$

Since $\hat{V} \in \mathfrak{o}(\mathfrak{V})$, by [1, Lemma 7.1] we have

$$\frac{1}{t}\mathbb{E}_x^{v^*}\left[\hat{V}(X_t)\right] \xrightarrow[t\to\infty]{} 0$$

and thus taking limits as $t \to \infty$ in (1.20), we obtain $\rho^* \ge \hat{\rho}$. Therefore, we must have $\hat{\rho} = \rho^*$.  □


## 1.5  Concluding Remarks

We have concentrated on the model of controlled diffusions with near-monotone running costs. The case of stable controls with a blanket Lyapunov condition is much simpler. If, for example, we impose the assumption that there exist a constant $k_0 > 0$ and a pair of nonnegative, inf-compact functions $(\mathscr{V}, h) \in \mathscr{C}^2(\mathbb{R}^d) \times \mathscr{C}(\mathbb{R}^d)$ satisfying $1 + c \in \mathfrak{o}(h)$ and such that

$$L^u\mathscr{V}(x) \le k_0 - h(x, u) \qquad \forall (x, u) \in \mathbb{R}^d \times \mathbb{U},$$

then the PIA always converges to the optimal solution.

**Acknowledgement** This work was supported in part by the Office of Naval Research through the Electric Ship Research and Development Consortium.

# References

1. Arapostathis, A., Borkar, V.S.: Uniform recurrence properties of controlled diffusions and applications to optimal control. SIAM J. Control Optim. **48**(7), 152–160 (2010)
2. Arapostathis, A., Borkar, V.S., Fernández-Gaucherand, E., Ghosh, M.K., Marcus, S.I.: Discrete-time controlled Markov processes with average cost criterion: a survey. SIAM J. Control Optim. **31**(2), 282–344 (1993)
3. Arapostathis, A., Borkar, V.S., Ghosh, M.K.: Ergodic control of diffusion processes, *Encyclopedia of Mathematics and its Applications*, vol. 143. Cambridge University Press, Cambridge (2011)
4. Bensoussan, A., Borkar, V.: Ergodic control problem for one-dimensional diffusions with near-monotone cost. Systems Control Lett. **5**(2), 127–133 (1984)
5. Bogachev, V.I., Krylov, N.V., Röckner, M.: On regularity of transition probabilities and invariant measures of singular diffusions under minimal conditions. Comm. Partial Differential Equations **26**(11–12), 2037–2080 (2001)
6. Costa, O.L.V., Dufour, F.: The policy iteration algorithm for average continuous control of piecewise deterministic Markov processes. Appl. Math. Optim. **62**(2), 185–204 (2010)
7. Gilbarg, D., Trudinger, N.S.: Elliptic partial differential equations of second order, *Grundlehren der Mathematischen Wissenschaften*, vol. 224, second edn. Springer-Verlag, Berlin (1983)
8. Gyöngy, I., Krylov, N.: Existence of strong solutions for Itô's stochastic equations via approximations. Probab. Theory Related Fields **105**(2), 143–158 (1996)
9. Hernández-Lerma, O., Lasserre, J.B.: Policy iteration for average cost Markov control processes on Borel spaces. Acta Appl. Math. **47**(2), 125–154 (1997)
10. Krylov, N.V.: Controlled diffusion processes, *Applications of Mathematics*, vol. 14. Springer-Verlag, New York (1980)
11. Meyn, S.P.: The policy iteration algorithm for average reward Markov decision processes with general state space. IEEE Trans. Automat. Control **42**(12), 1663–1680 (1997)
12. Stannat, W.: (Nonsymmetric) Dirichlet operators on $L^1$: existence, uniqueness and associated Markov processes. Ann. Scuola Norm. Sup. Pisa Cl. Sci. (4) **28**(1), 99–140 (1999)

# Chapter 2
# Discrete-Time Inventory Problems with Lead-Time and Order-Time Constraint

**Lankere Benkherouf and Alain Bensoussan**

## 2.1 Introduction

Demand uncertainty of the various items in the supply chain plays an important role in the control of material flow. Moreover, information on the suppliers mode of deliveries is key in selecting appropriate policies for inventory management. In some situations, orders for products cannot be placed while waiting for a delivery of previous orders. Thus placing an order-time constraint on deliveries. This applies where production capacity at the supplier side is limited. This situation may also arise in certain organizations where transportation capacity is limited: see Bensoussan, Çakanyidirim, and Moussaoui [4].

This chapter considers the basic inventory model of Scarf [16] and modifies it by including lead time with order-time constraints of the type mentioned above. It is well known that an $(s, S)$ policy is optimal for Scarf's model for the infinite horizon stationary model: see Igelhart [10], Veinott [18], Beyer and Sethi [6], and Benkherouf [3]. Also, $(s, S)$ policies remain optimal in the presence of lead time: see [18]. This chapter shows that $(s, S)$ policies are still optimal with the additional constraints on order time. A result, which seems to follow straightforwardly from existing results in the literature, turned out to be delicate and needing some care.

L. Benkherouf
Department of Statistics and Operations Research, Kuwait University,
Safat 13060, Kuwait
e-mail: lakdereb@kuc01.kuniv.edu.kw

A. Bensoussan (✉)
International Center for Decision and Risk Analysis,
School of Management, University of Texas at Dallas, Richardson, TX 75080, USA

The Hong Kong Polytechnic University, Hung Hom, Kowloon, Hong Kong
Ajou University, San S. Woncheon-Dong, Yeontong-Gu, Suwon 443–749, Korea
e-mail: alain.bensoussan@utdallas.edu

D. Hernández-Hernández and A. Minjárez-Sosa (eds.), *Optimization, Control, and Applications of Stochastic Systems*, Systems & Control: Foundations & Applications,
DOI 10.1007/978-0-8176-8337-5_2, © Springer Science+Business Media, LLC 2012

A related model to this chapter's is discussed in [4] for a discrete-state, continuous-time inventory model with infinite horizon where the demand process is assumed to be generated by some Poisson process.

The main ingredients for the problem under consideration in this chapter are given below:

The demand process is assumed to be composed of a sequence of i.i.d variables,

$$D_1, \ldots, D_n, \ldots,$$

constructed on a probability space $(\Omega, \mathscr{A}, P)$, where $D_n$ represents the demand at time $n$. Let $\mathscr{F}^n = \sigma(D_1, \ldots, D_n)$ be the $\sigma$-algebra generated by the demand process and $\mathscr{F}^0 = (\Omega, \emptyset)$.

An ordering policy $V$ is composed of a sequence of ordering times

$$\tau_1, \ldots, \tau_j, \ldots,$$

with $\tau_j$ being $\mathscr{F}^{\tau_j-1}$ measurable. To each ordering time $\tau_j$, one associates a quantity $v_{\tau_j}$ which represents the amount ordered. The new element here is that the usual condition $\tau_j \leq \tau_{j+1}$ is replaced with the constraint

$$\tau_{j+1} \geq \tau_j + L, \tag{2.1}$$

in which $L$ is the lead time, $L \geq 1$, and integer.

The inventory evolves as follows:

$$y_{n+1} = y_n + v_n - D_n,$$

with $y_1 = x$, $v_n = 0$, if $n \neq \tau_j$, for some $j \geq L$, and $y_n$ is the inventory level at time $n$.

To define the objective function, we introduce the cost in a single period. It is composed of a cost on the inventory

$$l(x) = hx^+ + px^-, \tag{2.2}$$

and an order cost

$$C(v) = K \mathbb{1}_{v>0} + cv, \tag{2.3}$$

where $h$, $p$, $K$, $c$ are strictly positive known constants and $x^+ = \max\{0, x\}$, $x^- = -\min\{0, x\}$, with

$$\mathbb{1}_{v>0} = \begin{cases} 1, & \text{If } v > 0 \\ 0, & \text{otherwise} \end{cases}.$$

Costs are assumed to be additive and discounted geometrically at a known rate $\alpha$, $0 < \alpha < 1$, and that unmet demand is completely backlogged.

The objective function is given by the formula

$$J_x(V) = \sum_{n=1}^{\infty} \alpha^{n-1} (E[l(y_n) + C(v_n)]), \tag{2.4}$$

and the value function is

$$u(x) = \inf_{V} J_x(V),\tag{2.5}$$

where $E$ is the usual expectation operator.

Inventory models with deterministic lead time can be found in [18], Archibald [1], Sahin [13, 14], and Fegergruen and Schechner [7]. These papers, but [18], are concerned with the computations of system performance measures for $(s, S)$ policies. The closest to the present work is [1], where at most one outstanding order is permitted. This is equivalent to constraint (2.1). However, no proof of the optimality of $(s, S)$ policies was provided. In fact the notion of policies with at most one outstanding order is not new. It can be found in Hadley and Within [9], where it was discussed in the context of lost sales models: see also Kim and Park [11]. This chapter, unlike those that deals with order-time constraint, but [4], provides a proof of the optimality of $(s, S)$ policies.

In the next section, we shall derive the Bellman equation associated to $u(x)$. Section 2.3 is concerned with ground preparation for showing the derivation of $(s, S)$ policies. Section 2.4 deals with existence and derivation of the pair $(s, S)$. The optimality of $(s, S)$ policies is shown, under some technical conditions, in Sect. 2.5. Section 2.6 treats the special case where the demand in each period is exponentially distributed. Further, this section contains some general remarks and a conclusion.

## 2.2  Dynamic Programming

It is easy to figure out the Bellman equation, considering the possibilities at the initial time. If there is no order at time 0, then (by (2.5) and (2.4)) the best which can be achieved over an infinite planning horizon is

$$l(x) + \alpha E u(x - D).$$

On the other hand if an order of size $v$ is made at time 0, which is possible since there is no pending order, then the inventory evolves as follows:

$$
\begin{aligned}
y_1 &= x,\\
y_2 &= x - D_1,\\
&\cdots\\
y_{L+1} &= x - (D_1 + \cdots + D_L) + v,
\end{aligned}
$$

and the best which can be achieved is

$$K + cv + l(x) + \sum_{j=2}^{L} \alpha^{j-1} E l(x - (D_1 + \cdots + D_{j-1}))$$

$$+ \alpha^L E u(x + v - (D_1 + \cdots + D_L)).$$

From these considerations, we can easily derive Bellman equation

$$u(x) = \min \left\{ l(x) + \alpha E u(x - D), \right.$$

$$\left. K - cx + \sum_{j=0}^{L-1} \alpha^j E l(x - D^{(j)}) + \inf_{\eta > x} [c\eta + \alpha^L E u(\eta - D^{(L)})] \right\}, \qquad (2.6)$$

in which we use the notation

$$D^{(j)} = \begin{cases} D_1 + \cdots + D_j, & \text{if } j \geq 1 \\ 0, & \text{if } j = 0 \end{cases}.$$

In the next section, we shall recast the function $u$ given in (2.6) in a form which we shall find useful later on in our analysis.

## 2.3  (s,S) Policy

We first transform equation (2.6), by introducing a constant $s \in R$, and changing $u(x)$ into $H_s(x)$, using the formula

$$u(x) = -cx + \sum_{j=0}^{L-1} \alpha^j E l(x - D^{(j)}) + C_s + H_s(x), \qquad (2.7)$$

where $C_s$ will be defined below. In fact, let

$$g(x) = (1 - \alpha)cx + \alpha^L E l(x - D^{(L)}), \qquad (2.8)$$

then we take

$$C_s = \frac{g(s) + \alpha c \bar{D}}{1 - \alpha},$$

where $\bar{D}$ is the expected value of $D$, with $0 < \bar{D} < \infty$.
We note the formula

$$\sum_{j=0}^{L-1} \alpha^j E g(x - D^{(j)})$$

$$= cx(1 - \alpha^L) - \alpha c \bar{D} \frac{1 - \alpha^L}{1 - \alpha} + \alpha^L c \bar{D} L + \sum_{j=0}^{L-1} \alpha^{j+L} E l(x - D^{(j+L)}).$$

We then check that $H_s(x)$ is the solution of

$$H_s(x) = \min \left\{ g(x) - g(s) + \alpha E H_s(x - D), \right.$$

$$\left. K + \inf_{\eta > x} \left[ \sum_{j=0}^{L-1} \alpha^j E(g(\eta - D^{(j)}) - g(s)) + \alpha^L E H_s(\eta - D^{(L)}) \right] \right\}. \quad (2.9)$$

On the other hand, for any $s$ we define

$$H_s(x) = \begin{cases} g(x) - g(s) + \alpha E H_s(x - D), & x \geq s \\ 0, & x < s \end{cases}. \quad (2.10)$$

A solution $H_s$ satisfying (2.10) exists, is unique, and is continuous in $\mathbb{R}$.

We have used the same notation $H_s(x)$ between (2.9) and (2.10), because we want to find $s$ so that the solution of (2.10) is also a solution of (2.9).

Let us set

$$g_s(x) = (g(x) - g(s)) \mathbb{1}_{x \geq s},$$

then the solution of (2.10) satisfies, for all $x$,

$$H_s(x) = g_s(x) + \alpha E H_s(x - D). \quad (2.11)$$

From this relation we deduce, after iterating that

$$H_s(x) = \sum_{j=0}^{L-1} \alpha^j E g_s(x - D^{(j)}) + \alpha^L E H_s(x - D^{(L)}),$$

therefore also

$$\sum_{j=0}^{L-1} \alpha^j E(g(x - D^{(j)}) - g(s)) + \alpha^L E H_s(x - D^{(L)})$$

$$= H_s(x) + \sum_{j=0}^{L-1} \alpha^j E(g(x - D^{(j)}) - g(s)) \mathbb{1}_{x - D^{(j)} < s}.$$

Define

$$\Psi_s(x) = H_s(x) + \sum_{j=0}^{L-1} \alpha^j E(g(x - D^{(j)}) - g(s)) \mathbb{1}_{x - D^{(j)} < s}, \quad (2.12)$$

then if we want $H_s(x)$ to satisfy (2.9), we must have

$$H_s(x) = \min\left\{ g(x) - g(s) + \alpha E H_s(x - D), K + \inf_{\eta > x} \Psi_s(\eta) \right\}. \tag{2.13}$$

This relation motivates the choice of $s$. We want to find $s$ so that

$$K + \inf_{\eta > s} \Psi_s(\eta) = 0. \tag{2.14}$$

If such an $s$ exists, we define $S = S(s)$ as the point where the infimum is attained.

$$\inf_{\eta > s} \Psi_s(\eta) = \Psi_s(S(s)). \tag{2.15}$$

Note, from (2.13), that $\Psi_s(x)$ coincides with $H_s(x)$, for $x > s$, when $L = 1$.

In the next section we shall show the existence of the pair $(s, S)$ satisfying (2.14) and (2.15).

## 2.4   Existence and Characterization of the Pair $(s, S)$

Let $F^{(L)}(x)$ and $f^{(L)}(x)$ be the cumulative distribution and the density function of $D^{(L)}$ respectively and $\bar{F}^{(L)}(x) = 1 - F^{(L)}(x)$, with $F^{(0)}(x) = 1$. Then (2.8) and (2.2) give

$$g(x) = (1 - \alpha)cx + \alpha^L h \int_0^x (x - \xi) f^{(L)}(\xi) d\xi + \alpha^L p \int_x^\infty (\xi - x) f^{(L)}(\xi) d\xi.$$

It follows that $\mu(x) = g'(x)$ is given by

$$\mu(x) = (1 - \alpha)c + \alpha^L h - \alpha^L (h + p) \bar{F}^{(L)}(x). \tag{2.16}$$

We assume that

$$(1 - \alpha)c - \alpha^L p < 0. \tag{2.17}$$

There exists a unique $\bar{s} > 0$ such that

$$\begin{aligned} \mu(x) \leq 0, & \quad \text{if } x \leq \bar{s} \\ \mu(x) \geq 0, & \quad \text{if } x \geq \bar{s} \end{aligned} \tag{2.18}$$

Note that if $L = 0$, (2.17) reduces to the classical condition of optimality of $(s, S)$ policies with no lead time.

**Lemma 2.4.1.** *The solution $H_s$ of (2.10) satisfies for $s \leq x$,*

$$H_s(x) + \sum_{j=0}^{L-1} \alpha^j \int_s^x \bar{F}^{(j)}(x-\xi)\mu(\xi)d\xi$$

$$= \frac{1-\alpha^L}{1-\alpha}(g(x) - g(s)) + \sum_{j=L}^{\infty} \alpha^j \int_s^x F^{(j)}(x-\xi)\mu(\xi)d\xi.$$

*Furthermore, the equation is still valid for $s > x$ by dropping the integral term on the right-hand side.*

*Proof.* The proof follows immediately by recalling the definition of $\mu$ in (2.16) and checking that $H_s$ given by

$$H_s(x) = \sum_{j=0}^{\infty} \alpha^j \int_s^x F^{(j)}(x-\xi)\mu(\xi)d\xi, \tag{2.19}$$

solves (2.10).                                                                                                      □

**Lemma 2.4.2.** *The function $\Psi_s$ defined in (2.12) is given by*

$$\Psi_s(x) = \frac{1-\alpha^L}{1-\alpha}(g(x) - g(s)) + \sum_{j=L}^{\infty} \alpha^j \int_s^x F^{(j)}(x-\xi)\mu(\xi)d\xi$$

$$-((1-\alpha)c + \alpha^L h)\bar{D} \sum_{j=1}^{L-1} j\alpha^j$$

$$+\alpha^L(h+p) \sum_{j=1}^{L-1} \alpha^j \int_0^{+\infty} F^{(j)}(\zeta)\bar{F}^{(L)}(x-\zeta)d\zeta. \tag{2.20}$$

*Proof.* Note that

$$E(g(x - D^{(j)}) - g(s))\mathbb{1}_{x-D^{(j)}<s}$$

$$= -E\mathbb{1}_{D^{(j)}>x-s}\left(\int_{x-D^{(j)}}^s \mu(\xi)d\xi\right)$$

$$= -\int_{x-s}^{+\infty}\left(\int_{x-\zeta}^s \mu(\xi)d\xi\right)dF^{(j)}(\zeta)$$

$$= -\int_{x-s}^{+\infty} \bar{F}^{(j)}(\zeta)\mu(x-\zeta)d\zeta$$

$$= -\int_{-\infty}^s \bar{F}^{(j)}(x-\xi)\mu(\xi)d\xi$$

$$= -\int_{-\infty}^x \bar{F}^{(j)}(x-\xi)\mu(\xi)d\xi + \int_s^x \bar{F}^{(j)}(x-\xi)\mu(\xi)d\xi.$$

The Lemma is now immediate from further direct computations on $\int_{-\infty}^{x} \bar{F}^{(j)}(x - \xi)\mu(\xi)d\xi$, (2.16), Lemma 2.4.1, and the definition of $\Psi_s$. This finishes the proof. $\square$

It is easy to deduce from Lemma 2.4.2 that $\Psi_s(x)$ is bounded below and tends to $+\infty$ as $x \to +\infty$. Therefore the infimum over $x \geq s$ is attained and we can define $S(s)$ as the smallest infimum.

**Proposition 2.4.1.** *We assume that (2.17) holds, then one has*

1.

$$\Psi_s(S(s)) \to -((1-\alpha)c + \alpha^L h)\bar{D} \sum_{j=1}^{L-1} j\alpha^j, \ as \ s \to +\infty. \tag{2.21}$$

2.

$$\max_s \Psi_s(S(s)) = \Psi_{\bar{s}}(S(\bar{s})) \geq 0, \tag{2.22}$$

3.

$$\Psi_s(S(s)) \to -\infty, \ as \ s \to -\infty, \tag{2.23}$$

4. *There exists one and only one solution of (2.14) such that $s \leq \bar{s}$.*

*Proof.* We compute, by Lemma 2.4.2,

$$\Psi_s'(x) = \frac{1-\alpha^L}{1-\alpha}\mu(x) + \sum_{j=L}^{\infty} \alpha^j \int_s^x f^{(j)}(x-\xi)\mu(\xi)d\xi$$

$$- \alpha^L(h+p) \sum_{j=1}^{L-1} \alpha^j \int_0^x \bar{F}^{(j)}(\zeta)f^{(L)}(x-\zeta)d\zeta, \tag{2.24}$$

which can also be written as

$$\Psi_s'(x) = \gamma(x) + \sum_{j=L}^{\infty} \alpha^j \int_s^x f^{(j)}(x-\xi)\mu(\xi)d\xi, \tag{2.25}$$

with

$$\gamma(x) = \frac{1-\alpha^L}{1-\alpha}((1-\alpha)c - \alpha^L p) + \alpha^L(h+p) \sum_{j=0}^{L-1} \alpha^j \int_0^x F^{(j)}(x-\xi)f^{(L)}(\xi)d\xi. \tag{2.26}$$

The function $\gamma$ is increasing in $x$ and

$$\gamma(x) = \frac{1-\alpha^L}{1-\alpha}((1-\alpha)c - \alpha^L p) < 0, \ \text{for} \ x \leq 0,$$

$$\gamma(\infty) = \frac{1-\alpha^L}{1-\alpha}((1-\alpha)c + \alpha^L h),$$

therefore there exists a unique $s^*$ such that

$$\gamma(x) < 0, \quad \text{for } x < s^*,$$
$$\gamma(x) > 0, \quad \text{for } x > s^*,$$
$$\gamma(s^*) = 0, \, s^* > 0.$$

Note that

$$\frac{1 - \alpha^L}{1 - \alpha} \mu(x) - \gamma(x) = \alpha^L (h + p) \sum_{j=0}^{L-1} \alpha^j \int_0^x \bar{F}^{(j)}(x - \xi) f^{(L)}(\xi) d\xi. \qquad (2.27)$$

The quantity on the right-hand side of the above equality vanishes for $x \leq 0$ or $L = 1$.
Otherwise

$$\frac{1 - \alpha^L}{1 - \alpha} \mu(x) - \gamma(x) > 0, \quad \text{for } x > 0, \, L \geq 2.$$

Since $\gamma(s^*) = 0$, we have using (2.18), $\mu(s^*) > 0$, hence $0 < \bar{s} < s^*$.

Next, if $s \geq s^*$, $\Psi_s'(x) \geq 0$; for $x \geq s$ by (2.25); hence $S(s) = s$. Therefore, we get by (2.20) that

$$\Psi_s(S(s)) = -((1 - \alpha)c + \alpha^L h)\bar{D} \sum_{j=1}^{L-1} j\alpha^j$$

$$+ \alpha^L (h + p) \sum_{j=1}^{L-1} \alpha^j \int_0^{+\infty} \bar{F}^{(j)}(\zeta) \bar{F}^{(L)}(s - \zeta) d\zeta, \qquad (2.28)$$

this function decreases in $s$ and converges to $-((1 - \alpha)c + \alpha^L h)\bar{D} \sum_{j=1}^{L-1} j\alpha^j$ as $s \to +\infty$. This proves part (1).

Consider now $\bar{s} < s < s^*$. We note that

$$\Psi_s'(s) = \gamma(s) < 0,$$
$$\Psi_s'(s^*) = \sum_{j=L}^{\infty} \alpha^j \int_s^{s^*} f^{(j)}(s^* - \xi) \mu(\xi) d\xi > 0,$$

hence in this case

$$\bar{s} < s < S(s) < s^*.$$

If $s < \bar{s}$, then we can claim that

$$s < \bar{s} < S(s).$$

Indeed, for $s < x < \bar{s}$, we can see from formula (2.25) and (2.18) that $\Psi_s'(x) < 0$.
However, we cannot compare $S(s)$ with $s^*$ in this case.

We then study the behavior of $\Psi_s(S(s))$. We have already seen that, for $s > s^*$ the function $\Psi_s(S(s))$ is decreasing to the negative constant $-((1-\alpha)c + \alpha^L h)\bar{D}\sum_{j=1}^{L-1} j\alpha^j$. In this case, it follows from (2.24) that

$$\frac{\mathrm{d}}{\mathrm{d}s}\Psi_s(S(s)) = -\alpha^L(h+p)\sum_{j=1}^{L-1}\alpha^j\int_0^s \bar{F}^{(j)}(\zeta)f^{(L)}(s-\zeta)\mathrm{d}\zeta, \quad s > s^*. \qquad (2.29)$$

Note that $s > 0$. If $s < s^*$, then $s < S(s)$; therefore

$$\frac{\mathrm{d}}{\mathrm{d}s}\Psi_s(S(s)) = \frac{\partial \Psi_s}{\partial s}(S(s)),$$

hence

$$\frac{\mathrm{d}}{\mathrm{d}s}\Psi_s(S(s)) = -\mu(s)\left[\frac{1-\alpha^L}{1-\alpha} + \sum_{j=L}^{\infty}\alpha^j F^{(j)}(S(s)-s)\right]. \qquad (2.30)$$

Note that at $s = s^*$ the two formulas in (2.29) and (2.30) coincide, since

$$-\frac{1-\alpha^L}{1-\alpha}\mu(s^*) = -\alpha^L(h+p)\sum_{j=1}^{L-1}\alpha^j\int_0^{s^*}\bar{F}^{(j)}(\zeta)f^{(L)}(s^*-\zeta)\mathrm{d}\zeta,$$

which can be easily checked by remembering the definition of $s^*(\gamma(s^*)=0)$, and (2.27). It follows clearly that $\Psi_s(S(s))$ decreases on $(\bar{s},s^*)$ and increases on $(-\infty,\bar{s})$. Finally $\Psi_s(S(s))$ decreases on $(\bar{s},+\infty)$ and increases on $(-\infty,\bar{s})$. So it attains its maximum at $\bar{s}$. From the formula for $\Psi_{\bar{s}}(x); H_{\bar{s}}(x)$; for $x > \bar{s}$ and the fact that $\bar{s}$ is the minimum of $g$, we get immediately that $\Psi_{\bar{s}}(S(\bar{s})) \geq 0$. This shows part (2) of the proposition.

Finally, for $s < 0$ we have

$$\Psi_s(S(s)) \leq \Psi_s(0)$$

$$= \frac{1-\alpha^L}{1-\alpha}(g(0)-g(s)) + \sum_{j=L}^{\infty}\alpha^j\int_s^0 F^{(j)}(-\xi)\mathrm{d}\xi$$

$$- ((1-\alpha)c - \alpha^L h)\bar{D}\sum_{j=1}^{L-1}j\alpha^j,$$

which tends to be $-\infty$ as $s \to -\infty$. This proves (3). Therefore $\Psi_s(S(s))$ is increasing from $-\infty$ to a positive number as $s$ grows from $-\infty$ to $\bar{s}$. Therefore there is one and only one $s < \bar{s}$ satisfying (2.14). The proof has been completed.                    □

## 2.5  Solution as an $(s,S)$ Policy

It remains to see whether the solution $H_s$ of equation (2.10) where $s$ is the solution of (2.14) is indeed a solution of (2.13). It is useful to use, instead of $\Psi_s(x)$ a different function, namely

$$\Phi_s(x) = H_s(x) + \sum_{j=1}^{L-1} \alpha^j E(g(x - D^{(j)}) - g(s)) \mathbb{1}_{x - D^{(j)} < s}, \qquad (2.31)$$

which differs from $\Psi_s(x)$ by simply deleting the term corresponding to $j = 0$.
  Clearly

$$\Phi_s(x) = \Psi_s(x), \quad \text{for } x \geq s.$$

However, when $x < s$,

$$\Psi_s(x) = \Phi_s(x) + g(x) - g(s),$$

Note that in finding $S$ it is indifferent to work with one or the other function. Note also that, when $L = 1$, $\Phi_s(x) = H_s(x)$, for all $x$. Now we claim that

$$\inf_{\eta > x} \Psi_s(\eta) = \inf_{\eta > x} \Phi_s(\eta).$$

This equality is obvious when $x > s$, since the functions are identical. If $x < s$ we have

$$\inf_{\eta > x} \Psi_s(\eta) = \inf_{\eta > x} \Phi_s(\eta) = \inf_{\eta > s} \Phi_s(\eta) = -K.$$

Indeed,

$$\inf_{\eta > x} \Psi_s(\eta) = \min \left[ \inf_{\eta > s} \Psi_s(\eta), \inf_{x < \eta < s} \Psi_s(\eta) \right],$$

and $\inf_{\eta > s} \Psi_s(\eta) = -K$, whereas (2.10) and (2.13) give

$$\inf_{x < \eta < s} \Psi_s(\eta) = \inf_{x < \eta < s} \sum_{j=0}^{L-1} \alpha^j E(g(\eta - D^{(j)}) - g(s)) \mathbb{1}_{x - D^{(j)} < s} > 0,$$

hence the claim is true. Therefore (2.13) becomes

$$H_s(x) = \min \left\{ g(x) - g(s) + \alpha E H_s(x - D), K + \inf_{\eta > x} \Phi_s(\eta) \right\}. \qquad (2.32)$$

For $x < s$, this relation reduces to

$$0 = \min[g(x) - g(s), 0],$$

which is true since $g(x)$ is decreasing for $x < s$. We then consider $x > s$. Since $H_s(x)$ is equal to the first term of the bracket. Therefore, what we have to prove is

$$H_s(x) \leq K + \inf_{\eta > x} \Phi_s(\eta), \quad \text{for } x > s. \tag{2.33}$$

In fact, we have not been able to prove (2.33) for all values of $L$. We know it is true for $L = 1$. We will prove it afterwards for exponential demands. For general demand distributions we have:

**Proposition 2.5.2.** *We assume*

$$\alpha^L((1-\alpha)c - \alpha^L p)(L-1)\bar{D} + (1-\alpha)K \geq 0, \tag{2.34}$$

*then property (2.33) is satisfied.*

*Proof.* We note that this result includes the case $L = 1$ in which the condition is automatically satisfied. The proof is similar to that of the case $L = 1$.

We recall from (2.10) that

$$H_s(x) - \alpha E H_s(x - D) = g_s(x), \quad \text{for all } x. \tag{2.35}$$

We then find a similar equation for $\Phi_s(x)$. This is where it is important to consider $\Phi_s(x)$ and not $\Psi_s(x)$, since we write the equation for any $x$ and not just for $x > s$. It is easy to verify, using (2.31), that $\Phi_s(x)$ is the solution of

$$\begin{aligned}
\Phi_s(x) &- \alpha E \Phi_s(x - D) \\
&= g_s(x) + \alpha E(g(x-D) - g(s))\mathbb{1}_{x-D<s} \\
&\quad - \alpha^L E(g(x-D^{(L)}) - g(s))\mathbb{1}_{x-D^{(L)}<s},
\end{aligned} \tag{2.36}$$

and again we check that this equation coincides with (2.35) when $L = 1$. Going back to (2.33) we recall that

$$K + \inf_{\eta > x} \Phi_s(\eta) \geq K + \inf_{\eta > s} \Phi_s(\eta) = 0, \quad \text{for all } x > s. \tag{2.37}$$

So it is sufficient to prove (2.33) for $x > x_0 > s$, where $x_0$ is the first value $x$ such that $H_s(x) \geq 0$. We necessarily have $H_s(x_0) = 0$. We have $s < \bar{s} < x_0$. Let us fix $\xi > 0$ and consider the domain $x \geq x_0 - \xi$. We can write, using (2.35), that for all $x$

$$H_s(x) - \alpha E H_s(x - D)\mathbb{1}_{x-D \geq x_0-\xi} \leq g_s(x), \tag{2.38}$$

using the fact that

$$E H_s(x - D)\mathbb{1}_{x-D<x_0-\xi} \leq 0.$$

Define next

$$M_s(x) = \Phi_s(x + \xi) + K.$$

We note that $M_s(x) > 0$ for all $x$. We then state, by (2.37), that

$$
\begin{aligned}
M_s(x) &- \alpha E M_s(x - D) \\
&= g_s(x + \xi) + \alpha E(g(x + \xi - D) - g(s))\mathbb{1}_{x+\xi-D<s} \\
&\quad - \alpha^L E(g(x + \xi - D^{(L)}) - g(s))\mathbb{1}_{x+\xi-D^{(L)}<s} + (1 - \alpha)K,
\end{aligned}
$$

and using the positivity of $M$ we can assert that

$$
\begin{aligned}
M_s(x) &- \alpha E M_s(x - D)\mathbb{1}_{x-D \geq x_0 - \xi} \\
&\geq g_s(x + \xi) + \alpha E(g(x + \xi - D) - g(s))\mathbb{1}_{x+\xi-D<s} \\
&\quad - \alpha^L E(g(x + \xi - D^{(L)}) - g(s))\mathbb{1}_{x+\xi-D^{(L)}<s} + (1 - \alpha)K. \qquad (2.39)
\end{aligned}
$$

We now consider the difference $Y_s(x) = H_s(x) - M_s(x)$, in the domain $x \geq x_0 - \xi$. We have by (2.38) and (2.39) that

$$
\begin{aligned}
Y_s(x) &- \alpha E Y_s(x - D)\mathbb{1}_{x-D \geq x_0 - \xi} \\
&\leq g_s(x) - g_s(x + \xi) - \alpha E(g(x + \xi - D) - g(s))\mathbb{1}_{x+\xi-D<s} \\
&\quad + \alpha^L E(g(x + \xi - D^{(L)}) - g(s))\mathbb{1}_{x+\xi-D^{(L)}<s} - (1 - \alpha)K. \qquad (2.40)
\end{aligned}
$$

We first check easily, that

$$g_s(x + \xi) - g_s(x) \geq 0, \quad \text{for all } x \geq x_0 - \xi. \qquad (2.41)$$

Consider next the function

$$\chi_s(y) = \alpha E(g(y - D) - g(s))\mathbb{1}_{y-D<s} - \alpha^L E(g(y - D^{(L)}) - g(s))\mathbb{1}_{y-D^{(L)}<s},$$

for $y \geq x_0$. We check that

$$\chi_s(y) = \int_{-\infty}^{s} \left( \int_{y-\zeta}^{+\infty} \left(-\alpha f(\eta) + \alpha^L f^{(L)}(\eta)\right) d\eta \right) \mu(\zeta)d\zeta,$$

so in fact

$$\chi_s(y) = \int_{-\infty}^{s} (-\alpha \bar{F}(y - \zeta) + \alpha^L \bar{F}^{(L)}(y - \zeta))\mu(\zeta)d\zeta. \qquad (2.42)$$

Note that in the integral, $\mu(\zeta) < 0$, since $s < \bar{s}$. We deduce that

$$\chi_s(y) \geq \alpha \int_{-\infty}^{s} (\bar{F}^{(L)}(y - \zeta) - \bar{F}(y - \zeta))\mu(\zeta)d\zeta.$$

Further, note that $\bar{F}^{(L)}(y - \zeta) - \bar{F}(y - \zeta) \geq 0$, and $\mu(\zeta) \geq ((1 - \alpha)c - \alpha^L p)$. Therefore

$$\chi_s(y) \geq \alpha^L((1 - \alpha)c - \alpha^L p) \int_{-\infty}^{s} (\bar{F}^{(L)}(y - \zeta) - \bar{F}(y - \zeta))d\zeta.$$

It follows that for $y \geq s$,

$$\begin{aligned}
\chi_s(y) &\geq \alpha^L((1 - \alpha)c - \alpha^L p) \int_{-\infty}^{y} (\bar{F}^{(L)}(y - \zeta) - \bar{F}(y - \zeta))d\zeta \\
&= \alpha^L((1 - \alpha)c - \alpha^L p) \int_{0}^{\infty} (\bar{F}^{(L)}(u) - \bar{F}(u))du \\
&= \alpha^L((1 - \alpha)c - \alpha^L p)(L - 1)\bar{D}.
\end{aligned}$$

Thanks to assumption (2.34) we can assert from (2.40) and (2.42) that

$$Y_s(x) - \alpha E Y_s(x - D)\mathbb{1}_{x-D\geq x_0-\xi} \leq 0, \quad \text{for all } x \geq x_0 - \xi.$$

Also

$$Y_s(x_0 - \xi) \leq -\Phi_s(x_0) - K \leq -K,$$

since $\Phi_s(x_0) \geq H_s(x_0) = 0$. It follows that

$$Y_s(x) \leq 0, \quad \text{for all } x \geq x_0 - \xi,$$

which is the desired result.                                                                    $\square$

## 2.6   The Exponential Case

In this section, we shall examine the special case when the demand $D$ is exponentially distributed. That is,

$$f(x) = \begin{cases} \beta \exp(-\beta x), & \text{if } x \geq 0, \\ 0, & \text{otherwise} \end{cases}, \tag{2.43}$$

for some $\beta > 0$. The next proposition shows that property (2.33) automatically holds in this case.

**Proposition 2.6.3.** *We assume that the demand is distributed according to an exponential distribution, then (2.33) holds.*

Before we proceed to the proof of Proposition 2.6.3, the following result is needed.

**Lemma 2.6.3.** *For $s \leq \bar{s}$, the solution $H_s$ of (2.10) satisfies the following properties:*

1. *Is constant on $(-\infty, s]$,*
2. *Strictly decreasing on $(s, \bar{s}]$,*
3. *$H_s(x) \to \infty$, as $x \to \infty$*

*Proof.* It clear that (1) follows from (2.10). Property (2) can be easily deduced from (2.18) and (2.19).
We claim that

$$H_s(x) \geq \frac{1}{1 - \alpha}(g(\bar{s}) - g(s)). \qquad (2.44)$$

Assume otherwise and define

$$A_s = \{x \in \mathbb{R}, \ s \leq x \leq \bar{s}\}.$$

Let

$$x^* = \min\left\{x \in A_s : H_s(x^*) = \frac{1}{1 - \alpha}(g(\bar{s}) - g(s))\right\}.$$

If follows from (2.10), the definitions of $g$, $x^*$ and (2) that

$$H_s(x^*) = g(x^*) - g(s) + \alpha E H_s(x^* - D) > (g(\bar{s}) - g(s)) + \alpha H_s(x^*).$$

Hence,

$$H_s(x^*) > \frac{1}{1 - \alpha}(g(\bar{s}) - g(s)).$$

This is in contradiction with the definition of $x^*$. Therefore (2.44) is true. This leads by (2.10) to

$$H_s(x) \geq g(x) - g(s)) + \alpha \frac{1}{1 - \alpha}(g(\bar{s}) - g(s)).$$

It is then immediate to see that (3) holds. This finishes the proof.  □

*Proof (Proposition 2.6.3).* We are going to show that

$$H_s'(x) \geq 0, \quad \text{if } x \geq x_0, \qquad (2.45)$$

then (2.33) will follow immediately, since for $x \geq x_0$

$$H_s(x) \leq H_s(x + \xi) \leq \Phi_s(x + \xi) \leq \Phi_s(x + \xi) + K, \quad \text{for } \xi \geq 0.$$

It can be shown using integration by parts that if $f$ is given (2.43), then $H_s$ is the solution of the differential equation

$$(1-\alpha)\beta H_s(x) + H_s'(x) = G(x), \qquad (2.46)$$

where $H_s(s) = 0$ and $G(x) = \beta(g(x) - g(s)) + \mu(x)$. In fact the function $H_s$ can be written explicitly. However, we content ourselves with (2.46) since this will be sufficient to proceed further in the proof.

We claim that $H_s$ has a unique minimum point.

Indeed, Lemma 2.6.3 implies $H_s$ attains a minimum belonging to the interval $(\bar{s}, \infty)$. Further, $H_s$ has a local minimum on $(s, x_0)$. Moreover, it is easy to show that $H_s$ is twice differentiable on $(s, \infty)$. Assume now that $H_s$ has two minima $x_1$ and $x_2$, with $x_1 < x_2$. It is clear that we can select $x_1$ such that $s < x_1 < x_0$. It follows that there exists an $x^* \in (x_1, x_2)$ such that $x^*$ is a local maximum. Therefore, $H'(x^*) = 0$, and $H''(x^*) \le 0$. Differentiating both sides of (2.46), we get that

$$H_s''(x^*) = G'(x^*) = \beta(\mu(x^*) + \mu'(x^*)).$$

The definition of $\mu$ in (2.16) with (2.18) and the fact $x^* > \bar{s}$ imply that $H_s''(x^*) > 0$. This leads to a contradiction. Therefore, $H_s$ has a unique minimum and this minimum belongs to the interval $(s, x_0)$. Consequently, (2.45) is true. The result has been proven. □

In this chapter a discrete-time continuous-state inventory model, with a fixed lead time of several periods, was considered. Orders for products cannot be placed while waiting for a delivery of previous orders. It was shown that the policy which minimizes the total expected inventory costs over an infinite planning horizon is of an $(s, S)$ type: see Propositions 2.5.2 and 2.6.3.

It seems natural to ask if $(s, S)$ policies are still optimal for continuous-state, continuous-time models for inventory models with constraint (2.1). A possible starting point for such investigation is the work of Bensoussan, Liu, and Sethi [5]: see also Benkherouf and Bensoussan [2]. The answer to this question remains open. Another interesting problem is to see if requirement (2.34) can be weakened. Moreover, since the present work seems to be among the first attempts at examining the optimality of $(s, S)$ policies for inventory models with lead-time and order-time constraints and has only dealt with the basic stationary model of Scarf, then it may be worthwhile to look at possible extensions of the present work to the large body of existing inventory models in the literature. These may include models with Markovian demand as discussed in Sethi and Feng [15], or demand dependent on the environment as found in Song and Zipkin [17], or models with two suppliers as treated in Fox, Metters, and Semple [8].

**Acknowledgments** Alain Bensoussan would like to acknowledge the support of WCU (World Class University) program through the Korea Science and Engineering Foundation funded by the Ministry of Education, Science and Technology (R31-20007).

The authors would also like to thank an anonymous referee for comments on an earlier version of the paper.

# References

1. Archibald, B.C.: Continuous review $(s,S)$ policies with lost sales. Management Science. **27**, 1171–1177 (1981).
2. Benkherouf, L., Bensoussan, A.: On the Optimality of $(s,S)$ inventory policies: A Quasi-Variational Inequalities Approach. SIAM Journal on Control and Optimization. **48**, 756–762 (2009).
3. Benkherouf, L.: On the optimality of $(s,S)$ policies: A Quasi-variational Approach. Journal of Applied Mathematics and Stochastic Analysis. doi 158193, 1–9 (2008).
4. Bensoussan, A., Çakanyidirim, M., Moussaoui, L.: Optimality of $(s,S)$ policies for continuous-time inventory problems with order constraint. Annals of Operations Research. (In Press).
5. Bensoussan, A., Liu, R.H, Sethi, S.P.: Optimality of an $(s,S)$ policy with compound Poisson and diffusion demand: A QVI approach. SIAM Journal on Control and Optimization. **44**, 1650–1676 (2006).
6. Beyer, D., Sethi, S.P.: The classical average-cost inventory models of Iglehart and Veinott-Wagner revisited. Journal of Optimization Theory and Applications. **101**, 523–555 (1999).
7. A. Federgruen, Schechner, Z.: Cost formulas for continuous review inventory models with fixed deliver lags. Operations Research. **31**, 957–965 (1983).
8. Fox, E.J., Cox,E.L., Metters, R.: Optimal Inventory Policy with Two Suppliers. Operations Research. **54**, 389–393 (2006).
9. Hadley, G.J., Within, T.M.: Analysis of Inventory Systems. Prentice-Hall, New-Jersey (1963).
10. Iglehart, D.L.: Optimality of $(s,S)$ policies in the infinite horizon dynamic inventory problem. Management Science. **9**,259–267 (1963).
11. Kim, D.E., Park, K.S.: $(Q,r)$ inventory model with a mixture of lost sales and time-weighted backorders. Journal of the Operational Research Society. **36**, 231–238 (1985).
12. Moinzadeh, K., Nahmias, S.: A continuous review for an inventory system with two supply modes. Management Science. **34**, 761–773 (1988).
13. Sahin, I.: On the stationary analysis of continuous review $(s,S)$ inventory systems with constant lead times Operations Research. **27**, 717–729 (1971).
14. Sahin, I.: On the objective function behavior in $(s,S)$ inventory models Operations Research. **30**, 709–724 (1982).
15. Sethi, S.P., Cheng, F.: Optimality of $(s,S)$ policies in inventory models with Markovian demand Operations Research. **45**, 931–939 (1997).
16. Scarf, H.: The optimality of $(s,S)$ policies in the dynamic inventory problem. Mathematical Methods in social Sciences, ed. by K. Arrow, S. Karlin, P. Suppes, Stanford University Press, Stanford, Chap.13, (1960).
17. Song, J.S., Zipkin, P.: Inventory Control in a Fluctuating Demand Environment. Operations Research. **41**, 351–370 (1993).
18. Veinott, A.F.: On the optimality of $(s,S)$ inventory policies: New conditions and a new proof. J. Siam. Appl. Math. **14**, 1067–1083 (1966).

# Chapter 3
# Sample-Path Optimality in Average Markov Decision Chains Under a Double Lyapunov Function Condition

**Rolando Cavazos-Cadena and Raúl Montes-de-Oca**

## 3.1 Introduction

This note concerns discrete-time Markov decision processes (MDPs) evolving on a denumerable state space. The performance index of a control policy is an (long run) average criterion, and besides standard continuity compactness conditions, the main structural assumption on the model is that (a) the (possibly unbounded) cost function has a Lyapunov function $\ell(\cdot)$ and (b) a power of order larger than 2 of $\ell$ also admits a Lyapunov function [14]. Within this context, the main purpose of the paper is to analyze the *sample-path* average optimality of some policies whose *expected* optimality is well known. More specifically, the main results in this direction are as follows:

(i) The stationary policy $f$ obtained by optimizing the right-hand side of the optimality equation is sample-path optimal in the strong sense, that is, under the action of $f$, the observed average costs in finite times converge almost surely to the optimal expected average cost $g$, whereas if the system is driven by any other policy, then with probability 1, the inferior limit of those averages is at least $g$.

(ii) The Markovian policies obtained from procedures frequently used to approximate a solution of the optimality equation, like the vanishing discount or the successive approximations methods, are sample-path average optimal in the strong sense.

R. Cavazos-Cadena (✉)
Departamento de Estadística y Cálculo, Universidad Autónoma Agraria Antonio Narro,
Buenavista, Saltillo, Coahuila 25315, México
e-mail: rcavazos@uaaan.mx

R. Montes-de-Oca
Departamento de Matemáticas, Universidad Autónoma Metropolitana–Iztapalapa,
México D.F. 09340, México
e-mail: momr@xanum.uam.mx

D. Hernández-Hernández and A. Minjárez-Sosa (eds.), *Optimization, Control, and Applications of Stochastic Systems*, Systems & Control: Foundations & Applications, DOI 10.1007/978-0-8176-8337-5_3, © Springer Science+Business Media, LLC 2012

The expected average criteria have been intensively studied, and a fairly complete account of the theory can be found in [9, 12, 13]; see also [1]. In this last paper, it was shown that, for a general MDP, if the optimality equation has a bounded solution, then the stationary policy $f$ referred to in the point (i) above is optimal in the sample-path sense. In [3, 4], a similar conclusion was obtained for models with denumerable state space if the cost function has an almost monotone (or penalized) structure, in the sense that the costs are sufficiently large outside a compact set; such a conclusion was extended to models on Borel spaces by [10, 16, 20]. More recently, for models with denumerable state space and finite actions sets, the sample-path average criterion was studied in [15] under the uniform ergodicity assumption. On the other hand, the first result described above is an extension of Theorem 4.1 in [6], where the sample-path optimality of the policy $f$ mentioned above was established in a weaker sense than the one used in the present work.

The *approach* of this note relies on basic probabilistic ideas, like Kolmogorov's inequality and the first Borel-Cantelli lemma, and was motivated by the elementary analysis of the strong law of large numbers as presented in [2].

The *organization* of the subsequent material is as follows: In Sect. 3.2, the decision model is presented and the conditions to obtain an (expected) average optimal stationary policy from a solution of the optimality equation are briefly described. Next, in Sect. 3.3, the idea of Lyapunov function is introduced and some of its elementary properties are established, whereas in Sect. 3.4, the basic structural restriction on the model, namely, the double Lyapunov function condition, is formulated as Assumption 3.4.1, and the main result of the chapter, solving problem (i) above, is stated as Theorem 3.4.1. The argument to establish this result relies on some properties of the sequence of innovations associated with the sequence of optimal relative costs, which are presented in Sect. 3.5, and then, the main theorem is proved in Sect. 3.6. Next, the result on the sample-path optimality of Markovian policies is stated as Theorem 3.7.1 in Sect. 3.7, and the necessary technical tools to prove that result, concerning tightness of the sequence of empirical measures and uniform integrability of the cost function, are established in Sect. 3.8. Finally, the exposition concludes with the proof of Theorem 3.7.1 in Sect. 3.9.

**Notation.** Throughout the remainder, $\mathbb{N}$ stands for the set of all nonnegative integers and the indicator function of a set $A$ is denoted by $I_A$, so that $I_A(x) = 1$ if $x \in A$ and $I_A(x) = 0$ when $x \notin A$. On the other hand, for a topological space $\mathbb{K}$, the class of all continuous functions defined on $\mathbb{K}$ and the Borel $\sigma$-field of $\mathbb{K}$ are denoted by $\mathscr{C}(\mathbb{K})$ and $\mathscr{B}(\mathbb{K})$, respectively, whereas $\mathbb{P}(\mathbb{K})$ stands for the class of all probability measures defined in $\mathscr{B}(\mathbb{K})$. Finally, for an event $G$, the corresponding indicator function is denoted by $I[G]$.

## 3.2  Decision Model

Let $\mathcal{M} = (S, A, \{A(x)\}_{x \in S}, C, P)$ be an MDP, where the state space $S$ is a denumerable set endowed with the discrete topology and the action set $A$ is a metric space. For each $x \in S$, $A(x) \subset A$ is the nonempty subset of admissible actions at $x$ and, defining the class of admissible pairs by $\mathbb{K} := \{(x, a) \mid a \in A(x), x \in S\}$, the mapping $C: \mathbb{K} \to \mathbb{R}$ is the cost function, whereas $P = [p_{xy}(\cdot)]$ is the controlled transition law on $S$ given $\mathbb{K}$, that is, for all $(x, a) \in \mathbb{K}$ and $y \in S$, the relations $p_{xy}(a) \geq 0$ and $\sum_{y \in S} p_{xy}(a) = 1$ are satisfied. This model $\mathcal{M}$ is interpreted as follows: At each time $t \in \mathbb{N}$, the decision maker observes the state of a dynamical system, say $X_t = x \in S$, and selects an action (control) $A_t = a \in A(x)$ incurring a cost $C(x, a)$. Then, regardless of the previous states and actions, the state at time $t + 1$ will be $X_{t+1} = y \in S$ with probability $p_{xy}(a)$; this is the Markov property of the decision process.

**Assumption 3.2.1**

 (i)  *For each $x \in S$, $A(x)$ is a compact subset of $A$.*
 (ii) *For every $x, y \in S$, the mappings $a \mapsto C(x, a)$ and $a \mapsto p_{xy}(a)$ are continuous in $a \in A(x)$.*

**Policies.** The space $\mathbb{H}_t$ of possible histories up to time $t \in \mathbb{N}$ is defined by $\mathbb{H}_0 := S$ and $\mathbb{H}_t := \mathbb{K}^t \times S$ for $t \geq 1$, whereas a generic element of $\mathbb{H}_t$ is denoted by $\mathbf{h}_t = (x_0, a_0, \ldots, x_i, a_i, \ldots, x_t)$, where $a_i \in A(x_i)$. A policy $\pi = \{\pi_t\}$ is a special sequence of stochastic kernels: For each $t \in \mathbb{N}$ and $\mathbf{h}_t \in \mathbb{H}_t$, $\pi_t(\cdot \mid \mathbf{h}_t)$ is a probability measure on $\mathscr{B}(A)$ concentrated on $A(x_t)$, and for each Borel subset $B \subset A$, the mapping $\mathbf{h}_t \mapsto \pi_t(B \mid \mathbf{h}_t)$, $\mathbf{h}_t \in \mathbb{H}_t$, is Borel measurable. The class of all policies is denoted by $\mathscr{P}$ and when the controller chooses actions according to $\pi$, the control $A_t$ applied at time $t$ belongs to $B \subset A$ with probability $\pi_t(B \mid \mathbf{h}_t)$, where $\mathbf{h}_t$ is the observed history of the process up to time $t$. Given the policy $\pi$ being used for choosing actions and the initial state $X_0 = x$, the distribution of the state-action process $\{(X_t, A_t)\}$ is uniquely determined [9], and such a distribution and the corresponding expectation operator are denoted by $P_x^\pi$ and $E_x^\pi$, respectively. Next, define $\mathscr{F} := \prod_{x \in S} A(x)$ and notice that $\mathscr{F}$ is a compact metric space, which consists of all functions $f: S \to A$ such that $f(x) \in A(x)$ for each $x \in S$. A policy $\pi$ is *Markovian* if there exists a sequence $\{f_t\} \subset \mathscr{F}$ such that the probability measure $\pi_t(\cdot \mid \mathbf{h}_t)$ is always concentrated at $f_t(x_t)$, and if $f_t \equiv f$ for every $t$, the Markovian policy $\pi$ is is referred to as *stationary*. The classes of stationary and Markovian policies are naturally identified with $\mathscr{F}$ and $\mathcal{M} := \prod_{t=0}^\infty \mathscr{F}$, respectively, and with these conventions $\mathscr{F} \subset \mathcal{M} \subset \mathscr{P}$.

**Performance Criteria.** Suppose that the cost function $C(\cdot, \cdot)$ is such that

$$E_x^\pi \left[ |C(X_t, A_t)| \right] < \infty, \quad x \in S, \quad \pi \in \mathscr{P}, \quad t \in \mathbb{N}. \tag{3.1}$$

In this case, the (long-run superior limit) average cost corresponding to $\pi \in \mathscr{P}$ at state $x \in S$ is defined by

$$J(x,\pi) := \limsup_{k \to \infty} \frac{1}{k} E_x^{\pi} \left[ \sum_{t=0}^{k-1} C(X_t, A_t) \right], \tag{3.2}$$

and the corresponding optimal value function is specified by

$$J^*(x) := \inf_{\pi \in \mathscr{P}} J(x,\pi), \quad x \in S; \tag{3.3}$$

a policy $\pi^* \in \mathscr{P}$ is (superior limit) average optimal if $J(x,\pi^*) = J^*(x)$ for every $x \in S$. The criterion (3.2) evaluates the performance of a policy in terms of the largest among the limit points of the expected average costs in finite times. In contrast, the following index assesses a policy in terms of the smallest of such limit points:

$$J_-(x,\pi) := \liminf_{k \to \infty} \frac{1}{k} E_x^{\pi} \left[ \sum_{t=0}^{k-1} C(X_t, A_t) \right] \tag{3.4}$$

is the (long run) inferior limit average criterion associated with $\pi \in \mathscr{P}$ at a state $x$, whereas the optimal value function associated with this criterion is given by

$$J_-^*(x) := \inf_{\pi \in \mathscr{P}} J_-(x,\pi), \quad x \in S. \tag{3.5}$$

From these specifications, it follows that

$$J_-^*(\cdot) \leq J^*(\cdot), \tag{3.6}$$

and within the context described below, it will be shown that the equality holds in this last relation.

**Optimality Equation.** A basic instrument to analyze the above average criteria is the following *optimality equation*:

$$g + h(x) = \inf_{a \in A(x)} \left[ C(x,a) + \sum_{y \in S} p_{xy}(a) h(y) \right], \quad x \in S, \tag{3.7}$$

where $g \in \mathbb{R}$ and $h \in \mathscr{C}(S)$ are given functions, and it is supposed that

$$E_x^{\pi}[|h(X_n)|] < \infty, \quad x \in S, \quad \pi \in \mathscr{P}, \quad n \in \mathbb{N}.$$

Under this condition and (3.1), a standard induction argument combining (3.7) and the Markov property yields that, for every nonnegative integer $n$,

$$(n+1)g + h(x) \leq E_x^\pi \left[ \sum_{t=0}^n C(X_t, A_t) + h(X_{n+1}) \right], \quad x \in S, \quad \pi \in \mathscr{P}.$$

Moreover, if $f \in \mathscr{F}$ satisfies that

$$g + h(x) = C(x, f(x)) + \sum_{y \in S} p_{xy}(f(x))h(y), \quad x \in S, \tag{3.8}$$

then

$$(n+1)g + h(x) = E_x^f \left[ \sum_{t=0}^n C(X_t, A_t) + h(X_{n+1}) \right], \quad x \in S.$$

Therefore, assuming that the condition

$$\lim_{n \to \infty} \frac{E_x^\pi [h(X_{n+1})]}{n+1} = 0 \tag{3.9}$$

holds for every $x \in S$ and $\pi \in \mathscr{P}$, it follows that the relation

$$\lim_{n \to \infty} \frac{1}{n+1} E_x^f \left[ \sum_{t=0}^n C(X_t, A_t) \right] = g \leq \liminf_{n \to \infty} \frac{1}{n+1} E_x^\pi \left[ \sum_{t=0}^n C(X_t, A_t) \right] \tag{3.10}$$

is always valid, and then, (3.2)–(3.6) immediately yield that

$$J_-^*(x) = J^*(x) = g = \lim_{n \to \infty} \frac{1}{n+1} E_x^f \left[ \sum_{t=0}^n C(X_t, A_t) \right], \quad x \in S, \tag{3.11}$$

so that:

(i) The superior and inferior limit average criteria render the same optimal value function,
(ii) A stationary policy $f$ satisfying (3.8) is average optimal, and
(iii) The optimal average cost is constant and is equal to $g$.

## 3.3  Lyapunov Functions

In this section, a structural condition on the model $\mathscr{M}$ will be introduced under which (a) the basic condition (3.1) holds, (b) the optimality equation (3.7) has a solution $(g, h(\cdot))$ such that the convergence (3.9) occurs, and (c) a policy $f \in \mathscr{F}$

satisfying (3.8) exists, so that the conclusions (i)–(iii) stated above hold. Throughout the rest of this chapter

$$z \in S \quad \text{is a fixed state}$$

and $T$ stands for the first return time to state $z$, that is,

$$T := \min\{n > 0 \mid X_n = z\}, \tag{3.12}$$

where, as usual, the minimum of the empty set is $\infty$. The following idea was introduced in [14] and was analyzed in [5]:

**Definition 3.3.1.** Let $D \in \mathscr{C}(\mathbb{K})$ and $\ell \colon S \to [1, \infty)$ be given functions. The function $\ell$ is a Lyapunov function for $D$, or "$D$ has the Lyapunov function $\ell$", if the following conditions (i)–(iii) hold:

(i)  $1 + |D(x,a)| + \sum_{y \neq z} p_{xy}(a)\ell(y) \leq \ell(x)$ for all $(x,a) \in \mathbb{K}$.
(ii) For each $x \in S$, the mapping $f \mapsto \sum_y p_{xy}(f(x))\ell(y) = E_x^f[\ell(X_1)]$ is continuous in $f \in \mathscr{F}$.
(iii) For each $f \in \mathscr{F}$ and $x \in S$, $E_x^f[\ell(X_n)I[T > n]] \to 0$ as $n \to \infty$.

The sentence "$D$ admits a Lyapunov function" means that there exists a function $\ell \colon S \to [1, \infty)$ such that conditions (i)–(iii) above hold.

The following simple lemma will be useful.

**Lemma 3.3.1.** *Suppose that C has the Lyapunov function $\ell(\cdot)$. In this case the assertions (i) and (ii) below are valid.*

*(i) For every $n \in \mathbb{N}$ and $\pi \in \mathscr{P}$,*

$$\frac{1}{n+1} E_x^\pi \left[ \sum_{t=0}^n (1 + |C(X_t, A_t)|) + \ell(X_{n+1}) \right] \leq B(x) := \ell(x) + \ell(z), \quad x \in S;$$

*in particular, the basic condition (3.1) holds.*

*(ii)* $\displaystyle \lim_{n \to \infty} \frac{1}{n} E_x^\pi[\ell(X_n)] \to 0.$

*Proof.* Notice that the inequality $1 + |C(x,a)| + \sum_{y \in S} p_{xy}(a)\ell(y) \leq \ell(x) + \ell(z)$ is always valid, by Definition 3.3.1(i), and then, an induction argument using the Markov property yields that for arbitrary $x \in S$ and $\pi \in \mathscr{P}$,

$$E_x^\pi \left[ \sum_{t=0}^n (1 + |C(X_t, A_t)|) + \ell(X_{n+1}) \right] \leq \ell(x) + (n+1)\ell(z), \quad n \in \mathbb{N},$$

a relation that immediately yields part (i); a proof of the second assertion can be found in Lemma 3.2 of [6]. $\qquad\square$

The following lemma, originally established by [14], shows that the existence of a Lyapunov function has important implications for the analysis of the average criteria in (3.2) and (3.4).

**Lemma 3.3.2.** *Suppose that the cost function C has a Lyapunov function $\ell$. In this case, there exists a unique pair $(g, h(\cdot)) \in \mathbb{R} \times \mathscr{C}(S)$ such that the following assertions (i)–(v) hold:*

(i) *$g = J^*(x)$ for each $x \in S$.*

(ii) *$h(z) = 0$ and $|h(x)| \leq (1 + \ell(z)) \cdot \ell(x)$ for all $x \in S$. Therefore, by Lemma 3.3.1 (ii), the convergence (3.9) holds.*

(iii) *The pair $(g, h(\cdot))$ satisfies the optimality equation (3.7).*

(iv) *For each $x \in S$, the mapping $a \mapsto \sum_{y \in S} p_{xy}(a) h(y)$ is continuous in $a \in A(x)$.*

(v) *An optimal stationary policy exists: For each $x \in S$, the term within brackets in the right-hand side of (3.7) has a minimizer $f(x) \in A(x)$, and the corresponding policy $f \in \mathscr{F}$ is optimal. Moreover, (3.11) holds.*

A proof of this result can be essentially found in Chapter 5 of [14]; see also Lemma 3.1 in [6] for a proof of the inequality in part (ii).

*Remark 3.3.1.* Notice that $g$ in Lemma 3.3.2 is uniquely determined, since it is *the* optimal (expected) average cost at every state. The function $h(\cdot)$ in the above lemma is also unique, as established in Lemma A.2(iv) in [7]. Indeed, defining the relative cost function as $C(\cdot, \cdot) - g$, the function $h(\cdot)$ is the optimal total relative cost incurred before the first return time to state $z$; more explicitly, $h(x) = \inf_{\pi \in \mathscr{P}} E_x^{\pi} \left[ \sum_{t=0}^{T-1} [C(X_t, A_t) - g] \right]$ for all $x \in S$.

This section concludes with some simple but useful properties of Lyapunov functions stated in Lemma 3.3.3 below, whose statement involves the following notation.

**Definition 3.3.2.** The class $\mathscr{L}(\ell)$ consists of all functions $D \in \mathscr{C}(\mathbb{K})$ such that a positive multiple of $\ell$ is a Lyapunov function for $D$, that is, $D \in \mathscr{L}(\ell)$ if and only if

$$\text{for some } c > 0, \quad 1 + |D(x,a)| + \sum_{y \neq z} p_{xy}(a)[c\ell(y)] \leq c\ell(x), \quad (x,a) \in \mathbb{K}.$$

Notice that the function $c\ell(\cdot)$ inherits the properties (ii) and (iii) of the function $\ell(\cdot)$ in Definition 3.3.1.

**Lemma 3.3.3.** *Suppose that $\ell \colon S \to [1, \infty)$ is a Lyapunov function for a function $\tilde{D} \in \mathscr{C}(\mathbb{K})$:*

(i) *If $D_0 \in \mathscr{C}(\mathbb{K})$ is such that $|D_0| \leq |\tilde{D}|$, then $\ell$ is also a Lyapunov function for $D_0$.*

(ii) *With the notation in Definition 3.3.2, the following properties (a) and (b) hold:*

(a) *$\mathscr{L}(\ell)$ is a vector space that contains the constant functions.*

(b) *If $D_1, D_2 \in \mathscr{L}(\ell)$, then $\max\{D_1, D_2\}$ and $\min\{D_1, D_2\}$ also belong to $\mathscr{L}(\ell)$.*

*Proof.* The first part follows directly from Definition 3.3.1. To establish part (ii), first notice that $\mathscr{L}(\ell)$ is nonempty since $\tilde{D} \in \mathscr{L}(\ell)$. Next, suppose that $D_1, D_2 \in \mathscr{L}(\ell)$ and observe that

$$1 + |D_i(x,a)| + \sum_{y \neq z} p_{xy}(a)[c_i \ell(y)] \leq c_i \ell(x), \quad (x,a) \in \mathbb{K}, \quad i = 1, 2,$$

where $c_1$ and $c_2$ are positive constants. If $d_1$ and $d_2$ are real numbers, multiplying both sides of the above equality by $1 + |d_i|$, it follows that, for every $(x,a) \in \mathbb{K}$,

$$([1 + |d_i|) + (1 + |d_i|)|D_i(x,a)| + \sum_{y \neq z} p_{xy}(a)[(1 + |d_i|)c_i \ell(y)] \leq (1 + |d_i|)c_i \ell(x),$$

and then,

$$1 + |d_i D_i(x,a)| + \sum_{y \neq z} p_{xy}(a)[(1 + |d_i|)c_i \ell(y)] \leq (1 + |d_i|)c_i \ell(x), \quad i = 1, 2.$$

these inequalities, it follows that

$$1 + [1 + |d_1 D_1(x,a)| + |d_2 D_2(x,a)|] + \sum_{y \neq z} p_{xy}(a)[(1 + |d_1|)c_1 + (1 + |d_2|)c_2] \ell(y)$$

$$\leq [(1 + |d_1|)c_1 + (1 + |d_2|)c_2] \ell(x),$$

showing that

$$1 + |d_1 D_1| + |d_2 D_2| \in \mathscr{L}(\ell),$$

and because the function in the left-hand side of this inclusion dominates $|d_1 D_1 + d_2 D_2|$ and 1, part (i) yields that (a) $d_1 D_1 + d_2 D_2 \in \mathscr{L}(\ell)$ and $1 \in \mathscr{L}(\ell)$, so that $\mathscr{L}(\ell)$ is a vector space that contains the constant functions. Finally, since $|\max\{D_1, D_2\}|, |\min\{D_1, D_2\}| \leq |D_1| + |D_2|$, the above displayed inclusion and part (i) together imply that (b) $\max\{D_1, D_2\}, \min\{D_1, D_2\} \in \mathscr{L}(\ell)$.      □

## 3.4   A Double Lyapunov Function Condition and Sample-Path Optimality

In this section, a notion of sample-path average optimality is introduced, and using the idea of Lyapunov function, a structural condition on the decision model $\mathscr{M}$ is formulated. Under such an assumption, it is stated in Theorem 3.4.1 below that, in addition to being expected average optimal, a stationary policy $f$ satisfying the (3.8) is also average optimal in the sample-path sense.

**Definition 3.4.1.** A policy $\pi^* \in \mathscr{P}$ is sample-path average optimal with optimal value $g^* \in \mathbb{R}$ if the following conditions (i) and (ii) are valid:

(i)  For each state $x \in S$, $\displaystyle \lim_{n \to \infty} \frac{1}{n} \sum_{t=0}^{n-1} C(X_t, A_t) = g^*$ $\quad P_x^{\pi^*}$-a.s.;

(ii)  For every $\pi \in \mathscr{P}$ and $x \in S$, $\displaystyle \liminf_{n \to \infty} \frac{1}{n} \sum_{t=0}^{n-1} C(X_t, A_t) \geq g^*$ $\quad P_x^{\pi}$-a.s..

The existence and construction of sample-path optimal policies will be studied under the following structural condition on the model $\mathscr{M}$.

**Assumption 3.4.1 [Double Lyapunov function condition]**

(i)  *The cost function $C(\cdot, \cdot)$ has a Lyapunov function $\ell$.*
(ii)  *For some $\beta > 2$, the mapping $\ell^\beta$ admits a Lyapunov function.*

A simple class of models satisfying this assumption is presented below.

*Example 3.4.1.* For each $t \in \mathbb{N}$, let $X_t \in S = \mathbb{N}$ be the number of customers waiting for a service at a time $t$ in a single-server station. To describe the evolution of $\{X_t\}$, let the action set $A$ be a compact metric space and set $A(x) = A$ for every $x$. Next, let $\{\Delta_t(a) \mid t \in \mathbb{N}, a \in A\}$ and $\{\xi_t(a) \mid t \in \mathbb{N}, a \in A\}$ be two families of independent and identically distributed random variables taking values in the set $\mathbb{N}$, and suppose that the following conditions hold:

(i)  For each $k \in \mathbb{N}$, the mappings $a \mapsto P[\Delta_t(a) = k]$, $a \mapsto P[\xi_t(a) = k]$, and $a \mapsto E[\xi_t(a)^r] \in (0, \infty)$, $1 \leq r \leq 2m+3$, are continuous, where $m \in \mathbb{N} \setminus \{0\}$ is fixed.
(ii)  $P[\Delta_t(a) = 1] = \mu(a) = 1 - P[\Delta_t(a) = 0]$ and $E[\xi_t(a)] - \mu(a) \leq -\rho < 0$ for all $a \in A$.

When the action $a \in A$ is chosen, (the Bernoulli variable) $\Delta_t(a)$ and $\xi_t(a)$ represent the number of service completions and arrivals in $[t, t+1)$, respectively, and the evolution of the state process is determined by

$$X_{t+1} = X_t + \xi_t(a) - \Delta_t(a) I_{\mathbb{N} \setminus \{0\}}(X_t) \quad \text{if } A_t = a, \quad t \in \mathbb{N}, \tag{3.13}$$

an equation that allows to obtain the transition law and to show that $p_{xy}(a)$ is a continuous function of $a \in A$, by the first of the conditions presented above. In the third part of the following proposition, a class of cost functions satisfying Assumption 3.4.1 is identified. □

**Proposition 3.4.1.** *In the context of Example 3.4.1, the following assertions hold when $z = 0$ and $T$ is the first return time in (3.12):*

(i)  *For each $r = 1, 2, \ldots, 2m+2$, there exist positive constants $b_r$ and $c_r$ such that the function $\ell_{r+1} \in \mathscr{C}(S)$ given by*

$$\ell_{r+1}(x) = x^{r+1} + b_r x + c_r, \quad x \in S, \tag{3.14}$$

*satisfies*

$$\rho x^r + 1 + E[\ell_{r+1}(X_{t+1}) \mid X_t = x, A_t = a] \leq \ell_{r+1}(x), \quad x \in S, \qquad (3.15)$$

*where $\rho > 0$ is the number in condition (ii) stated in Example 3.4.1. Moreover, for each $x \in S$,*

$$a \mapsto E[\ell_{r+1}(X_{t+1}) \mid X_t = x, A_t = a], \quad a \in A, \quad \text{is continuous}, \qquad (3.16)$$

*and for every $\pi \in \mathscr{P}$,*

$$\lim_{n \to \infty} E_x^\pi[(1 + \rho X_n^r) I[T > n]] = 0. \qquad (3.17)$$

*Consequently,*
(ii) *For $j = 1, 2, \dots, 2m + 1$, the above mapping $\ell_{j+1}(\cdot)$ is a Lyapunov function for the cost function*

$$C_j(x) \colon = \rho x^j, \quad x \in S.$$

(iii) *Suppose that, for some integer $j = 1, 2, \dots, m - 1$, the cost function $C \in \mathscr{C}(\mathbb{K})$ is such that*

$$\max_{a \in A} |C(x, a)| \leq b_1 x^j + b_0, \quad x \in S,$$

*where $b_0$ and $b_1$ are positive constants. In this case, the function $C$ satisfies Assumption 3.4.1.*

*Proof.* (i) For $X_t = x \neq 0$ and $1 \leq r \leq 2m + 2$, the evolution equation (3.13) yields that

$$E[(X_{t+1})^{r+1} \mid X_t = x, A_t = a] = E[(x + \xi_t(a) - \Delta_t(a))^{r+1}]$$
$$\leq x^{r+1} - (r+1)\rho x^r + R(x, a), \qquad (3.18)$$

where $\sup_{a \in A} |R(x, a)| = O(x^{r-1})$; thus, there exists a constant $b > 0$ such that $R(x, a) \leq \rho x^r + b$ for every $x \neq 0$, and it follows that

$$\rho x^r - b + E[(X_{t+1})^{r+1} \mid X_t = x, A_t = a] \leq x^{r+1}.$$

When $r = 0$, the term $R(x, a)$ in (3.18) is null, so that $\rho + E[X_{t+1} \mid X_t = x, A_t = a] \leq x$; multiplying both sides of this relation by a sufficiently large constant $b_r$ such that $\rho b_r > b + 1$ and combining the resulting inequality with the one displayed above, it follows that

$$\rho x^r + 1 + E[(X_{t+1})^{r+1} + b_r X_{t+1} \mid X_t = x, A_t = a] \leq x^{r+1} + b_r x, \quad x \in S \setminus \{0\}.$$

Defining $c_r \colon = 1 + \max_{a \in A} E[\xi_t(a)^{r+1} + b_r \xi_t(a)]$, it follows that the function $\ell_{r+1}$ in (3.14) satisfies the inequality (3.15), whereas (3.16) follows from the continuity properties of the distributions of the departure and arrival streams in Example 3.4.1. On the other hand, (3.15) yields that the inequality $E_x^\pi[\rho X_0^r + 1 + \ell_{r+1}(X_1)I[T > 1]] \le \ell_{r+1}(x)$ is always valid, and an induction argument using the Markov property leads to

$$E_x^\pi\left[\sum_{t=0}^{n-1}(\rho X_t^r + 1)I[T > t] + \ell_{r+1}(X_n)I[T > n]\right] \le \ell_{r+1}(x), \quad n \in \mathbb{N};$$

taking the limit as $n$ goes to $+\infty$, this implies that $E_x^\pi[\sum_{t=0}^\infty (\rho X_t^r + 1)I[T > t]] \le \ell_{r+1}(x)$, and (3.17) follows.

(ii) The relations (3.15) and (3.16) immediately show that $\ell_{j+1}$ satisfies the requirements (i) and (ii) in Definition 3.3.1 of a Lyapunov function for $C_j$. Next, observe that $\ell_{j+1} \le c_0 x^{j+1} + c_1$ for some constants $c_0$ and $c_1$, and then, (3.17) with $j + 1(\le m+2)$ instead of $r$ implies that $\ell_{j+1}$ also satisfies the third property in Definition 3.3.1.

(iii) Observe that the condition on the function $C(\cdot, \cdot)$ can be written as $|C| \le b_1 C_j + b_0$, and then, it is sufficient to show that $C_j$ satisfies Assumption 3.4.1, since in this case the corresponding conclusion for $C$ follows from Lemma 3.3.3. Let the integer $j$ between 1 and $m-1$ be arbitrary and notice that part (ii) yields that $\ell_{j+1}$ is a Lyapunov function for $C_j$. Next, set $\beta = (2j+3)/(j+1) > 2$ and observe that (3.14) implies that there exist positive constants $c_0$ and $c_1$ such that $\ell_{j+1}^\beta(x) \le c_1 x^{2j+3} + c_0 = c_1 C_{2j+3}(x) + c_0$; since $2j + 3 \le 2m + 1$, part (ii) shows that $C_{2j+3}$ has a Lyapunov function, and then, $\ell_{j+1}^\beta$ also admits a Lyapunov function, by Lemma 3.3.3.

□

The following result establishes the existence of sample-path average optimal stationary polices.

**Theorem 3.4.1.** *Suppose that Assumptions 3.2.1 and 3.4.1 hold, and let $(g, h(\cdot))$ be the solution of the optimality equation guaranteed by Lemma 3.3.2. In this case, if the stationary policy $f$ satisfies (3.8), then $f$ is sample-path average optimal with the optimal value $g$. More explicitly, for each $x \in S$ and $\pi \in \mathscr{P}$,*

$$\lim_{n\to\infty}\frac{1}{n}\sum_{t=0}^{n-1}C(X_t, A_t) = g \quad P_x^f\text{-}a.s. \tag{3.19}$$

*and*

$$\liminf_{n\to\infty}\frac{1}{n}\sum_{t=0}^{n-1}C(X_t, A_t) \ge g \quad P_x^\pi\text{-}a.s. \tag{3.20}$$

*Remark 3.4.1.* This theorem is related to Theorem 4.1 in [6] where MDPs with average reward criteria were considered. In the context of the present work, Theorem 4.1 in that paper establishes that, under Assumption 3.2.1, if the cost function has the Lyapunov function $\ell$, then $\limsup_{n\to\infty} n^{-1}\sum_{t=0}^{n-1} C(X_t, A_t) \ge g$  $P_x^\pi$-a.s. for each

$\pi \in \mathscr{P}$ and $x \in S$ and that the equality holds if $\pi = f$ satisfies (3.8). In the present Theorem 3.4.1, the additional condition in Assumption 3.4.1(ii) is incorporated, and in this context, the stronger conclusions (3.19) and (3.20) are obtained.

A proof of Theorem 3.4.1 will be given after presenting the necessary preliminaries in the following section:

## 3.5  Innovations of the Sequence of Optimal Relative Costs

This section contains the main technical tool that will be used to establish Theorem 3.4.1. The necessary result concerns properties of the sequence of innovations associated to $\{h(X_t)\}$ which is introduced below. Throughout the remainder of this chapter Assumptions 3.2.1 and 3.4.1 are supposed to be valid even without explicit reference, and $(g, h(\cdot)) \in \mathbb{R} \times \mathscr{C}(S)$ stands for the pair satisfying the optimality equation (3.7) as described in Lemma 3.3.2. Next, for each positive integer $n$, let $\mathscr{F}_n$ be the $\sigma$-field generated by the states observed and actions applied up to time $n$:

$$\mathscr{F}_n: \ = \sigma(X_t, A_t, 0 \le t \le n), \quad n = 1, 2, 3, \dots, \tag{3.21}$$

and observe that for each initial state $x$ and $\pi \in \mathscr{P}$, the Markov property of the decision process yields that

$$E_x^\pi[h(X_n) \,|\, \mathscr{F}_{n-1}] = \sum_{y \in S} p_{X_{n-1}y}(A_{n-1})h(y). \tag{3.22}$$

**Definition 3.5.1.** The process of $\{Y_k, k \ge 1\}$ of *innovations* associated to the sequence of observed optimal relative costs $\{h(X_k), k \ge 1\}$ is given by

$$Y_n = h(X_n) - \sum_{y \in S} p_{X_{n-1}, y}(A_{n-1})h(y), \quad n = 1, 2, 3, \dots .$$

Now, let $x \in S$ and $\pi \in \mathscr{P}$ be arbitrary but fixed, and notice that combining the definition above with (3.21) and (3.22), it follows that (i) $Y_n$ is $\mathscr{F}_n$ measurable, and (ii) the innovations $Y_n$ can be written as

$$Y_n = h(X_n) - E_x^\pi[h(X_n) \,|\, \mathscr{F}_{n-1}], \tag{3.23}$$

and then, $Y_n$ is uncorrelated with the $\sigma$-field $\mathscr{F}_{n-1}$ with respect to $P_x^\pi$, that is,

$$E_x^\pi[Y_n W] = 0 \text{ if } W \text{ is } \mathscr{F}_{n-1} \text{ measurable and } Y_n W \text{ is } P_x^\pi \text{ integrable} \tag{3.24}$$

([2]). The following is the main result of this section.

**Theorem 3.5.1.** *Suppose that Assumptions 3.2.1 and 3.4.1 hold and let the pair* $(g, h(\cdot)) \in \mathbb{R} \times \mathscr{C}(S)$ *be as in Lemma 3.3.2. In this context, for each initial state* $x \in S$ *and* $\pi \in \mathscr{P}$, *the following convergences hold:*

$$\lim_{n\to\infty} \frac{h(X_n)}{n} = 0 \quad P_x^\pi\text{-}a.s.$$  (3.25)

*and*

$$\lim_{n\to\infty} \frac{1}{n}\sum_{k=1}^{n} Y_k = 0 \quad P_x^\pi\text{-}a.s.$$  (3.26)

This theorem will be established using two elementary facts stated in the following lemmas. The first one is a criterion for almost sure convergence, which is a consequence of the first Borel-Cantelli lemma [2].

**Lemma 3.5.1.** *Let $\{W_n\}$ be a sequence of random variables defined on a probability space $(\Omega, \mathscr{F}, P)$. In this case, if $\sum_{n=1}^\infty P[|W_n| > \varepsilon] < \infty$ for each $\varepsilon > 0$, then $\lim_{n\to\infty} W_n = 0$ $P$-a.s..*

The second result involved in the proof of Theorem 3.5.1 is the following inequality by Kolmogorov.

**Lemma 3.5.2.** *If $n$ and $k$ are two positive integers such that $n > k$, then for every $\alpha > 0$*

$$P_x^\pi\left[ \max_{r:k\le r\le n} \left| \sum_{t=k}^{r} Y_t \right| \ge \alpha \right] \le \frac{1}{\alpha^2} \sum_{t=k}^{n} E_x^\pi[Y_t^2].$$

This classical result is established as Theorem 22.4 in [2] for the case in which the $Y_n$'s are independent. In the present context, from the relations (3.27)–(3.29) below, it follows that $E_x^\pi[Y_n^2]$ is always finite, and then, (3.24) yields that, if $n > k$, then $E_x^\pi[Y_n Y_k I[G]] = 0$ for every $G \in \mathscr{F}_k$; from this last observation, the same arguments in the aforementioned book allow to establish Lemma 3.5.2. Now, let $\ell$ be a Lyapunov function for the cost function $C$ such that $\ell^\beta$ admits a Lyapunov function for some $\beta > 2$, as ensured by Assumption 3.4.1. Applying Lemma 3.3.1 to the cost function $\ell^\beta$, it follows that there exists a function $b: S \to (0,\infty)$ such that

$$\frac{1}{n+1}E_x^\pi\left[ \sum_{t=0}^{n} \ell^2(X_t) \right] \le \frac{1}{n+1}E_x^\pi\left[ \sum_{t=0}^{n} \ell^\beta(X_t) \right] \le b(x), \quad x \in S,$$  (3.27)

where the first inequality is due to the fact that $\ell(\cdot) \ge 1$.

*Proof of Theorem 3.5.1.* Let $\varepsilon > 0$ be arbitrary and notice that, by Lemma 3.3.2(ii),

$$|h(\cdot)| \le c\ell(\cdot),$$  (3.28)

where $c = 1 + \ell(z)$. Combining this relation with Markov's inequality, it follows that, for each positive integer $n$,

$$P_x^\pi[|h(X_n)/n| > \varepsilon] \le \frac{E_x^\pi[|h(X_n)|^\beta]}{n^\beta \varepsilon^\beta} \le \frac{c^\beta E_x^\pi[\ell(X_n)^\beta]}{n^\beta \varepsilon^\beta}$$

and then, (3.27) yields that

$$P_x^\pi[|h(X_n)/n| > \varepsilon] \le \frac{c^\beta(n+1)b(x)}{n^\beta \varepsilon^\beta} \le 2\frac{c^\beta b(x)}{n^{\beta-1}\varepsilon^\beta};$$

since $\beta > 2$, it follows that $\sum_{n=1}^\infty P_x^\pi[|h(X_n)/n| > \varepsilon] < \infty$, and the convergence (3.25) follows from Lemma 3.5.1. Next, using that the (unconditional) variance of a random variable is an upper bound for the expectation of its conditional variance [19], from (3.23), it follows that

$$
\begin{aligned}
E_x^\pi[Y_t^2] &= E_x^\pi[(h(X_t) - E_x^\pi[h(X_t)|\mathscr{F}_{t-1}])^2]\\
&\le E_x^\pi[(h(X_t) - E_x^\pi[h(X_t)])^2]\\
&\le E_x^\pi[h(X_t)^2]\\
&\le c^2 E_x^\pi[\ell(X_t)^2];
\end{aligned}
\tag{3.29}
$$

see (3.28) for the last inequality. This fact and Lemma 3.5.2 together lead to

$$P_x^\pi\left[\max_{r:k\le r\le n}\left|\sum_{t=k}^r Y_t\right| > \alpha\right] \le \frac{c^2}{\alpha^2}\sum_{t=k}^n E_x^\pi[\ell(X_t)^2],$$

and then, (3.27) yields that

$$P_x^\pi\left[\max_{r:k\le r\le n}\left|\sum_{t=k}^r Y_t\right| > \alpha\right] \le \frac{c^2(n+1)b(x)}{\alpha^2}, \qquad \alpha > 0, \quad n > k \ge 1. \tag{3.30}$$

Using this relation with $k = 1$, $n = m^2$, and $\alpha = \varepsilon m^2$, it follows that

$$
q_m := P_x^\pi\left[m^{-2}\left|\sum_{t=1}^{m^2} Y_t\right| > \varepsilon\right] \le P_x^\pi\left[\max_{r:1\le r\le m^2}\left|\sum_{t=1}^r Y_t\right| > m^2\varepsilon\right]
$$
$$
\le \frac{c^2(m^2+1)b(x)}{\varepsilon^2 m^4}.
$$

Therefore, $\sum_{m=1}^\infty q_m < \infty$, and recalling that $\varepsilon > 0$ is arbitrary, an application of Lemma 3.5.1 implies that

$$\lim_{m\to\infty}\frac{1}{m^2}\sum_{t=1}^{m^2} Y_t = 0 \quad P_x^\pi\text{-a.s.} \tag{3.31}$$

On the other hand, given a positive integer $m$, from the inclusion

$$\left[\max_{j:0\le j\le 2m}\left|(m^2+j)^{-1}\sum_{t=m^2}^{m^2+j}Y_t\right|\ge\varepsilon\right]\subset\left[\max_{j:0\le j\le 2m}\left|\sum_{t=m^2}^{m^2+j}Y_t\right|\ge m^2\varepsilon\right],$$

it follows that

$$p_m:\ =P_x^\pi\left[\max_{j:0\le j\le 2m}\left|(m^2+j)^{-1}\sum_{t=m^2}^{m^2+j}Y_t\right|\ge\varepsilon\right]$$

$$\le P_x^\pi\left[\max_{j:0\le j\le 2m}\left|\sum_{t=m^2}^{m^2+j}Y_t\right|\ge m^2\varepsilon\right]$$

$$=P_x^\pi\left[\max_{r:m^2\le r\le(m+1)^2-1}\left|\sum_{t=m^2}^{r}Y_t\right|\ge m^2\varepsilon\right]$$

$$\le\frac{c^2(m+1)^2 b(x)}{\varepsilon^2 m^4,}$$

where the last inequality was obtained from (3.30) with $n=(m+1)^2-1$, $k=m^2$, and $\alpha=m^2\varepsilon$. Thus, $\sum_{m=1}^{\infty}p_m<\infty$, and then, Lemma 3.5.1 implies that

$$\lim_{m\to\infty}\left\{\max_{j:0\le j\le 2m}\left|(m^2+j)^{-1}\sum_{t=m^2}^{m^2+j}Y_t\right|\right\}=0\quad P_x^\pi\text{-a.s.}\qquad(3.32)$$

To conclude, let $n$ be a positive integer and let $m$ be the integral part of $\sqrt{n}$, so that

$$n=m^2+i,\quad 0\le i\le 2m.$$

Assume that $i$ is positive and notice that in this case

$$\left|\frac{1}{n}\sum_{t=1}^{n}Y_t\right|\le\frac{m^2}{n}\left|\frac{1}{m^2}\sum_{t=1}^{m^2}Y_t\right|+\frac{1}{m^2+i}\left|\sum_{t=m^2+1}^{m^2+i}Y_t\right|,$$

and then,

$$\left|\frac{1}{n}\sum_{t=1}^{n}Y_t\right|\le\left|\frac{1}{m^2}\sum_{t=1}^{m^2}Y_t\right|+\max_{j:0\le j\le 2m}\left\{\frac{1}{m^2+j}\left|\sum_{t=m^2+1}^{m^2+j}Y_t\right|\right\},$$

a relation that is also valid when $i=0$, that is, if $n=m^2$. Taking the limit when $m$ goes to $+\infty$ in both sides of this last inequality, the convergences in (3.31) and (3.32) together imply that (3.26) holds.                                     $\square$

## 3.6   Proof of Theorem 3.4.1

In this section, a criterion for the sample-path average optimality of a policy will be derived from Theorem 3.5.1 and that result will be used to establish Theorem 3.4.1. The arguments use the following notation:

**Definition 3.6.1.** The discrepancy function $\Phi\colon \mathbb{K} \to \mathbb{R}$ associated to the pair $(g, h(\cdot))$ in Lemma 3.3.2 is defined by

$$\Phi(x,a)\colon = C(x,a) + \sum_{y \in S} p_{xy}(a)h(y) - h(x) - g.$$

Notice that $\Phi$ is a continuous mapping, by Assumption 3.2.1 and Lemma 3.3.2(iv). Also, observe that the optimality equation (3.7) yields that

$$\Phi(x,a) \geq 0, \quad (x,a) \in \mathbb{K}.$$

**Lemma 3.6.1.** *Suppose that Assumptions 3.2.1 and 3.4.1 hold. In this context, a policy $\pi^* \in \mathscr{P}$ is sample-path average optimal if and only if*

$$\lim_{n \to \infty} \frac{1}{n} \sum_{t=0}^{n-1} \Phi(X_t, A_t) = 0 \quad P_x^{\pi^*}\text{-a.s.} \tag{3.33}$$

*Proof.* It will be verified that for all $x \in S$ and $\pi \in \mathscr{P}$,

$$\lim_{n \to \infty} \frac{1}{n} \sum_{t=0}^{n-1} [C(X_t, A_t) - g - \Phi(X_t, A_t)] = 0 \quad P_x^{\pi}\text{-a.s.} \tag{3.34}$$

Assuming that this relation holds, the desired conclusion can be established as follows: Observing that

$$\frac{1}{n} \sum_{t=0}^{n-1} C(X_t, A_t) = \frac{1}{n} \sum_{t=0}^{n-1} [C(X_t, A_t) - g - \Phi(X_t, A_t)] + g + \frac{1}{n} \sum_{t=0}^{n-1} \Phi(X_t, A_t),$$

and taking the inferior limit as $n$ goes to $\infty$ in both sides of this equality, the nonnegativity of the discrepancy function and (3.34) together imply that the relation

$$\liminf_{n \to \infty} \frac{1}{n} \sum_{t=0}^{n-1} C(X_t, A_t) \geq g, \quad P_x^{\pi}\text{-a.s.}$$

is always valid. Thus, by Definition 3.4.1, $\pi^* \in \mathscr{P}$ is sample-path average optimal if and only if

$$\lim_{n \to \infty} \frac{1}{n} \sum_{t=0}^{n-1} C(X_t, A_t) = g \quad P_x^{\pi^*}\text{-a.s.}, \quad x \in S,$$

a property that, by (3.34) with $\pi^*$ instead of $\pi$, is equivalent to the criterion (3.33).

Thus, to conclude the argument, it is sufficient to verify the statement (3.34). To achieve this goal, notice that the definition of the discrepancy function yields that following equality is always valid for $t \geq 1$:

$$C(X_{t-1}) - g - \Phi(X_{t-1}, A_{t-1}) = h(X_{t-1}) - \sum_{y \in S} p_{X_{t-1}, y}(A_{t-1}) h(y),$$

a relation that, *via* the specification of the innovation $Y_t$ in Definition 3.5.1, leads to

$$C(X_{t-1}) - g - \Phi(X_{t-1}, A_{t-1}) = h(X_{t-1}) - h(X_t) + Y_t.$$

Therefore,

$$\sum_{t=1}^{n} [C(X_{t-1}) - g - \Phi(X_{t-1}, A_{t-1})] = h(X_0) - h(X_n) + \sum_{t=1}^{n} Y_t,$$

and then, for every initial state $X_0 = x$ and $\pi \in \mathscr{P}$,

$$\frac{1}{n} \sum_{t=1}^{n} [C(X_{t-1}) - g - \Phi(X_{t-1}, A_{t-1})] = \frac{h(x)}{n} - \frac{h(X_n)}{n} + \frac{1}{n} \sum_{t=1}^{n} Y_t, \quad P_x^{\pi}\text{-a.s.},$$

where the equality $P_x^{\pi}[X_0 = x] = 1$ was used; from this point, (3.34) follows directly from Theorem 3.5.1. □

*Proof of Theorem 3.4.1.* From Definitions 3.6.1 and (3.8), it follows that $\Phi(x, f(x)) = 0$ for every state $x$. Thus, using that $A_t = f(X_t)$ when the system is running under the policy $f$, it follows that, for every initial state $x$ and $t \in \mathbb{N}$, the equality $\Phi(X_t, A_t) = \Phi(X_t, f(X_t)) = 0$ holds with probability 1 with respect to $P_x^f$. Therefore, the criterion (3.33) is satisfied by $f$, and then, Lemma 3.6.1 yields that the policy $f$ is sample-path average optimal. □

## 3.7  Approximations Schemes and Sample-Path Optimality

In the remainder of this chapter the sample-path optimality of Markovian policies is analyzed. The interest in this problem stems from the fact that an explicit solution $(g, h(\cdot))$ of the optimality equation (3.7) is seldom available, and in this case, the sample-path average optimal policy $f$ in (3.8) cannot be determined. When a solution of the optimality equation is not at hand, an iterative approximation procedure is implemented and (i) approximations $\{(g_n, h_n(\cdot))\}_{n \in \mathbb{N}}$ for $(g, h(\cdot))$ are generated, and (ii) a stationary policy $f_n$ is obtained from $(g_n, h_n(\cdot))$. Such a policy $f_n$ is '"nearly optimal"' in the sense that, for each fixed $x$, the convergence $\Phi(x, f_n(x)) \to 0$ occurs as $n \to \infty$, and the next objective is to establish the sample-path average optimality of the Markovian policy $\{f_n\}$.

*Remark 3.7.1.* Two procedures that can be used to approximate the solution of the optimality equation and to generate a Markovian policy $\{f_n\}$ such that the $f_n$'s are nearly optimal are briefly described below; for details see, for instance, [1,9,11–13], or [17].

(i) The discounted method. For each $\alpha \in (0,1)$, the total expected $\alpha$-discounted cost at the state $x$ under $\pi \in \mathscr{P}$ is given by $V_\alpha(x,\pi) := E_x^\pi [\sum_{t=0}^\infty \alpha^t C(X_t, A_t)]$, whereas $V_\alpha^*(x) := \inf_{\pi \in \mathscr{P}} V_\alpha(x,\pi)$, $x \in S$, is the $\alpha$-optimal value function, which satisfies the optimality equation

$$V_\alpha(x) = \inf_{a \in A(x)} \left[ C(x,a) + \alpha \sum_{y \in S} p_{xy}(a)V_\alpha(y) \right], \quad x \in S.$$

Now let $\{\alpha_n\} \subset (0,1)$ be a sequence increasing to 1, and define

$$(g_n, h_n) := ((1 - \alpha_n)V_{\alpha_n}(z), V_{\alpha_n}(\cdot) - V_{\alpha_n}(z))$$

and let the policy $f_n$ be such that

$$V_{\alpha_n}(x) = C(x, f_n(x)) + \alpha_n \sum_{y \in S} p_{xy}(f_n(x))V_{\alpha_n}(y), \quad x \in S. \tag{3.35}$$

(ii) Value iteration. This procedure approximates the solution $(g, h(\cdot))$ of the optimality equation (3.7) using the total cost criterion over a finite horizon. For each $n \in \mathbb{N} \setminus \{0\}$ let $J_n(x, \pi)$ be the total cost incurred when the system runs during $n$ steps under policy $\pi$ starting at $x$, that is, $J_n(x, \pi) := E_x^\pi \left[ \sum_{t=0}^{n-1} C(X_t, A_t) \right]$, and let $J_n^*(x) := \inf_{\pi \in \mathscr{P}} J_n(x, \pi)$ be the corresponding optimal value function; the sequence $\{J_n^*(\cdot)\}$ satisfies the relation

$$J_n^*(x) = \inf_{a \in A(x)} \left[ C(x,a) + \sum_{y \in S} p_{xy}(a)J_{n-1}^*(y) \right], \quad x \in S, \quad n = 1,2,3,\ldots, \tag{3.36}$$

where $J_0^*(\cdot) = 0$, so that the functions $J_n^*(\cdot)$ are determined recursively, which is an important feature of the method. The approximations to $(g, h(\cdot))$ are given by

$$(g_n, h_n(\cdot)) := (J_n^*(z) - J_{n-1}^*(z), J_n^*(\cdot) - J_n^*(z)),$$

whereas the policy $f_n$ is such that $f_n(x)$ is a minimizer of the term within brackets in (3.36):

$$J_n^*(x) = C(x, f_n(x)) + \sum_{y \in S} p_{xy}(f_n(x))J_{n-1}^*(y), \quad x \in S, \quad n = 1,2,3,\ldots$$

Under Assumptions 3.2.1 and 3.4.1(i), the approximations $(g_n, h_n(\cdot))$ generated by the discounted method converge pointwise to $(g, h(\cdot))$, and the policies $f_n$ in (3.35) are nearly optimal in the sense that $\Phi(x, f_n(x)) \to 0$ as $n \to \infty$. Similar conclusions hold for the value iteration scheme if, additionally, the transition law satisfies that

$$p_{xx}(a) > 0, \quad a \in A(x), \quad x \in S;$$

this requirement can be avoided if the transformation by Schewitzer (1971) is applied to the transition law and the value iteration method is applied to the transformed model; see, for instance, [7] or [8].

The following theorem establishes a sufficient condition for the sample-path optimality of a Markovian policy.

**Theorem 3.7.1.** *Suppose that Assumptions 3.2.1 and 3.4.1 hold, and let* $\mathbf{f} = \{f_t\}$ *be a Markov policy such that*

$$\lim_{n \to \infty} \Phi(x, f_n(x)) = 0, \quad x \in S, \tag{3.37}$$

*where* $\Phi$ *is the discrepancy function introduced in Definition 3.6.1. In this case, the policy* $\mathbf{f}$ *is sample-path average optimal; see Definition 3.4.1.*

The proof of this result relies on some consequences of Theorem 3.4.1 which will be analyzed below.

## 3.8  Tightness and Uniform Integrability

This section contains the technical preliminaries that will be used to establish Theorem 3.7.1. The necessary results are concerned with properties of the sequence of empirical measures, which is now introduced.

**Definition 3.8.1.** The random sequence $\{v_n\}$ of empirical measures associated with the state-action process $\{(X_t, A_t)\}$ is defined by

$$v_n(B): \ = \frac{1}{n} \sum_{t=0}^{n-1} \delta_{(X_t, A_t)}(B), \quad B \in \mathscr{B}(\mathbb{K}), \quad n = 1, 2, 3, \dots,$$

where $\delta_{\mathbf{k}}$ stands for the Dirac measure concentrated at $\mathbf{k}$, that is, $\delta_{\mathbf{k}}(B) = 1$ if $\mathbf{k} \in B$ and $\delta_{\mathbf{k}}(B) = 0$ when $\mathbf{k} \notin B$.

Notice that this specification yields that, for each positive integer $n$ and $D \in \mathscr{C}(\mathbb{K})$,

$$v_n(D): \ = \int_{\mathbb{K}} D(\mathbf{k}) v_n(d\mathbf{k}) = \frac{1}{n} \sum_{t=1}^{n} D(X_t, A_t) .$$

The main result of this section concerns the asymptotic behavior of $\{v_n\}$ and involves the following notation: Given a set $\tilde{S} \subset S$, for each $D \in \mathscr{C}(\mathbb{K})$, defines the new function $D_{\tilde{S}} \in \mathscr{C}(\mathbb{K})$ as follows:

$$D_{\tilde{S}}(x,a): = \max\{|D(x,a)|, 1\} I_{\tilde{S}}(x), \quad (x,a) \in \mathbb{K}. \tag{3.38}$$

**Theorem 3.8.1.** *Suppose that Assumptions 3.2.1 and 3.4.1 hold and let $x \in S$ and $\pi \in \mathscr{P}$ be arbitrary but fixed. In this context, for each $\varepsilon > 0$, there exists a finite set $F_\varepsilon \subset S$ such that*

$$\limsup_{n \to \infty} v_n(C_{S \setminus F_\varepsilon}) \le \varepsilon \quad P_x^\pi\text{-a.s.;} \tag{3.39}$$

*see (3.38).*

*Remark 3.8.1.* For a positive integer $r$, let $F_{1/r}$ be the set corresponding to $\varepsilon = 1/r$ in the above theorem and define the event $\Omega^*$ by

$$\Omega^*: = \bigcap_{r=1}^\infty \left[ \limsup_{n \to \infty} v_n(C_{S \setminus F_{1/r}}) \le 1/r \right].$$

(i)  Let the set $\mathbb{K}_r \subset \mathbb{K}$ be given by

$$\mathbb{K}_r = \{(x,a) \in \mathbb{K} \,|\, x \in F_{1/r}\}. \tag{3.40}$$

With this notation, $\mathbb{K}_r$ is a compact set, since $F_{1/r}$ is finite, and (3.38) yields that $C_{S \setminus F_{1/r}}(x,a) \ge 1$ for $(x,a) \in \mathbb{K} \setminus \mathbb{K}_r$, so that

$$v_n(\mathbb{K} \setminus \mathbb{K}_r) \le \int_{\mathbb{K}} C_{S \setminus F_{1/r}}(\mathbf{k}) v_n(d\mathbf{k}) = v_n(C_{S \setminus F_{1/r}}),$$

and then,

$$\limsup_{n \to \infty} v_n(\mathbb{K} \setminus \mathbb{K}_r) \le \limsup_{n \to \infty} v_n(C_{S \setminus F_{1/r}}).$$

Consequently, along a sample trajectory $\{(X_t, A_t)\}$ in $\Omega^*$ the corresponding sequence $\{v_n\}$ satisfies $\limsup_{n \to \infty} v_n(\mathbb{K} \setminus \mathbb{K}_r) \le 1/r$ for every $r > 0$ so that $\{v_n\}$ is tight.

(ii)  Given a probability measure $\mu$ defined in $\mathscr{B}(\mathbb{K})$, a function $D \in \mathscr{C}(\mathbb{K})$ is integrable with respect to $\mu$ if, and only if, for each positive integer $r$, there exists a compact set $\tilde{\mathbb{K}}_r$ such that

$$\int_{\mathbb{K} \setminus \tilde{\mathbb{K}}_r} |D(\mathbf{k})| \mu(d\mathbf{k}) \le 1/r.$$

Notice now that (3.38) and (3.40) yield that $C_{S \setminus F_{1/r}}(x,a) \geq |C(x,a)|$ for $(x,a) \in \mathbb{K} \setminus \mathbb{K}_r$, so that

$$\int_{\mathbf{K} \setminus \mathbf{K}_r} |C(\mathbf{k})| v_n(d\mathbf{k}) \leq \int_{\mathbf{K} \setminus \mathbf{K}_r} C_{S \setminus F_{1/r}}(\mathbf{k}) v_n(d\mathbf{k}) = v_n(C_{S \setminus F_{1/r}}).$$

Thus,

$$\limsup_{n \to \infty} \int_{\mathbf{K} \setminus \mathbf{K}_r} |C(\mathbf{k})| v_n(d\mathbf{k}) \leq \limsup_{n \to \infty} v_n(C_{S \setminus F_{1/r}}),$$

and then, along a sample trajectory in $\Omega^*$,

$$\limsup_{n \to \infty} \int_{\mathbf{K} \setminus \mathbf{K}_r} |C(\mathbf{k})| v_n(d\mathbf{k}) \leq 1/r, \quad r = 1,2,3,\ldots,$$

showing that the cost function $C$ is uniformly integrable with respect to the family $\{v_n\}$ of empirical measures.

(iii) Since Theorem 3.8.1 yields that $P_x^\pi[\Omega^*] = 1$ for every $x \in S$ and $\pi \in \mathscr{P}$, the previous discussion can be summarized as follows: Regardless of the initial state and the policy used to drive the system, the following assertions hold with probability 1: (a) The sequence $\{v_n\}$ is tight and (b) the cost function is uniformly integrable with respect to $\{v_n\}$.

The proof of Theorem 3.8.1 relies on the following lemma.

**Lemma 3.8.1.** *Let $\varepsilon > 0$ be arbitrary, and suppose that Assumptions 3.2.1 and 3.4.1 hold, and let $g_{S \setminus F}$ be the optimal expected average cost associated with the cost function $-C_{S \setminus F}$, that is,*

$$g_{S \setminus F} := \inf_{\pi \in \mathscr{P}} J(x, \pi, -C_{S \setminus F}), \tag{3.41}$$

*where $J(x, \pi, -C_{S \setminus F})$ is given by the right-hand side of (3.2) with the function $-C_{S \setminus F}$ instead of $C$. With this notation, there exists a finite set $F \subset S$ such that*

$$g_{S \setminus F} \geq -\varepsilon. \tag{3.42}$$

*Proof.* Let $\{F_k\}$ be a sequence of finite subsets of $S$ such that

$$F_k \subset F_{k+1}, \quad k = 1,2,3,\ldots, \quad \text{and} \quad \bigcup_{k=1}^{\infty} F_k = S; \tag{3.43}$$

from (3.38), it follows that $-C_{S \setminus F_k} \nearrow 0$ as $k \nearrow \infty$, a property that *via* (3.41) immediately yields that

$$g_{S \setminus F_k} \leq g_{S \setminus F_{k+1}} \leq 0, \quad k = 1,2,3,\ldots,$$

so that $\{g_{S\setminus F_k}\}$ is a convergent sequence; set

$$\bar{g}: = \lim_{k\to\infty} g_{S\setminus F_k}. \tag{3.44}$$

To establish the desired conclusion, it is sufficient to show that

$$\bar{g} = 0,$$

since in this case (3.42) occurs when $F$ is replaced by $F_k$ with $k$ large enough. To verify the above equality, let $\ell$ be a Lyapunov function for the function $C$ and notice that Lemma 3.3.3 yields that, for some constant $c > 0$, the mapping $c\ell$ is a Lyapunov function for $-C_{S\setminus F_k}$, and then, this last function also satisfies Assumption 3.4.1. Thus, from Lemma 3.3.2 applied to the cost function $C_{S\setminus F_k}$, it follows that there exists a function $h_k: S \to \mathbb{R}$ as well as a policy $f_k \in \mathscr{F}$ such that

$$g_{S\setminus F_k} + h_k(x) = -C_{S\setminus F_k}(x, f_k(x)) + \sum_{y\in S} p_{xy}(f_k(x))h_k(y), \quad x \in S, \tag{3.45}$$

where $h_k(\cdot) \le c\ell(\cdot)$, that is,

$$h_k(\cdot) \in \prod_{x\in S}[c\ell(x), c\ell(x)]. \tag{3.46}$$

Using the fact that the right-hand side of this inclusion as well as $\mathscr{F}$ are compact metric spaces, it follows that there exists a sequence $\{k_r\}$ of positive integers increasing to $\infty$ such that the following limits exist:

$$\bar{f}(x): = \lim_{r\to\infty} f_{k_r}(x), \quad \bar{h}(x): = \lim_{r\to\infty} h_{k_r}(x), \quad x \in S.$$

Next, observe that (3.38) and (3.43) together yield that for each state $x$,

$$C_{S\setminus F_k}(x, f_k(x)) = 0 \quad \text{when } k \text{ is large enough,}$$

whereas, *via* Proposition 2.18 in p. 232 of [18], the continuity property in Definition 3.3.1(ii) and (3.46) leads to

$$\lim_{r\to\infty} \sum_{y\in S} p_{xy}(f_{k_r}(x))h_{k_r}(y) = \sum_{y\in S} p_{xy}(\bar{f}(x))\bar{h}(y).$$

Replacing $k$ by $k_r$ in (3.45) and taking the limit as $r$ goes to $\infty$ in both sides of the resulting equation, (3.44) and the three last displays allow to write that

$$\bar{g} + \bar{h}(x) = \sum_{y\in S} p_{xy}(\bar{f}(x))\bar{h}(y), \quad x \in S.$$

Starting from this relation, an induction argument yields that $(n+1)\bar{g} + \bar{h}(x) = E_x^{\bar{f}}[\bar{h}(X_{n+1})]$ for every $x \in S$ and $n \in \mathbb{N}$, that is,

$$\bar{g} = \frac{1}{n+1} E_x^{\bar{f}}[\bar{h}(X_{n+1})] - \frac{\bar{h}(x)}{n+1},$$

and taking the limit as $n$ goes to $\infty$, the inclusion (3.46) and Lemma 3.3.1(ii) together imply that $\bar{g} = 0$; as already mentioned, this completes the proof of the lemma.  $\square$

*Proof of Theorem 3.8.1.* Recalling that Assumption 3.4.1 is in force, let $\ell$ be a Lyapunov function for the cost function $C$ such that $\ell^{\beta}$ admits a Lyapunov function for some $\beta > 2$. As already noted, for some constant $c > 0$, the function $c\ell$ is a Lyapunov function for $-C_{S\backslash F}$, a fact that immediately implies that this last function also satisfies Assumption 3.4.1. Therefore, applying Theorem 3.4.1 with the cost function $-C_{S\backslash F}$ instead of $C$, it follows that for every $x \in S$ and $\pi \in \mathscr{P}$,

$$\liminf_{n\to\infty} v_n(-C_{S\backslash F}) \geq g_{S\backslash F}, \quad P_x^{\pi}\text{-a. s.},$$

and selecting $F$ as the finite set in Lemma 3.8.1, it follows that

$$\liminf_{n\to\infty} v_n(-C_{S\backslash F}) \geq -\varepsilon, \quad P_x^{\pi}\text{-a. s.},$$

a statement that is equivalent to (3.39).                                      $\square$

## 3.9  Proof of Theorem 3.7.1

In this section a proof of the sample-path average optimality of a Markovian policy satisfying condition (3.37) will be presented. The argument combines Theorem 3.8.1 with the following lemma.

**Lemma 3.9.1.** *Let the Markovian policy $\mathbf{f} = \{f_t\}$ be such that (3.37) holds, and consider a fixed sample trajectory $\{(X_t, A_t)\}$ along which properties (i) and (ii) below hold:*

*(i) The sequence $\{v_n\}$ of empirical measures is tight.*
*(ii) $A_t = f(X_t)$ for all $t \in N$.*

*In this context, if $v^*$ is a limit point of the sequence $\{v_n\}$ in the weak convergence topology, then $v^*$ is supported on*

$$\mathbb{K}^* = \{(x,a) \in \mathbb{K} \mid \Phi(x,a) = 0\}, \tag{3.47}$$

*that is,*

$$v^*(\mathbb{K}^*) = 1.$$

*Proof.* For each $x \in S$ and $\varepsilon > 0$ defines the set

$$\mathbb{K}(x, \varepsilon) = \{(x, a) \mid a \in A(x) \text{ and } \Phi(x, a) > \varepsilon\},$$

which is an open subset of $\mathbb{K}$, since the function $\Phi$ is continuous and $S$ is endowed with the discrete topology. In this case,

$$\delta_{(X_t, A_t)}(\mathbb{K}(x, \varepsilon)) = \delta_{(X_t, f_t(X_t))}(\mathbb{K}(x, \varepsilon)) = 1 \iff X_t = x \text{ and } \Phi(x, f_t(x)) > \varepsilon.$$

Therefore, using condition (3.37), it follows that there exists an integer $N > 0$ such that

$$\delta_{(X_t, A_t)}(\mathbb{K}(x, \varepsilon)) = 0, \quad t > N,$$

so that

$$v_n(\mathbb{K}(x, \varepsilon)) = \frac{1}{n} \sum_{t=0}^{n-1} \delta_{(X_t, A_t)}(\mathbb{K}(x, \varepsilon)) = \frac{1}{n} \sum_{t=0}^{N} \delta_{(X_t, f_t(X_t))}(\mathbb{K}(x, \varepsilon)), \quad n > N,$$

by Definition 3.8.1, and it follows that $v_n(\mathbb{K}(x, \varepsilon)) \to 0$ as $n \to \infty$. Therefore, recalling that $\mathbb{K}(x, \varepsilon)$ is an open subset of $\mathbb{K}$, the fact that $v^*$ is a limit point of the sequence $\{v_n\}$ implies that

$$v^*(\mathbb{K}(x, \varepsilon)) \leq \limsup_{n \to \infty} v_n(\mathbb{K}(x, \varepsilon)) = 0, \quad x \in S, \quad \varepsilon > 0$$

[2]. Finally, using that $\mathbb{K} \setminus \mathbb{K}^* = \bigcup_{x \in S, \, r \in \mathbb{N}} \mathbb{K}(x, 1/r)$ (because of the nonnegativity of $\Phi$), the above inequality leads to $v^*(\mathbb{K} \setminus \mathbb{K}^*) = 0$.  $\square$

*Proof of Theorem 3.7.1.* It will be proved that

$$\text{the discrepancy function } \Phi \text{ satisfies Assumption 3.4.1.} \tag{3.48}$$

Assuming that this assertion holds, the conclusion of Theorem 3.7.1 can be obtained as follows: Let $\mathbf{f} = \{f_t\}$ be a Markovian policy satisfying the property (3.37) and, for $\tilde{S} \subset S$, define

$$\Phi_{(\tilde{S})}(x, a) := \Phi(x, a) I_{\tilde{S}}(x), \quad (x, a) \in \mathbb{K}, \tag{3.49}$$

so that the equality

$$\Phi = \Phi_{(\tilde{S})} + \Phi_{(S \setminus \tilde{S})} \tag{3.50}$$

is always valid. Next, given $\varepsilon > 0$, observe the following facts (a) and (b):

(a) The property (3.48) allows to apply Theorem 3.8.1 with the cost function $C$ replaced by $\Phi$ to conclude that there exists a finite set $F_\varepsilon \subset S$ such that for each $x \in S$, the relation $\limsup_{n \to \infty} v_n(\Phi_{S \setminus F_\varepsilon}) \leq \varepsilon$ occurs almost surely with respect to $P_x^{\mathbf{f}}$; since (3.49) and (3.38) together imply that $\Phi_{(S \setminus F_\varepsilon)} \leq \Phi_{S \setminus F_\varepsilon}$, it follows that

$$\limsup_{n \to \infty} V_n(\Phi_{(S \setminus F_\varepsilon)}) \leq \varepsilon \quad P_x^{\mathbf{f}}\text{-a.s.} \tag{3.51}$$

(b) Let $\{(X_t, A_t)\}$ be a fixed sample trajectory along which

    (i) $\{v_n\}$ is tight, and

    (ii) $A_t = f_t(X_t)$.

Now select a sequence $\{n_k\}$ of positive integers such that $\lim_{k \to \infty} n_k = \infty$ and

$$\lim_{k \to \infty} V_{n_k}(\Phi_{(F_\varepsilon)}) = \limsup_{n \to \infty} V_n(\Phi_{(F_\varepsilon)}).$$

Because of the tightness of $\{v_n\}$, taking a subsequence—if necessary—it can be assumed that $\{v_{n_k}\}$ converges weakly to some $v^* \in \mathbb{P}(\mathbb{K})$, and in this case, observing that $\Phi_{(F_\varepsilon)}$ is continuous and has compact support (since $F_\varepsilon \subset S$ is finite), it follows that

$$\lim_{k \to \infty} V_{n_k}(\Phi_{(F_\varepsilon)}) = v^*(\Phi_{(F_\varepsilon)});$$

on the other hand, using that $v^*$ is supported in the set $\mathbb{K}^*$ specified in (3.47), by Lemma 3.9.1, and that $\Phi_{F_\varepsilon}$ is null on that set (see (3.47) and (3.49)), it follows that $v^*(\Phi_{F_\varepsilon}) = 0$. Combining this equality with the two last displays, it follows that $\limsup_{n \to \infty} V_n(\Phi_{(F_\varepsilon)}) = 0$ along a sample-path satisfying conditions (i) and (ii) above. Observing that condition (i) holds almost surely with respect to $P_x^{\mathbf{f}}$, by Remark 3.8.1, and that $P_x^{\mathbf{f}}[A_t = f_t(X_t)] = 1$ for all $t$, it follows that

$$\limsup_{n \to \infty} V_n(\Phi_{(F_\varepsilon)}) = 0 \quad P_x^{\mathbf{f}}\text{-a.s.} ,$$

a relation that, combined with (3.50), (3.51) and the nonnegativity of $\Phi$, yields that the convergence $\lim_{n \to \infty} V_n(\Phi) = 0$ occurs with probability 1 with respect to $P_x^{\mathbf{f}}$, and then, the policy $\mathbf{f}$ is sample-path average optimal, by Lemma 3.6.1. Thus, to conclude the argument, it is sufficient to verify (3.48). To achieve this goal, let $\ell$ be a Lyapunov function for the cost function $C$ such that $\ell^\beta$ admits a Lyapunov function for some $\beta > 2$, and recall that the solution $(g, h(\cdot))$ of the optimality Equation (3.7) satisfies

$$|h(\cdot)| \leq c\ell(\cdot) \tag{3.52}$$

for some $c > 0$, as well as $h(z) = 0$. Combining this last equality with the specification of the discrepancy function, it follows that, for all $(x, a) \in \mathbb{K}$,

$$h(x) = C(x, a) - g - \Phi(x, a) + \sum_{y \neq z} p_{xy}(a) h(y). \tag{3.53}$$

On the other hand, Lemma 3.3.3 yields that $|C(\cdot) - g|$ admits a Lyapunov function of the form $c_1 \ell$ where $c_1 > 0$ so that

$$c_1 \ell(x) \geq |C(x,a) - g| + 1 + \sum_{y \in S, \, y \neq z} p_{xy}(a) c_1 \ell(y);$$

multiplying both sides of this relation by a constant $c_2$ satisfying

$$c_2 c_1 > c \text{ and } c_2 > 1, \tag{3.54}$$

it follows that

$$c_2 c_1 \ell(x) \geq c_2 |C(x,a) - g| + c_2 + \sum_{y \in S, \, y \neq z} p_{xy}(a) c_2 c_1 \ell(y),$$

and then,

$$c_2 c_1 \ell(x) \geq |C(x,a) - g| + 1 + \sum_{y \in S, \, y \neq z} p_{xy}(a) c_2 c_1 \ell(y).$$

Combining this inequality with (3.53), it is not difficult to obtain that

$$\tilde{\ell}(x) \geq 1 + \Phi(x,a) + \sum_{y \in S, \, y \neq z} p_{xy}(a) \tilde{\ell}(y), \quad (x,a) \in \mathbb{K}, \tag{3.55}$$

where $\tilde{\ell}(\cdot): \ = c_1 c_2 \ell(\cdot) - h(\cdot) \geq 0$ and the inequality follows from (3.52) and (3.54). Recalling that $\Phi$ is nonnegative, the above display yields that that $\tilde{\ell}$ takes values in $[1, \infty)$ and that $\tilde{\ell}$ satisfies the first requirement for being a Lyapunov function for $\Phi$; setting $\tilde{c}: \ = c_1 c_2 + c > 0$, (3.53) implies that $\tilde{\ell}(\cdot) \leq \tilde{c}\, \ell$, and then, $\tilde{\ell}$ inherits the second and third properties in Definition 3.3.1 from the corresponding ones of $\ell$. Thus, $\tilde{\ell}$ is a Lyapunov function for $\Phi$, and using that $\tilde{\ell}^\beta \leq \tilde{c}^\beta \ell^\beta$, it follows that $\tilde{\ell}^\beta$ also admits a Lyapunov function, by Lemma 3.3.3. Thus, the statement (3.48) holds, and as already mentioned, this concludes the proof of Theorem 3.7.1. $\qquad\square$

**Acknowledgment** With sincere gratitude and appreciation, the authors dedicate this work to Professor Onésimo Hernández-Lerma on the occasion of his 65th anniversary, for his friendly and generous support and clever guidance.

# References

1. Arapostathis A., Borkar V.K., Fernández-Gaucherand E., Ghosh M.K., Marcus S.I.: Discrete time controlled Markov processes with average cost criterion: a survey. SIAM J. Control Optim. **31**, 282–344 (1993)
2. Billingsley P. : Probability and Measure, 3rd edn. Wiley, New York (1995)

3. Borkar V.K: On minimum cost per unit of time control of Markov chains, SIAM J. Control Optim. **21**, 652–666 (1984)
4. Borkar V.K.: Topics in Controlled Markov Chains, Longman, Harlow (1991)
5. Cavazos-Cadena R., Hernández-Lerma O.: Equivalence of Lyapunov stability criteria in a class of Markov decision processes, Appl. Math. Optim. **26**, 113–137 (1992)
6. Cavazos-Cadena R., Fernández-Gaucherand E.: Denumerable controlled Markov chains with average reward criterion: sample path optimality, Math. Method. Oper. Res. **41**, 89–108 (1995)
7. Cavazos-Cadena R., Fernández-Gaucherand E.: Value iteration in a class of average controlled Markov chains with unbounded costs: Necessary and sufficient conditions for pointwise convergence, J. App. Prob. **33**, 986–1002 (1996)
8. Cavazos-Cadena R.: Adaptive control of average Markov decision chains under the Lyapunov stability condition, Math. Method. Oper. Res. **54**, 63–99 (2001)
9. Hernández-Lerma O.: Adaptive Markov Control Processes, Springer, New York (1989)
10. Hernández-Lerma O.: Existence of average optimal policies in Markov control processes with strictly unbounded costs, Kybernetika, **29**, 1–17 (1993)
11. Hernández-Lerma O., Lasserre J.B.: Value iteration and rolling horizon plans for Markov control processes with unbounded rewards, J. Math. Anal. Appl. **177**, 38–55 (1993)
12. Hernández-Lerma O., Lasserre J.B.: Discrete-time Markov control processes: Basic optimality criteria, Springer, New York (1996)
13. Hernández-Lerma O., Lasserre J.B.: Further Topics on Discrete-time Markov Control Processes, Springer, New York (1999)
14. Hordijk A.: Dynamic Programming and Potential Theory (Mathematical Centre Tract 51.) Mathematisch Centrum, Amsterdam (1974)
15. Hunt F.Y.: Sample path optimality for a Markov optimization problem, Stoch. Proc. Appl. **115**, 769–779 (2005)
16. Lasserre J.B:: Sample-Path average optimality for Markov control processes, IEEE T. Automat. Contr. **44**, 1966–1971 (1999)
17. Montes-de-Oca R., Hernández-Lerma O.: Value iteration in average cost Markov control processes on Borel spaces, Acta App. Math. **42**, 203–221 (1994)
18. Royden H.L.: Real Analysis, 2nd edn. MacMillan, New York (1968)
19. Shao J.: Mathematical Statistics, Springer, New York (1999)
20. Vega-Amaya O.: Sample path average optimality of Markov control processes with strictly unbounded costs, Applicationes Mathematicae, **26**, 363–381 (1999)

# Chapter 4
# Approximation of Infinite Horizon Discounted Cost Markov Decision Processes

François Dufour and Tomás Prieto-Rumeau

## 4.1 Introduction

Markov decision processes (MDPs) constitute a general family of controlled
stochastic processes suitable for the modeling of sequential decision-making problems. They appear in many fields such as, for instance, engineering, computer
science, economics, operations research, etc. A significant list of references on
discrete-time MDPs may be found in the survey [2] and the books [1,4,8,10,11,17,
18]. The analysis of MDPs leads to a large variety of interesting mathematical and
computational problems. The corresponding theory has reached a rather high degree
of maturity although the classical tools, such as value iteration, policy iteration,
linear programming, and their various extensions, are generally hardly applicable in
practice. Hence, solving MDPs numerically is an awkward and important problem
mainly because MDPs are generally *very large* due to their inherent structure, and
so, solving the associated dynamic programming equation leads to the well known
curse of dimensionality. In order to meet this challenge, different approaches have
emerged, which can be roughly classified in three classes of techniques: they are
based on:

(i) The approximation of the control model via state aggregation or discretization
    so as to provide a simpler model
(ii) The approximation of the value function through adaptation of classical
     techniques (such as dynamic programming or value iteration)

F. Dufour
Institut de Mathématiques de Bordeaux, Université Bordeaux I, Talence 33405, France

INRIA Bordeaux Sud Ouest, Team CQFD, Bordeaux 78153, France
e-mail: dufour@math.u-bordeaux1.fr

T. Prieto-Rumeau (✉)
Department of Statistics, UNED, Madrid 28040, Spain
e-mail: tprieto@ccia.uned.es

D. Hernández-Hernández and A. Minjárez-Sosa (eds.), *Optimization, Control, and
Applications of Stochastic Systems*, Systems & Control: Foundations & Applications,
DOI 10.1007/978-0-8176-8337-5_4, © Springer Science+Business Media, LLC 2012

(iii) The approximation of the optimal policy by developing efficient policy improvement steps

This chapter belongs to class (i) of the methods described above. Our objective is to present an approximation procedure which transforms an infinite horizon discounted cost MDP with general state and action spaces, and possibly unbounded cost function, into a simpler model (with finite state and action spaces, and, consequently, bounded cost function). We show that the optimal value function and the optimal policies of the MDP under consideration can be approximated by the corresponding optimal value function and the optimal policies of the finite MDP. Most interestingly, *explicit bounds* on the approximation errors are derived. Moreover, it is well known that for infinite horizon discounted cost MDPs, the optimal policy is deterministic and stationary. It must be pointed out that our approximation procedure preserves this important property by providing a suboptimal deterministic stationary policy with guaranteed approximation error.

There exists a huge literature regarding the approaches related to items (ii) and (iii), for example, reinforcement learning, neuro-dynamic programming, approximate dynamic programs, and simulation-based techniques, to name just a few. Without attempting to present an exhaustive panorama on numerical methods for MDPs, we suggest that the interested reader consults the survey [20] and the books [5, 6, 16, 19] and the references therein, to get a rather complete view of this research field.

It is important to stress the fact that the technique developed herein is rather different to those described in (ii) and (iii) above. Indeed, the approaches related to items (ii) and (iii) are related to probabilistic approximation techniques, whereas those of item (i) are related to numerical approximation methods. Moreover, the methods of items (ii) and (iii) mainly deal with MDPs with finite state and/or finite action spaces, and, more importantly, with bounded cost functions; see, for example, the books [5, 6, 16, 19]. These two classes of approximation techniques can be roughly described as follows:

(a) Numerical approximation techniques, which give *actual* bounds or approximations. To fix ideas, suppose that $V^*$ is the optimal value that we want to approximate. In a numerical approximation scheme, given some $\varepsilon > 0$, we find some (number) $\widehat{V}$ such that $|V^* - \widehat{V}| < \varepsilon$. Such techniques can be found in, for example, [9, 12, 14, 21].

(b) Probabilistic approximation techniques, which provide approximations that converge in a suitable probabilistic sense, or bounds that are satisfied with probability "close" to 1. They are usually based on simulating sample paths of the controlled process. In this setting, for given $\varepsilon > 0$, we find a random variable $\overline{V}$ such that $E|V^* - \overline{V}| < \varepsilon$ (convergence in $L_1$), or such that $P\{|V^* - \overline{V}| > \varepsilon\} < \delta$ for some small $\delta$ (convergence in probability). We note that, in practice, we do not obtain $\overline{V}$, but, rather, we observe a realization of the random variable $\overline{V}$.

The probabilistic techniques in (b) are usually more efficient—in computational terms—than the numerical methods in (a), and they even succeed to break the curse of dimensionality. The price to pay, however, is that we approximate a given value, say $V^*$, by a random observation $\bar{V}$. Depending on the application under consideration, the fact that the methods in (a) provide a deterministic estimate with guaranteed bounds can be more attractive than the stochastic estimates with bounds valid in *mean* given by the approaches (b). Clearly, the techniques in (a) and (b) are of a different nature and, so, hard to compare mainly because the approximation is deterministic in case (a) and stochastic in (b).

The discretization procedure (or state aggregation) to convert an MDP into a simpler optimization problem can be traced back to the middle of the 1970s; see [15, Sect. 2.3] for a rather complete account of the works developed at that time. In particular, the discretization procedure has been analyzed in a general context in [12,14,21] and for specific MDPs in [3,9]. It is important to point out the differences between the results obtained in our work and those presented in the literature. In [14], the author studies the approximation of the original decision model by means of a sequence of MDPs. Under mild hypotheses, such as continuous convergence of the data, a convergence result is established. However, as pointed in [14, p. 494], no rates of convergence are given mainly due to the fact that the problem under study is fairly general. The approximation of dynamic programs in the general framework of monotone contraction operator models is developed in [12, 21]; the analysis is then particularized to a classical MDP model. In [21], the author focuses on infinite horizon problems, while in [12], the author addresses the problem of finite-stage models. This approach is based on the construction of partitions of the state and action spaces. The corresponding convergence results and the error bounds are established under general hypotheses. As mentioned in [21, p. 236], however, this approach is "of limited practical value because the optimal return function is used" for the corresponding construction. Moreover, and more importantly, the convergence of the optimal value functions associated to the approximating models depends heavily on an appropriate and relevant choice of the partitions. In [12], the problem of how to construct the approximating decision model (and, therefore, the associated partitions) is not discussed. Nevertheless, it is mentioned that this is an important problem and that a theoretical guideline should be provided (see the introduction of [12, Sect. 4]). In [21], this problem is emphasized too, and under continuity and compactness assumptions, a convergence result is obtained (Theorem 5.2). In [3,9], the discretization procedure of the classical MDP model is addressed under the hypothesis that the state and action spaces are compact, and the cost function is bounded. Their approaches are constructive in the sense that they are applicable in practice.

Our framework here is clearly more general than the one studied in [3,9]. Indeed, we deal with an MDP with general state space (namely, a locally compact Borel space) and compact action sets, while the cost function is allowed to be unbounded. Such extensions are clearly not straightforward to obtain, and they need a careful and sharp analysis of the partitioning of the state and action spaces. Compared to [12, 14, 21], we impose stronger conditions though, as opposed to [14], we are able to derive explicit error bounds for our approximation scheme. Furthermore, rather than embedding our MDP into the general theory of monotone contraction operators as in [12, 21], our idea is to take into account the characteristic features of the MDP under consideration so as to provide a more constructive solution (with the important property that it is applicable in practice). In particular, our approach provides an explicit construction of the partitions of the state and action spaces. As mentioned before, this important problem is not discussed in [12] and it is just briefly studied in [21] for a very specific model (with compact state and action spaces). Consequently, our results are different, and complementary, to those obtained in [12,21]. Moreover, the analysis in [12, 21] is done by assuming that the cost function is bounded, whereas it is briefly explained how the results could be extended to the unbounded case. We claim, however, that these extensions are not straightforward and they depend again on an appropriate and relevant choice of the partitions of the state and action spaces.

The rest of the chapter is organized as follows. In Sect. 4.2, we introduce the MDP model we are interested in and state our main assumptions. Some useful preliminary results are proved in Sect. 4.3. We give our first main result in Sect. 4.4, on the approximation of the optimal value function. An approximation of a discount optimal policy is presented in Sect. 4.5. Our conclusions are stated in Sect. 4.6.

## 4.2 Definition of the Control Model

In this section, we introduce the discrete-time Markov control model we are concerned with and state our main assumptions. We follow closely the notation of Chap. 2 in [10]. Let us consider the five tuple for a Markov control model $\mathcal{M} = (X, A, (A(x), x \in X), P, c)$, where:

(a) The state space $X$ is a locally compact Borel space. The metric on $X$ and the family of the Borel subsets of $X$ will be respectively denoted by $d_X$ and $\mathcal{B}(X)$.

(b) The action space $A$ is a locally compact Borel space. The metric on $A$ will be denoted by $d_A$. In the family of closed subsets of $A$, we will consider the Hausdorff metric, defined as

$$d_H(C_1, C_2) := \sup_{a \in C_1} \inf_{b \in C_2} \{d_A(a, b)\} \vee \sup_{b \in C_2} \inf_{a \in C_1} \{d_A(a, b)\}$$

$$= \sup_{a \in C_1} \{d_A(a, C_2)\} \vee \sup_{b \in C_2} \{d_A(C_1, b)\}$$

for closed $C_1, C_2 \subseteq A$ (where $\vee$ stands for "maximum"). We know that $d_H$ is a metric, except that it might not be finite. By $\mathcal{B}(A)$, we will denote the family of Borel subsets of $A$.

(c) The set of feasible actions at state $x \in X$ is $A(x)$, assumed to be a nonempty measurable subset of $X$. We also suppose that

$$\mathbb{K} := \{(x, a) \in X \times A : a \in A(x)\}$$

is a measurable subset of $X \times A$, which contains the graph of a measurable function from $X$ to $A$.

(d) The transition probability function, which is given by the stochastic kernel $P$ on $X$ given $\mathbb{K}$. This means that $B \mapsto P(B|x,a)$ is a probability measure on $(X, \mathscr{B}(X))$ for every $(x,a) \in \mathbb{K}$ and, in addition, that $(x,a) \mapsto P(B|x,a)$ is measurable on $\mathbb{K}$ for every $B \in \mathscr{B}(X)$.

(e) The measurable cost-per-stage function $c : \mathbb{K} \to \mathbb{R}$.

The family of stochastic kernels $\varphi$ on $A$ given $X$ such that $\varphi(A(x)|x) = 1$ for all $x \in X$ is denoted by $\Phi$. Let $\mathscr{F}$ be the family of measurable functions $f : X \to A$ satisfying that $f(x) \in A(x)$ for all $x \in X$. (By (c) above, the set $\mathscr{F}$ is nonempty.)

In our next definition, we introduce the class of admissible policies for the decision-maker. We use the notation $\mathbb{N}$, which stands for the set of nonnegative integers.

**Definition 4.2.1.** Let $H_0 := X$ and $H_n := \mathbb{K} \times H_{n-1}$ for $n \geq 1$. A *control policy* is a sequence $\pi = \{\pi_n\}_{n \in \mathbb{N}}$ of stochastic kernels $\pi_n$ on $A$ given $H_n$ satisfying that $\pi_n(A(x_n)|h_n) = 1$ for all $h_n \in H_n$ and $n \in \mathbb{N}$, where $h_n := (x_0, a_0, \ldots, x_{n-1}, a_{n-1}, x_n)$ (note that, on $H_n$, we are considering the product $\sigma$-algebra). Let $\Pi$ be the class of all policies.

A policy $\pi = \{\pi_n\}_{n \in \mathbb{N}}$ is said to be *deterministic* if there exists a sequence $\{f_n\}$ in $\mathscr{F}$ such that $\pi_n(\cdot|h_n) = \delta_{f_n(x_n)}(\cdot)$ for all $n \in \mathbb{N}$ and $h_n \in H_n$, where $\delta_a$ denotes the Dirac probability measure supported on $a \in A$.

Finally, if the deterministic policy $\pi = \{\pi_n\}_{n \in \mathbb{N}}$ is such that $\pi_n(\cdot|h_n) = \delta_{f(x_n)}(\cdot)$ for some $f \in \mathscr{F}$, then we say that $\pi$ is a *deterministic stationary policy*. The set of such policies is identified with $\mathscr{F}$.

Let $(\Omega, \mathscr{F})$ be the canonical space consisting of the set of all sample paths $\Omega = (X \times A)^\infty = \{(x_t, a_t)\}_{t \in \mathbb{N}}$ and the associated product $\sigma$-algebra $\mathscr{F}$. Here, $\{x_t\}_{t \in \mathbb{N}}$ is the state process and $\{a_t\}_{t \in \mathbb{N}}$ stands for the action process. For every policy $\pi \in \Pi$ and any initial distribution $v$ on $X$, there exists a unique probability measure $P_v^\pi$ on $(\Omega, \mathscr{F})$ such that, for any $B \in \mathscr{B}(X)$, $C \in \mathscr{B}(A)$, and $h_t \in H_t$ with $t \in \mathbb{N}$,

$$P_v^\pi(x_0 \in B) = v(B), \quad P_v^\pi(a_t \in C|h_t) = \pi_t(C|h_t), \quad P_v^\pi(x_{t+1} \in B|h_t, a_t) = P(B|x_t, a_t).$$

(For such a construction, see, e.g., [10, Chap. 2].) The expectation with respect to $P_v^\pi$ is denoted by $E_v^\pi$. If $v = \delta_x$ for some $x \in X$, then we will write $P_x^\pi$ and $E_x^\pi$ in lieu of $P_v^\pi$ and $E_v^\pi$, respectively.

Next, we define our infinite horizon discounted cost Markov control problem. Suppose that $0 < \alpha < 1$ is a given *discount factor*. For any initial state $x \in X$, the total expected discounted cost (or discounted cost, for short) of a control policy $\pi \in \Pi$ is

$$J(x, \pi) := E_x^\pi \left[ \sum_{t=0}^\infty \alpha^t c(x_t, a_t) \right].$$

(In the sequel, we will impose conditions ensuring that this discounted cost is finite.) The optimal discounted cost function is then defined as

$$J(x) := \inf_{\pi \in \Pi} J(x, \pi) \quad \text{for } x \in X,$$

and we say that a policy $\pi^* \in \Pi$ is *discount optimal* if $J(x, \pi) = J(x)$ for every $x \in X$. We introduce some more notation. Given $t \geq 1$, an initial state $x \in X$, and a control policy $\pi \in \Pi$, let

$$J_t(x, \pi) := E_x^\pi \left[ \sum_{k=0}^{t-1} \alpha^k c(x_k, a_k) \right],$$

that is, $J_t(x, \pi)$ is the discounted cost of the policy $\pi$ on the first $t$ decision epochs. Also, let

$$J_t(x) := \inf_{\pi \in \Pi} J_t(x, \pi) \quad \text{for } x \in X.$$

Our assumptions on the control model use the notion of Lipschitz continuity. Given two metric spaces $(X_1, \rho_1)$ and $(X_2, \rho_2)$, and a function $g : X_1 \to X_2$, we say that $g$ is *L-Lipschitz continuous* on $X_1$ (for some so-called Lipschitz constant $L > 0$) if

$$\rho_2(g(x), g(y)) \leq L \cdot \rho_1(x, y) \quad \text{for every } x \text{ and } y \text{ in } X_1.$$

Typically, we will denote the Lipschitz constant of a Lipschitz continuous function $g$ by $L_g$. If $X_1$ is the product of two metric spaces: $(X_{11}, \rho_{11})$ and $(X_{12}, \rho_{12})$, then, when referring to Lipschitz continuity on $X_1 := X_{11} \times X_{12}$, the metric considered in $X_1$ is the sum of $\rho_{11}$ and $\rho_{12}$.

We state our assumptions on the control model.

**Assumption A.**

*(A1)   as:4:A1For each state $x \in X$, the action set $A(x)$ is compact.*

*(A2)   The multifunction $\Psi$ from $X$ to $A$, defined as $\Psi(x) := A(x)$, is $L_\Psi$-Lipschitz continuous with respect to the Hausdorff metric, i.e., for some constant $L_\Psi > 0$ and every $x, y \in X$,*

$$d_H(A(x), A(y)) \leq L_\Psi \cdot d_X(x, y).$$

*(The sets $A(x)$ being closed—Assumption (A1)—,the Hausdorff metric is indeed well defined.)*

*(A3)   There exists a continuous function $w : X \to [1, \infty)$ and a positive constant $\bar{c}$ such that $|c(x, a)| \leq \bar{c} w(x)$ for all $(x, a) \in \mathbb{K}$. Moreover, the cost function $c$ is $L_c$-Lipschitz continuous on $\mathbb{K}$.*

Before proceeding with our assumptions, we introduce some more notation. Given a function $v : X \to \mathbb{R}$, we define $Pv : \mathbb{K} \to \mathbb{R}$ as

$$Pv(x, a) := \int_X v(y) P(dy|x, a) \quad \text{for } (x, a) \in \mathbb{K},$$

provided that the corresponding integrals are well defined and finite. The class of measurable functions $v : X \to \mathbb{R}$ such that $||v||_w := \sup_{x \in X} \{|v(x)|/w(x)\}$ is finite, with $w$ as in Assumption (A3), denoted by $\mathbb{B}_w(X)$, is a Banach space and $||\cdot||_w$ is indeed a norm (called the $w$-norm). Let $\mathbb{L}_w(X)$ be the family of functions in $\mathbb{B}_w(X)$ which, in addition, are Lipschitz continuous. We proceed with Assumption A.

  *(A4.i)* *The function $Pw$ is continuous on $\mathbb{K}$. In addition, there exists $\overline{d} > 0$ such that $Pw(x,a) \leq \overline{d}w(x)$ for all $(x,a) \in \mathbb{K}$.*

  *(A4.ii)* *The stochastic kernel $P$ is weakly continuous, meaning that $Pv$ is continuous on $\mathbb{K}$ for every bounded and continuous function $v$ on $X$.*

  *(A4.iii)* *There exists a constant $L_P > 0$ such that for every $(x,a)$ and $(y,b)$ in $\mathbb{K}$, and $v \in \mathbb{L}_w(X)$, with Lipschitz constant $L_v$,*

$$|Pv(x,a) - Pv(y,b)| \leq L_P L_v [d_X(x,y) + d_A(a,b)].$$

  *In this case, the stochastic kernel $P$ is said to be $L_P$-Lipschitz continuous.*

  *(A5)* *The discount factor $\alpha$ satisfies $\alpha\overline{d} < 1$ and $\alpha L_P(1 + L_\Psi) < 1$.*

*Remark 4.2.1.* Lipschitz continuity of $\Psi$ in Assumption (A2) and the fact that the action sets are compact (Assumption (A1)) imply that $\Psi$ is a continuous multifunction from $X$ to $A$; see Lemma 2.6 in [7].

Given $x \in X$ and $f \in \mathscr{F}$, the notation

$$c(x,f) := c(x, f(x)), \ P(\cdot|x,f) := P(\cdot|x, f(x)), \text{ and } Pv(x,f) := Pv(x, f(x))$$

will be used in the sequel.

## 4.3 Preliminary Results

In this section, we state some technical lemmas that will be useful in the rest of the chapter. First, we derive some properties of the Bellman operator.

**Lemma 4.3.1.** *Suppose that Assumption A is satisfied, and define the operator $T$ as*

$$Tu(x) := \min_{a \in A(x)} \left[ c(x,a) + \alpha \int_X u(y)P(dy|x,a) \right] \quad \text{for } x \in X,$$

*where $u \in \mathbb{L}_w(X)$.*

  *(i)* *If $u \in \mathbb{L}_w(X)$, then $Tu \in \mathbb{L}_w(X)$ and $L_{Tu} = (L_c + \alpha L_P L_u)(1 + L_\Psi)$.*

  *(ii)* *The operator $T$ is a contraction mapping with modulus $\alpha\overline{d}$.*

  *(iii)* *$T$ has a unique fixed point $u^* \in \mathbb{L}_w(X)$, with $\|u^*\|_w \leq \overline{c}/(1 - \alpha\overline{d})$. The fixed point $u^*$ is the limit (in the $w$-norm) of $\{T^n u_0\}_{n \geq 1}$ for every $u_0 \in \mathbb{L}_w(X)$.*

*Proof.* *Part (i).* Given $x$ and $y$ in $X$, we have that

$$|Tu(x) - Tu(y)| \leq \sup_{b \in B(y)} \inf_{a \in A(x)} |\Delta(x,a,y,b)| \vee \sup_{a \in A(x)} \inf_{b \in B(y)} |\Delta(x,a,y,b)|,$$

where $\Delta(x,a,y,b) := c(x,a) - c(y,b) + \alpha(Pu(x,a) - Pu(y,b))$ for $(x,a)$ and $(y,b)$ in $\mathbb{K}$. By Lipschitz continuity of the cost function $c$ and the stochastic kernel $P$, we obtain

$$|\Delta(x,a,y,b)| \leq (L_c + \alpha L_P L_u) \cdot (d_X(x,y) + d_A(a,b)).$$

Finally, recalling the definition of the Hausdorff metric, we conclude that

$$|Tu(x) - Tu(y)| \leq (L_c + \alpha L_P L_u) \cdot (d_X(x,y) + d_H(A(x),A(y)))$$
$$\leq [(L_c + \alpha L_P L_u)(1 + L_\Psi)] \cdot d_X(x,y).$$

Hence, $Tu \in \mathbb{L}_w(X)$ and its Lipschitz constant is $(L_c + \alpha L_P L_u)(1 + L_\Psi)$.

*Part (ii).* Proceeding as in Part (i), given $u,v \in \mathbb{L}_w(X)$, and $x \in X$, we have

$$|Tu(x) - Tv(x)| \leq \alpha \sup_{a \in A(x)} P|u - v|(x,a)$$
$$\leq \alpha \|u - v\|_w \sup_{a \in A(x)} Pw(x,a) \leq \alpha \overline{d} \|u - v\|_w w(x),$$

by Assumption (A4.i), and the stated result follows.

*Part (iii).* Based on Banach's fixed point theorem, it is easily seen that, given arbitrary $u_0 \in \mathbb{L}_w(X)$, the sequence $\{T^n u_0\}_{n \in \mathbb{N}}$ is Cauchy in the $w$-norm. Hence, it converges to some $u^* \in \mathbb{B}_w(X)$. Observe now that, as a consequence of Part (i) and Assumption (A5), the Lipschitz constant $L_n$ of $T^n u_0$ verifies

$$\lim_{n \to \infty} L_n = \frac{L_c(1 + L_\Psi)}{1 - \alpha L_P(1 + L_\Psi)}.$$

Hence, since for every $n \in \mathbb{N}$ and $x,y \in X$ we have $|T^n u_0(x) - T^n u_0(y)| \leq L_n d_X(x,y)$, it follows that

$$|u^*(x) - u^*(y)| \leq \frac{L_c(1 + L_\Psi)}{1 - \alpha L_P(1 + L_\Psi)} \cdot d_X(x,y),$$

so that $u^* \in \mathbb{L}_w(X)$. The fact that $u^*$ is the unique fixed point of $T$ in $\mathbb{B}_w(X)$ follows from standard arguments. Finally, letting $u_0 \equiv 0$, by Assumption (A3), we have that $\|Tu_0\|_w \leq \overline{c}$, and so we obtain

$$\|u^*\|_w \leq \|u^* - Tu_0\|_w + \|Tu_0\|_w \leq \alpha \overline{d} \|u^*\|_w + \overline{c},$$

and $\|u^*\|_w \leq \overline{c}/(1 - \alpha \overline{d})$ follows.                                          $\square$

Under Assumption A, the Assumptions 8.5.1, 8.5.2, and 8.5.3 in [11] are satisfied. Then, as a direct consequence of Lemma 4.3.1 and the results in [11, Sect. 8.3.B], we derive the following facts. The Lipschitz continuity of the optimal discounted cost functions, established below, can be found also in [13, Theorem 4.1].

**Theorem 4.3.1.** *If Assumption A is satisfied, then the following statements hold.*

(i) *For every $t \geq 1$, the optimal discounted cost $J_t$ in the first $t$ decision epochs is $T^t 0$ (that is, the $t$-times composition of $T$ with itself starting from $J_0 \equiv 0$). In particular, $J_t \in \mathbb{L}_w(X)$ with*

$$L_{J_t} = \frac{L_c(1+L_\Psi)(1-(\alpha L_P(1+L_\Psi))^t)}{1-\alpha L_P(1+L_\Psi)} \leq \frac{L_c(1+L_\Psi)}{1-\alpha L_P(1+L_\Psi)}$$

*and $||J_t||_w \leq \bar{c}(1-(\alpha\bar{d})^t)/(1-\alpha\bar{d})$.*

(ii) *The optimal discounted cost function $J$ is in $\mathbb{L}_w(X)$, and it is the unique fixed point of the operator $T$ in $\mathbb{B}_w(X)$; its Lipschitz constant is*

$$L_J = \frac{L_c(1+L_\Psi)}{1-\alpha L_P(1+L_\Psi)} \tag{4.1}$$

*and $||J||_w \leq \bar{c}/(1-\alpha\bar{d})$.*

(iii) *The sequence $\{J_t\}_{t\in\mathbb{N}}$ converges to $J$ in the $w$-norm. In addition,*

$$||J_t - J||_w \leq \bar{c}(\alpha\bar{d})^t/(1-\alpha\bar{d}) \quad \textit{for every } t \geq 0.$$

Our next result is a well-known fact, and it will be useful in the forthcoming.

**Lemma 4.3.2.** *Suppose that Assumption A is verified, and let $f \in \mathscr{F}$ and $v \in \mathbb{B}_w(X)$ be such that*

$$v(x) \geq c(x,f) + \alpha \int_X v(y)P(dy|x,f) \quad \textit{for all } x \in X. \tag{4.2}$$

*Then $v(x) \geq J(x,f)$ for every $x \in X$.*

*Proof.* Integrating, recursively, the inequality (4.2) with respect to $P(dy|x,f)$, we obtain, for every $n \geq 1$,

$$v(x) \geq J_n(x,f) + \alpha^n E_x^f[v(x_n)] \quad \text{for } x = x_0 \in X. \tag{4.3}$$

Now, on the one hand, $|E_x^f[v(x_n)]| \leq ||v||_w \bar{d}^n w(x)$ by Assumption (A4.i), and thus, since $\alpha\bar{d} < 1$,

$$\lim_{n\to\infty} \alpha^n E_x^f[v(x_n)] = 0.$$

On the other hand,

$$\left| E_x^f \left[ \sum_{t \geq n} \alpha^t c(x_t, a_t) \right] \right| \leq \bar{c}w(x) \sum_{t \geq n} (\alpha\bar{d})^t,$$

and consequently, $\lim_{n\to\infty} J_n(x,f) = J(x,f)$. Taking the limit as $n \to \infty$ in (4.3), we obtain the desired result. □

The proof of our next result can be found in Lemma 2.9 in [7].

**Lemma 4.3.3.** *We suppose that Assumption A holds. Let $K_0$ be an arbitrary compact subset of $X$ and fix $\varepsilon > 0$. There exists a sequence $\{K_t\}_{t \in \mathbb{N}}$ of compact subsets of $X$ such that, for all $t \in \mathbb{N}$,*

$$\sup_{(x,a) \in \mathbb{K}_t} \int_{K_{t+1}^c} w(y) P(dy|x,a) \leq \varepsilon,$$

*where $\mathbb{K}_t := \{(x,a) \in X \times A : x \in K_t, a \in A(x)\}$.*

## 4.4 Approximation of the Optimal Discounted Cost Function

In this section, we propose an approximation of the optimal discounted cost function $J$ by using the fact that $J$ is the fixed point of the operator $T$. Our main result is presented in Theorem 4.4.2.

Suppose that $(Z, \rho)$ is a compact metric space and fix a constant $\eta > 0$. A finite subset $\Gamma := \{z_1, \ldots, z_r\}$ of $Z$ is said to be associated to an $\eta$-partition of $Z$ if there exists a finite measurable partition $\{Z_1, \ldots, Z_r\}$ of $Z$ which satisfies the two conditions below.

(i)  We have $z_i \in Z_i$ for all $1 \leq i \leq r$. The projection operator $p_\Gamma : Z \to \Gamma$ is then defined as $p_\Gamma(z) := z_i$ if $z \in Z_i$.
(ii) For every $z \in Z$, it is $\rho(z, p_\Gamma(z)) \leq \eta$.

Clearly, such partitions exist because $Z$ is compact.

Fix now arbitrary positive constants $\varepsilon$, $\beta$, and $\delta$. *In what follows, the constants $\varepsilon$, $\beta$, and $\delta$ remain fixed and, so, they will not be explicit in the notation.*

For $\varepsilon > 0$, consider the sequence of compact sets $\{K_t\}_{t \in \mathbb{N}}$ that was constructed in Lemma 4.3.3. For every $t \in \mathbb{N}$, let $\Gamma_t$ be a finite subset of $K_t$ associated to a $\beta$-partition of $K_t$. Now, given $x \in \Gamma_t$ for $t \in \mathbb{N}$, let $\Theta_t(x)$ be a finite subset of $A(x)$ associated to a $\delta$-partition of the compact metric action space $A(x)$.

Fix a time horizon $N \geq 1$. The functions $\widehat{J}_{N,t}$ on $\Gamma_t$, for $0 \leq t \leq N$, are given by $\widehat{J}_{N,N}(x) := 0$ for $x \in \Gamma_N$ and, for $0 \leq t < N$, by

$$\widehat{J}_{N,t}(x) := \min_{a \in \Theta_t(x)} \left[ c(x,a) + \alpha \sum_{y \in \Gamma_{t+1}} \widehat{J}_{N,t+1}(y) P(p_{\Gamma_{t+1}}^{-1}\{y\}|x,a) \right] \quad \text{for } x \in \Gamma_t. \quad (4.4)$$

The functions $\widehat{J}_{N,t}$ are extended to $K_t$ by letting $\widehat{J}_{N,t}(x) := \widehat{J}_{N,t}(p_{\Gamma_t}(x))$ for $x \in K_t$. Therefore, $\widehat{J}_{N,N}(x) = 0$ for all $x \in K_N$, and (4.4) becomes

$$\widehat{J}_{N,t}(x) = \min_{a \in \Theta_t(x)} \left[ c(x,a) + \alpha \int_{K_{t+1}} \widehat{J}_{N,t+1}(y) P(dy|x,a) \right] \quad \text{for } x \in \Gamma_t.$$

**Lemma 4.4.4.** *If Assumption A holds, then, for every $N \geq 1$ and $0 \leq t \leq N$,*

$$\sup_{x \in K_t} |\widehat{J}_{N,t}(x) - J_{N-t}(x)| \leq \frac{1}{1-\alpha} \cdot \left( \frac{L_c(1+L_\Psi)}{1-\alpha L_P(1+L_\Psi)} \cdot (\beta + \delta) + \frac{\alpha \overline{c}}{1-\alpha \overline{d}} \cdot \varepsilon \right).$$

*Proof.* Throughout this proof, the integer $N \geq 1$ remains fixed. To simplify the notation, in what follows, we will write $H_{N,t} := \sup_{x \in K_t} |\widehat{J}_{N,t}(x) - J_{N-t}(x)|$ for $0 \leq t \leq N$ and we note that $H_{N,N} = 0$.

For $0 \leq t < N$ and $z \in \Gamma_t$, we have

$$\widehat{J}_{N,t}(z) = \min_{a \in \Theta_t(z)} \left[ c(z,a) + \alpha \int_{K_{t+1}} \widehat{J}_{N,t+1}(y)P(dy|z,a) \right],$$

$$J_{N-t}(z) = \min_{b \in A(z)} \left[ c(z,b) + \alpha \int_X J_{N-t-1}(y)P(dy|z,b) \right].$$

Consequently, $|\widehat{J}_{N,t}(z) - J_{N-t}(z)|$ is bounded above by

$$\max_{a \in \Theta_t(z)} \inf_{b \in A(z)} |\Delta(z,a,b)| \vee \sup_{b \in A(z)} \min_{a \in \Theta_t(z)} |\Delta(z,a,b)|,$$

where

$$\Delta(z,a,b) = c(z,a) - c(z,b) + \alpha \int_{K_{t+1}} \widehat{J}_{N,t+1}(y)P(dy|z,a) - \alpha P J_{N-t-1}(z,b).$$

Now, given $z \in \Gamma_t$, $a \in \Theta_t(z)$ and $b \in A(z)$, we observe the following: Firstly,

$$|c(z,a) - c(z,b)| \leq L_c \cdot d_A(a,b). \tag{4.5}$$

Secondly,

$$|P J_{N-t-1}(z,b) - P J_{N-t-1}(z,a)| \leq L_P L_{J_{N-t-1}} \cdot d_A(a,b). \tag{4.6}$$

Finally, $| \int_{K_{t+1}} \widehat{J}_{N,t+1}(y)P(dy|z,a) - P J_{N-t-1}(z,a)|$ is less than or equal to

$$\|J_{N-t-1}\|_w \int_{K_{t+1}^c} w(y)P(dy|z,a) + \int_{K_{t+1}} |\widehat{J}_{N,t+1}(y) - J_{N-t-1}(y)|P(dy|z,a),$$

which, by Lemma 4.3.3 and the fact that $(z,a) \in \mathbb{K}_t$, is in turn bounded above by

$$\|J_{N-t-1}\|_w \cdot \varepsilon + \sup_{y \in K_{t+1}} |\widehat{J}_{N,t+1}(y) - J_{N-t-1}(y)| = \|J_{N-t-1}\|_w \cdot \varepsilon + H_{N,t+1}. \tag{4.7}$$

Combining (4.5)–(4.7) yields that $|\Delta(z,a,b)|$ is bounded above by

$$(L_c + \alpha L_P L_{J_{N-t-1}}) \cdot d_A(a,b) + \frac{\alpha \overline{c}}{1-\alpha \overline{d}} \cdot \varepsilon + \alpha H_{N,t+1},$$

where we have used the fact that $||J_{N-t-1}||_w \leq \overline{c}/(1 - \alpha\overline{d})$; see Theorem 4.3.1(i). On the other hand, also by Theorem 4.3.1 and recalling Lemma 4.3.1(i) and (4.1),

$$(L_c + \alpha L_P L_{J_{N-t-1}}) \leq (L_c + \alpha L_P L_{J_{N-t-1}})(1 + L_\Psi) = L_{J_{N-t}} \leq L_J.$$

Finally, by the definition of $\Theta_t(z)$,

$$\max_{a \in \Theta_t(z)} \inf_{b \in A(z)} d_A(a,b) \vee \sup_{b \in A(z)} \min_{a \in \Theta_t(z)} d_A(a,b) \leq \delta,$$

so that, for all $z \in \Gamma_t$,

$$|\widehat{J}_{N,t}(z) - J_{N-t}(z)| \leq L_J \cdot \delta + \frac{\alpha\overline{c}}{1 - \alpha\overline{d}} \cdot \varepsilon + \alpha H_{N,t+1}.$$

Suppose now that $x \in K_t$, and let $z = p_{\Gamma_t}(x) \in \Gamma_t$. We have that

$$|\widehat{J}_{N,t}(x) - J_{N-t}(x)| = |\widehat{J}_{N,t}(z) - J_{N-t}(x)|$$
$$\leq |\widehat{J}_{N,t}(z) - J_{N-t}(z)| + |J_{N-t}(z) - J_{N-t}(x)|,$$

where

$$|J_{N-t}(z) - J_{N-t}(x)| \leq L_{J_{N-t}} d_X(z,x) \leq L_J \cdot \beta.$$

We conclude that, for $0 \leq t < N$,

$$H_{N,t} \leq L_J \cdot (\delta + \beta) + \frac{\alpha\overline{c}}{1 - \alpha\overline{d}} \cdot \varepsilon + \alpha H_{N,t+1},$$

with $H_{N,N} = 0$. The bound on $H_{N,t}$ readily follows by induction: indeed, it holds for $t = N$, and we can prove recursively that it is satisfied for $t = N - 1, \ldots, 1, 0$. $\square$

To alleviate the notation, we will write

$$H := \frac{1}{1 - \alpha} \cdot \left( \frac{L_c(1 + L_\Psi)}{1 - \alpha L_P(1 + L_\Psi)} \cdot (\beta + \delta) + \frac{\alpha\overline{c}}{1 - \alpha\overline{d}} \cdot \varepsilon \right). \tag{4.8}$$

(As already mentioned, the constants $\beta$, $\delta$, and $\varepsilon$ are assumed to be fixed and, hence, we do not make them explicit in the notation $H$ above.)

**Theorem 4.4.2.** *Suppose that the control model $\mathcal{M}$ satisfies Assumption A. Given arbitrary positive constants $\beta$, $\delta$, and $\varepsilon$, consider the functions $\widehat{J}_{N,t}$ constructed above. For every $N \geq 1$ and $x \in K_0$,*

$$|\widehat{J}_{N,0}(x) - J(x)| \leq H + \frac{\overline{c}(\alpha\overline{d})^N}{1 - \alpha\overline{d}} \cdot w(x).$$

*Proof.* By Theorem 4.3.1(iii), for every $x \in K_0$, we have

$$|J_N(x) - J(x)| \le \frac{\overline{c}(\alpha\overline{d})^N}{1 - \alpha\overline{d}} \cdot w(x).$$

The stated result is now a direct consequence of Lemma 4.4.4, with $t = 0$.   □

As a consequence of Theorem 4.4.2, we have the following: Suppose that $x_0 \in X$ is an arbitrary initial state $x_0 \in X$, and let $\eta > 0$ be a given precision. Letting $K_0 = \{x_0\}$, we can, *a priori and explicitly*, determine $\beta$, $\delta$, $\varepsilon$, and $N$ such that

$$\frac{1}{1-\alpha} \cdot \left( \frac{L_c(1+L_\Psi)}{1-\alpha L_P(1+L_\Psi)} \cdot (\beta + \delta) + \frac{\alpha\overline{c}}{1-\alpha\overline{d}} \cdot \varepsilon \right) + \frac{\overline{c}(\alpha\overline{d})^N}{1-\alpha\overline{d}} \cdot w(x_0) < \eta.$$

Indeed, *all the constants involved in the above inequality are known* (they depend on the Lipschitz constants of the control model $\mathcal{M}$ and on other constants introduced in Assumption A). Once we determine $\beta$, $\delta$, $\varepsilon$, and $N$, we proceed with the constructive procedure described in this section and we compute explicitly $\widehat{J}_{N,0}(x_0)$ such that $|\widehat{J}_{N,0}(x_0) - J(x_0)| < \eta$.

## 4.5   Approximation of a Discount Optimal Policy

In the previous section, we provided an approximation of the optimal discounted cost function $J$. In this section, we are concerned with the approximation of a discount optimal policy. There is a straightforward naive approach (that we discard): for $N$ large enough, consider an approximation of the optimal policy of the finite horizon discounted control problem; see [7]. Then, for the first $N$ periods, use this policy, and from time $N$, choose arbitrary actions. This policy is close to discount optimality, but it is not stationary.

We know, however, that there exists a deterministic stationary policy that is discount optimal. Such a policy is obtained, loosely speaking, as the minimizer in the fixed point equation $J = TJ$. From the point of view of the decision-maker, it is more interesting to have at hand a deterministic stationary policy that is close to discount optimality, instead of a nonstationary one. Hence, our goal here is to provide such a deterministic stationary policy, based on our finite approximations scheme.

Given $N \ge 1$, $0 \le t < N$, and $z \in \Gamma_t$, we have

$$\widehat{J}_{N,t}(z) = \min_{a \in \Theta_t(z)} \left[ c(z,a) + \alpha \int_{K_{t+1}} \widehat{J}_{N,t+1}(y) P(dy|z,a) \right]$$

$$= c(z, \widetilde{a}_{N,t}(z)) + \alpha \int_{K_{t+1}} \widehat{J}_{N,t+1}(y) P(dy|z, \widetilde{a}_{N,t}(z))$$

for some $\widetilde{a}_{N,t}(z) \in \Theta_t(z)$. We define $\widetilde{f}_{N,t} \in \mathcal{F}$ as follows: Suppose that $f_\Delta \in \mathcal{F}$ is a fixed deterministic stationary policy. Let

$$\widetilde{f}_{N,t}(x) := \begin{cases} \arg\min_{a \in A(x)} d_A(a, \widetilde{a}_{N,t}(p_{\Gamma_t}(x))), & \text{if } x \in K_t; \\ f_\Delta(x), & \text{if } x \notin K_t. \end{cases}$$

(In particular, $\widetilde{f}_{N,t}(z) = \widetilde{a}_{N,t}(z)$ if $z \in \Gamma_t$.) We derive from Proposition D.5 in [10] that such $\widetilde{f}_{N,t}$ exists and that it is indeed measurable.

**Lemma 4.5.5.** *Suppose Assumption A holds. For every $N \geq 1$, $0 \leq t < N$, and $x \in K_t$,*

$$J(x) + \frac{2\overline{c}(\alpha\overline{d})^{N-t}}{1 - \alpha\overline{d}} \cdot w(x) + \frac{2\varepsilon\overline{c}\alpha}{1 - \alpha\overline{d}} + L_J\beta + 2H$$

$$\geq c(x, \widetilde{f}_{N,t}) + \alpha \int_{K_{t+1}} J(y)P(dy|x, \widetilde{f}_{N,t}).$$

*Proof.* Throughout this proof, $N \geq 1$ and $0 \leq t < N$ remain fixed. Given $x \in K_t$, by Theorem 4.3.1(iii) and Lemma 4.4.4,

$$J(x) \geq J_{N-t}(x) - \frac{\overline{c}(\alpha\overline{d})^{N-t}}{1 - \alpha\overline{d}} \cdot w(x) \geq \widehat{J}_{N,t}(x) - \frac{\overline{c}(\alpha\overline{d})^{N-t}}{1 - \alpha\overline{d}} \cdot w(x) - H, \qquad (4.9)$$

where we use the notation introduced in (4.8). Letting $z = p_{\Gamma_t}(x)$, observe that $\widehat{J}_{N,t}(x) = \widehat{J}_{N,t}(z)$ and also that

$$\widehat{J}_{N,t}(z) = c(z, \widetilde{f}_{N,t}) + \alpha \int_{K_{t+1}} \widehat{J}_{N,t+1}(y)P(dy|z, \widetilde{f}_{N,t}),$$

where, for $y \in K_{t+1}$, it is $\widehat{J}_{N,t+1}(y) \geq J_{N-t-1}(y) - H$, so that

$$\widehat{J}_{N,t}(z) \geq c(z, \widetilde{f}_{N,t}) + \alpha \int_{K_{t+1}} J_{N-t-1}(y)P(dy|z, \widetilde{f}_{N,t}) - H. \qquad (4.10)$$

We define the function $G$ on $X$ as $G(x) := c(x, \widetilde{f}_{N,t}(x)) + \alpha P J_{N-t-1}(x, \widetilde{f}_{N,t}(x))$. By the Lipschitz continuity of $c$, $J_{N-t-1}$, and the stochastic kernel $P$, given $x \in K_t$ and letting $z = p_{\Gamma_t}(x)$, we obtain

$$|G(x) - G(z)| \leq (L_c + \alpha L_P L_{J_{N-t-1}}) \cdot (d_X(x, z) + d_A(\widetilde{f}_{N,t}(x), \widetilde{f}_{N,t}(z))).$$

By definition,

$$d_A(\widetilde{f}_{N,t}(x), \widetilde{f}_{N,t}(z)) = \inf_{a \in A(x)} d_A(a, \widetilde{f}_{N,t}(z))$$

$$\leq \sup_{b \in A(z)} \inf_{a \in A(x)} d_A(a,b)$$

$$\leq d_H(A(x), A(z)) \leq L_\Psi \cdot d_X(x,z).$$

Hence, recalling that $d_X(x,z) \leq \beta$, this yields that

$$|G(x) - G(z)| \leq (L_c + \alpha L_P L_{J_{N-t-1}})(1 + L_\Psi)\beta$$

$$\leq (L_c + \alpha L_P L_J)(1 + L_\Psi)\beta = L_J\beta. \qquad (4.11)$$

Note also that, for all $x \in K_t$,

$$\left| c(x, \widetilde{f}_{N,t}) + \alpha \int_{K_{t+1}} J_{N-t-1}(y)P(dy|x, \widetilde{f}_{N,t}) - G(x) \right| \leq \frac{\varepsilon \overline{c}\alpha}{1 - \alpha\overline{d}} \qquad (4.12)$$

because $\|J_{N-t-1}\|_w \leq \overline{c}/(1 - \alpha\overline{d})$ (Theorem 4.3.1(i)). Adding and subtracting $G(x)$ and $G(z)$ in (4.10), we obtain from (4.11) and (4.12) that

$$\widehat{J}_{N,t}(z) + L_J\beta + \frac{2\varepsilon\overline{c}\alpha}{1 - \alpha\overline{d}} + H \geq c(x, \widetilde{f}_{N,t}) + \alpha \int_{K_{t+1}} J_{N-t-1}(y)P(dy|x, \widetilde{f}_{N,t}). \qquad (4.13)$$

Recalling that $\|J_{N-t-1} - J\|_w \leq \overline{c}(\alpha\overline{d})^{N-t-1}/(1 - \alpha\overline{d})$, the stated result is now derived from (4.9).                                                    $\square$

Let $C_0 := K_0$, and for $t \geq 1$, let

$$C_t := K_0^c \cap \cdots \cap K_{t-1}^c \cap K_t.$$

Also, we define $C_\infty := \bigcap_{t \in \mathbb{N}} K_t^c$. In plain words, if $x \in \bigcup_{t \in \mathbb{N}} K_t$, then $x \in C_t$ when $t$ is the first index for which $x \in K_t$. We have $K_t \subseteq C_0 \cup \cdots \cup C_t$. Then, $C_\infty$ is the complement of $\bigcup_{t \in \mathbb{N}} C_t = \bigcup_{t \in \mathbb{N}} K_t$. Clearly, $C_\infty$ and $\{C_t\}_{n \in \mathbb{N}}$ form a measurable partition of the state space $X$.

Given $N \geq 1$, we define the following deterministic stationary policy $\widehat{f}_N \in \mathscr{F}$. If $x \in C_t$ for some $t \in \mathbb{N}$, then $\widehat{f}_N(x) := \widetilde{f}_{N+t,t}(x)$, while if $x \in C_\infty$, then $\widehat{f}_N(x) := f_\Delta(x)$, where $f_\Delta \in \mathscr{F}$ is a fixed policy. Also, we define the function $\overline{J} \in \mathbb{B}_w(X)$ as

$$\overline{J}(x) := \begin{cases} J(x), & \text{if } x \in \bigcup_{t \in \mathbb{N}} C_t; \\ \overline{c}w(x)/(1 - \alpha\overline{d}), & \text{if } x \in C_\infty. \end{cases}$$

Since, by Theorem 4.3.1(ii), $\|J\|_w \leq \overline{c}/(1 - \alpha\overline{d})$, it should be clear that

$$J(x) \leq \overline{J}(x) \leq \overline{c}w(x)/(1 - \alpha\overline{d}) \quad \text{for every } x \in X, \qquad (4.14)$$

and also that

$$\int_{K_{t+1}^c} \bar{J}(y)P(dy|x,a) \leq \frac{\varepsilon\bar{c}}{1-\alpha\bar{d}} \quad \text{for } x \in K_t \text{ and } a \in A(x). \tag{4.15}$$

**Lemma 4.5.6.** *Suppose that Assumption A holds and fix $N \geq 1$. If $x \in \bigcup_{t\in\mathbb{N}} C_t$, then*

$$\bar{J}(x) + \frac{2\bar{c}(\alpha\bar{d})^N}{1-\alpha\bar{d}} \cdot w(x) + \frac{3\varepsilon\bar{c}\alpha}{1-\alpha\bar{d}} + L_J\beta + 2H \geq c(x,\tilde{f}_N) + \alpha\int_X \bar{J}(y)P(dy|x,\tilde{f}_N),$$

*while if $x \in C_\infty$, then*

$$\bar{J}(x) \geq c(x,\tilde{f}_N) + \alpha\int_X \bar{J}(y)P(dy|x,\tilde{f}_N).$$

*Proof.* Suppose that $x \in C_t$ for some $t \in \mathbb{N}$. Since $x \in K_t$, it follows from Lemma 4.5.5 applied to the pair $N+t,t$ that

$$J(x) + \frac{2\bar{c}(\alpha\bar{d})^N}{1-\alpha\bar{d}} \cdot w(x) + \frac{2\varepsilon\bar{c}\alpha}{1-\alpha\bar{d}} + L_J\beta + 2H$$

$$\geq c(x,\tilde{f}_{N+t,t}) + \alpha\int_{K_{t+1}} J(y)P(dy|x,\tilde{f}_{N+t,t}).$$

Recall now that, by definition, $\tilde{f}_{N+t,t}(x) = \tilde{f}_N(x)$. Note also that $J(x) = \bar{J}(x)$ for $x \in C_t$ and $J(y) = \bar{J}(y)$ when $y \in K_{t+1} \subseteq C_0 \cup \cdots \cup C_{t+1}$. Consequently,

$$\bar{J}(x) + \frac{2\bar{c}(\alpha\bar{d})^N}{1-\alpha\bar{d}} \cdot w(x) + \frac{2\varepsilon\bar{c}\alpha}{1-\alpha\bar{d}} + L_J\beta + 2H \geq c(x,\tilde{f}_N) + \alpha\int_{K_{t+1}} \bar{J}(y)P(dy|x,\tilde{f}_N).$$

Finally, the first statement of the lemma follows from (4.14) and (4.15).

The inequality when $x \in C_\infty$ is derived from the bound (4.14) after some straightforward calculations. $\qquad\qquad\square$

**Theorem 4.5.3.** *Suppose that the control model $\mathcal{M}$ satisfies Assumption A. Given arbitrary positive constants $\beta$, $\delta$, and $\varepsilon$, and $N \geq 1$, consider the policy $\tilde{f}_N$ constructed above. For every $x \in \bigcup_{t\in\mathbb{N}} C_t = \bigcup_{t\in\mathbb{N}} K_t$, we have*

$$0 \leq J(x,\tilde{f}_N) - J(x) \leq \frac{2\bar{c}(\alpha\bar{d})^N}{(1-\alpha\bar{d})^2} \cdot w(x) + \frac{3\varepsilon\bar{c}\alpha}{(1-\alpha\bar{d})(1-\alpha)} + \frac{L_J\beta + 2H}{1-\alpha}.$$

*Proof.* As a consequence of Lemma 4.5.6,

$$\bar{J}(x) + \frac{2\bar{c}(\alpha\bar{d})^N}{1-\alpha\bar{d}} \cdot w(x) + \frac{3\varepsilon\bar{c}\alpha}{1-\alpha\bar{d}} + L_J\beta + 2H \geq c(x,\tilde{f}_N) + \alpha\int_X \bar{J}(y)P(dy|x,\tilde{f}_N)$$

for every $x \in X$. Hence, letting

$$\widetilde{J}(x) := \overline{J}(x) + \frac{2\overline{c}(\alpha\overline{d})^N}{(1-\alpha\overline{d})^2} \cdot w(x) + \frac{3\varepsilon\overline{c}\alpha}{(1-\alpha\overline{d})(1-\alpha)} + \frac{L_J\beta + 2H}{1-\alpha} \quad \text{for } x \in X,$$

a direct calculation shows that the function $\widetilde{J} \in \mathbb{B}_w(X)$ satisfies $\widetilde{J}(x) \geq c(x,\widetilde{f}_N) + \alpha P\widetilde{J}(x,\widetilde{f}_N)$ for all $x \in X$. The stated result now follows from Lemma 4.3.2.    □

As for Theorem 4.4.2, given some precision $\eta > 0$, we can—explicitly and a priori—determine constants $\beta$, $\delta$, and $\varepsilon$, as well as $N \geq 1$, such that

$$\frac{2\overline{c}(\alpha\overline{d})^N}{(1-\alpha\overline{d})^2} \cdot w(x_0) + \frac{3\varepsilon\overline{c}\alpha}{(1-\alpha\overline{d})(1-\alpha)} + \frac{L_J\beta + 2H}{1-\alpha} < \eta$$

for some fixed initial state $x_0 \in X$. In this case, the deterministic stationary policy $\widetilde{f}_N$ is such that $|J(x_0,\widetilde{f}_N) - J(x_0)| < \eta$.


## 4.6  Conclusions

In this chapter, we have proposed approximations of the optimal discounted cost and a discount optimal policy of an infinite horizon MDP. Our approach uses a state and action discretization procedure, taking advantage of the Lipschitz continuity of the elements of the control model.

An important feature of our approach is that the approximation errors are given explicit bounds, so that it can be used to solve numerically an infinite horizon discounted MDP, as has been done in [7] for finite horizon MDPs by following a similar approach.

There remain, however, some interesting open issues. On the one hand, the approximation of the optimal discounted cost in Theorem 4.4.2 indeed allows the explicit calculation of $\widehat{J}_{N,0}(x)$, which is close to $J(x)$ for all $x$ in a given compact subset of the state space. On the other hand, to compute in practice the policy $\widetilde{f}_N$, one has to determine the policies $\widetilde{f}_{N+t,t}$ for all $t \geq 0$. The reason is that, when the system is in state $x_0 \in K_0$, we use the policy $\widetilde{f}_{N,0}$ (which has been obtained by computation of $\widehat{J}_{N,N},\ldots,\widehat{J}_{N,0}$). Then, with a "high" probability, the system makes a transition to some state $x_1 \in K_1$. The policy that we use in $x_1$ is not $\widetilde{f}_{N,1}$ but $\widetilde{f}_{N+1,1}$, which requires the computation of $\widehat{J}_{N+1,N+1},\ldots,\widehat{J}_{N+1,0}$. This makes that the approximation of a discount optimal policy poses some computational problems. It would be very interesting to find an approximation of an optimal policy which requires less computational effort.

Finally, after having studied finite horizon MDPs in [7], and infinite horizon discounted MDPs in this chapter, it is natural to wonder whether our techniques herein can be used to obtain approximations of long-run average reward MDPs.

# References

1. Altman, E.: Constrained Markov Decision Processes. Chapman & Hall/CRC, Boca Raton FL (1999)
2. Arapostathis, A., Borkar, V.S., Fernández-Gaucherand, E., Ghosh, M.K., Marcus, S.I.: Discrete-time controlled Markov processes with average cost criterion: a survey. SIAM J. Control Optim. **31**, 282–344 (1993)
3. Bertsekas, D.P.: Convergence of discretization procedures in dynamic programming. IEEE Trans. Automat. Control **20**, 415–419 (1975)
4. Bertsekas, D.P., Shreve, S.E.: Stochastic Optimal Control: the Discrete Time Case. Academic Press, New York (1978)
5. Bertsekas, D.P., Tsitsiklis, J.N.: Neuro-Dynamic Programming. Athena Scientific, Belmont MA (1996)
6. Chang, H.S., Fu, M.C., Hu, J.Q., Marcus, S.I.: Simulation-Based Algorithms for Markov Decision Processes. Springer, London (2007)
7. Dufour, F., Prieto-Rumeau, T.: Approximation of Markov decision processes with general state space. J. Math. Anal. Appl. **388**, 1254–1267 (2012)
8. Filar, J., Vrieze, K.: Competitive Markov Decision Processes. Springer, New York (1997)
9. Hernández-Lerma, O.: Adaptive Markov Control Processes. Springer, New York (1989)
10. Hernández-Lerma, O., Lasserre, J.B.: Discrete-Time Markov Control Processes: Basic Optimality Criteria. Springer, New York (1996)
11. Hernández-Lerma, O., Lasserre, J.B.: Further Topics on Discrete-Time Markov Control Processes. Springer, New York (1999)
12. Hinderer, K.: On approximate solutions of finite-stage dynamic programs, in *Dynamic Programming and Its Applications. Proc. Conf. Univ. British Columbia, Vancouver BC, 1977* (Academic Press, New York, 1978), pp. 289–317
13. Hinderer, K.: Lipschitz continuity of value functions in Markovian decision processes. Math. Methods Oper. Res. **62**, 3–22 (2005)
14. Langen, H.J.: Convergence of dynamic programming models. Math. Oper. Res. **6**, 493–512 (1981)
15. Morin, T.L.: Computational advances in dynamic programming, in *Dynamic Programming and Its Applications. Proc. Conf. Univ. British Columbia, Vancouver BC, 1977* (Academic Press, New York, 1978), pp. 53–90
16. Powell, W.B.: Approximate Dynamic Programming. Wiley, Hoboken NJ (2007)
17. Puterman, M.L.: Markov Decision Processes: Discrete Stochastic Dynamic Programming. Wiley, New York (1994)
18. Sennott, L.I.: Stochastic Dynamic Programming and the Control of Queueing Systems. Wiley, New York (1999)
19. Sutton, R.S., Barto, A.G.: Reinforcement Learning: an Introduction. MIT Press, Cambridge MA (1998)
20. Van Roy, B.: Neuro-dynamic programming: overview and recent trends, in *Handbook of Markov Decision Processes. Internat. Ser. Oper. Res. Management Sci* (Kluwer, Boston MA, 2002), pp. 431–459
21. Whitt, W.: Approximations of dynamic programs, I. Math. Oper. Res. **3**, 231–243 (1978)

# Chapter 5
# Reduction of Discounted Continuous-Time MDPs with Unbounded Jump and Reward Rates to Discrete-Time Total-Reward MDPs

Eugene A. Feinberg

## 5.1 Introduction

It is a great pleasure to devote this chapter to Professor Hernández-Lerma, an outstanding scholar and one of the major contributors to the theory of Markov decision processes (MDPs). Among many wonderful studies and discoveries, Professor Hernández-Lerma made profound contributions to the theory of continuous-time Markov decision processes (CTMDPs). He wrote two monographs [12, 15], a survey [13], and a large number of research papers on CTMDPs. References to most of these papers can be found in [12].

The first studies of CTMDPs were conducted by Bellman [2, Chap. 11], Howard [18], Zachrisson [36], Rykov [30], and Marin-Löf [26]. Miller [27, 28] studied CTMDPs with finite state and action sets controlled by Markov policies, and Kakumanu [20] studied such processes with countable state spaces.

In general, an important basic question for MDPs is how to define policies and, what is more important, how to define stochastic processes associated with policies and initial distributions? For discrete time, this is usually done via the Ionescu Tulcea theorem (see Hernández-Lerma and Lasserre [16, Appendix C]) that sequentially defines the probability measure on a sequence of states and actions. For CTMDPs, Miller [27, 28] used the Kolmogorov forward equation to define a stochastic process corresponding to a Markov policy and initial distribution of states. Since then, most of the publications on CTMDPs considered Markov policies as the most general class of policies.

E.A. Feinberg (✉)
Department of Applied Mathematics and Statistics, Stony Brook University,
Stony Brook, NY 11794, USA
e-mail: Eugene.Feinberg@sunysb.edu

D. Hernández-Hernández and A. Minjárez-Sosa (eds.), *Optimization, Control, and Applications of Stochastic Systems*, Systems & Control: Foundations & Applications, DOI 10.1007/978-0-8176-8337-5_5, © Springer Science+Business Media, LLC 2012

The following two natural questions have been addressed in the literature: is it possible to consider (i) randomized policies and (ii) more general history-dependent policies than Markov policies? Let us start with randomized policies.

In discrete time, it is possible to implement randomization procedures at each time epoch. This is not the case in continuous time. Let us consider the simplest case: there is only one state, no dynamics, and two actions, $a$ and $b$, are available in this state. The decision maker wants to use these actions independently at each time instance with probabilities 0.5. In this case, Kolmogorov's consistency theorem implies that there is a stochastic process $a(t)$ with independent values such that $P\{a(t) = a\} = P\{a(t) = b\} = 0.5$. However, as explained in Kallianpur [22, Example 1.2.5], almost all trajectories of $a(t)$ are not measurable. This implies that the corresponding expected rewards cannot be defined even if the reward rates $r(a)$ and $r(b)$ are given.

Hordijk and van der Duyn Schouten [17] introduced the notion of randomized Markov policies. Such policies are defined as convex extensions of original actions. Let an action $a$ define transition rates $q_a$ and reward rates $r_a$. The notion of randomized Markov policies is based on using convex hulls of all $q_a$ and $r_a$ for all actions $a$ at each state. Intuitively, this means that, for any actions $a$ and $b$ and for any constant $\lambda \in (0,1)$, there is an action $c = c(\lambda)$ with $q_c = \lambda q_a + (1 - \lambda)q_b$ and $r(c) = \lambda r(a) + (1 - \lambda)r(b)$. Markov policies with respect to such a convex extension were called randomized Markov in [17]. This term is broadly used in the literature, and it appears to be confusing, because this construction does not implement any randomization procedures. It simply relaxes the control sets. We think that the term "relaxed" is more appropriate, and we shall call such policies relaxed. In particular, the policies introduced by Hordijk and van der Duyn Schouten [17] will be called relaxed Markov in this chapter.

Yushkevich [35] introduced general policies with decisions depending on the history up to time $t$, not only on the current time and state as for Markov policies. Yushkevich [35] defined a trajectory as a sequence $(t_n, x_n)_{n=0,1,\ldots}$, where $t_n$ is the time of $n$th jump ($t_0 = 0$) and $x_n$ is the state immediately after the jump. Yushkevich [35] defined a policy as a function of the previous states, previous jump epochs, and the time passed since the last jump. By using the Ionescu Tulcea theorem, for a given policy and a given initial state distribution, Yushkevich [35] constructed the corresponding probability distribution on the set of trajectories.

For marked point processes, Jacod [19] introduced the notion of a compensator, originally called in [19] a predictable projection and sometimes called a dual predictable projection. In particular, any marked point process defines its compensator; [19, Theorem 2.1]. If the sample space is the set of sequences $x_0, t_1, x_1, t_2, \ldots$, the compensator and the initial state distribution uniquely define the unique marked point process $(t_n, x_n)$, $n = 0, 1, \ldots$ ; [19, Theorem 3.6].

Kitaev [23, 24] observed that a policy for a CTMDP defines in a natural way a compensator. Though this construction is equivalent to the construction introduced by Yushkevich [35], it leads to the direct definition of a stochastic process defined by a policy without using the Ionescu Tulcea theorem. Though past-dependent policies were introduced long ago, most of the studies of CTMDPs are still limited to Markov policies.

Early studies of CTMDPs, except Bather [1], were limited to problems with uniformly bounded jump rates. Such problems can be reduced to discrete time [21, 25, 33]. One of such reductions is uniformization. Another reduction method is based on the equivalence of optimality equations in discrete and continuous time. Neither of these methods is directly applicable to CTMDPs with unbounded jump rates. This topic is discussed in detail in the survey by Guo, Hernández-Lerma, and Prieto-Rumeau [13] and in the comments to this survey published in the same journal issue.

In order to study constrained discounted CTMDPs, Feinberg [10] introduced another reduction of discounted CTMDPs to discounted MDPs. This reduction is based on the fact that policies that control the process between jumps are equivalent to policies that change actions only at jump epochs. Continuous-time controlled stochastic models that allow action selection only at jump epochs are well-studied in the literature. They are known under the name of semi-Markov decision processes (SMDPs). If jump rates of a CTMDP are uniformly bounded, the equivalent SMDP can be easily reduced to the discounted discrete-time MDP [10, Appendix].

If jump rates of a CTMDP are not bounded, the reduction to an SMDP still takes place. Then, the corresponding SMDP can be reduced to a total-reward MDP as described in [10, Appendix]. Though the total rewards for this MDP do not have a representation in the form of total discounted rewards with some discount factor that is less than 1, the analysis of this MDP leads to the description of optimal policies of the original CTMDP with unbounded jump rates and to the methods of their computation. For example, the value function satisfies the discrete-time optimality equation, and can be computed by value iterations, and stationary optimal policies exist under natural assumptions.

This chapter applies the reduction results from [10] to CTMDPs with unbounded jump rates. The boundness of jump rates was not used in [10] to reduce a CTMDP to an SMDP. It was used to show that the corresponding SMDP can be reduced to a discounted MDP. This chapter describes in detail the reduction to a total-reward MDP.

Now we present the general idea of the reduction of a CTMDP to an SMDP from [10]. It is based on the property [8] of a nonstationary exponential distribution that generalizes $E[\mathbb{X}] = \lambda^{-1}$ for an exponential random variable $\mathbb{X}$ with intensity $\lambda > 0$.

Let $(A, \mathscr{A})$ be a measurable space and $\mathbb{X}$ be a random variable with

$$P\{\mathbb{X} > t\} = \exp\left(-\int_0^t \lambda(\phi(s))\mathrm{d}s\right),$$

where $\lambda(a)$ is a nonnegative measurable function on $A$ and $\phi$ is a measurable mapping of $[0,\infty)$ to $(A, \mathscr{A})$. The interpretation of this formula is that we deal with a fixed state, $\mathbb{X}$ is the time the process spends in this state until the jump $(A, \mathscr{A})$ is the action set, and $\phi$ is a policy that selects an action $\phi(t)$, when the process spends time $t$ in the state since the last jump in this state.

Define $p(B) = P\{\phi(\mathbb{X}) \in B\}$, $B \in \mathscr{A}$, the probability that an action from set $B$ was used at jump epoch. Let

$$m(B) = E\left[\int_0^{\mathbb{X}} \mathbf{1}\{\phi(s) \in B\}\mathrm{d}s\right], \qquad B \in \mathscr{A},$$

the expected time that actions from set $B$ were used until the jump. According to Feinberg [8, Theorem 1],

$$m(B) = \int_B \frac{p(\mathrm{d}a)}{\lambda(a)}, \qquad B \in \mathscr{A}. \tag{5.1}$$

The measure $m$ defines the expected total reward up to time $\mathbb{X}$. If the reward rate under an action $a$ is $r(a)$, where $r$ is a measurable function, then the expected total reward during the time interval $[0, \mathbb{X}]$ is

$$R = \int_A r(a)m(\mathrm{d}a). \tag{5.2}$$

If, instead of selecting actions $\phi(t)$ at time $t \in [0, \mathbb{X}]$, an action is selected randomly at time 0 according to the probability $p$ and is followed until the jump epoch $\mathbb{X}$, the expected total time $m(B)$, during which that actions $a \in B$ are used, is also defined by (5.1). Since the expected total rewards until jump and the distribution of an action selected at jump epoch are the same for $\phi$ and for the policy that selects actions only at time 0 according to the probability $p$, these two policies yield the same expected total rewards. Of course, for multiple jumps, the analysis is more involved, and it is carried in [10], where it is shown that the corresponding policies have the same occupancy measures. This is done there without the assumption that the jump rates are uniformly bounded.

We conclude the introduction with a terminological remark. We write "occupancy measure," instead of "occupation measure" frequently used in the literature on MDPs, because the former provides a more adequate description of the corresponding mathematical object.

## 5.2 Definition of a CTMDP

A CTMDP is defined by the multiplet $\{X, A, A(x), q(\cdot|x,a), r(x,a), R(x,a,y)\}$, where:

(i) $X$ is the state space endowed with a $\sigma$-field $\mathscr{X}$ such that $(X, \mathscr{X})$ is a standard Borel space; that is, $(X, \mathscr{X})$ is isomorphic to a Borel subset of a Polish space.
(ii) $A$ is the action space endowed with a $\sigma$-field $\mathscr{A}$ such that $(A, \mathscr{A})$ is a standard Borel space.

(iii) $A(x)$ are sets of actions available at $x \in X$. It is assumed that $A(x) \in \mathscr{A}$ for all $x \in X$ and the set of feasible state-action pairs $Gr(A) = \{(x,a) : x \in X, a \in A(x)\}$ is a Borel subset of $X \times A$ containing the graph of a Borel mapping from $X$ to $A$.

(iv) $q(\cdot|x,a)$ is a signed measure on $(X, \mathscr{X})$ taking nonnegative finite values on measurable subsets of $X \setminus \{x\}$ and satisfying $q(X|x,a) = 0$, where $(x,a) \in Gr(A)$. In addition, $q(\Gamma|x,a)$ is a measurable function on $Gr(A)$ for any $\Gamma \in \mathscr{X}$. Let $q(x,a) \triangleq -q(\{x\}|x,a)$. Since $q(X|x,a) = 0$, then $q(X \setminus \{x\}|x,a) = q(x,a)$ and $0 \le q(x,a) < \infty$ for all $(x,a) \in Gr(A)$ because $0 \le q(X \setminus \{x\}|x,a) < \infty$.

(v) $r(x,a)$ is a real-valued reward rate function that is assumed to be measurable on $Gr(A)$; $-\infty \le r(x,a) \le \infty$.

(vi) $R(x,a,y)$ is the instantaneous reward collected if the process jumps from state $x$ to state $y$, where $y \ne x$, and an action $a$ is chosen in state $x$ at the jump epoch. The function $R(x,a,y)$ is assumed to be measurable; $-\infty \le R(x,a,y) \le \infty$.

A signed measure is also called a kernel. A kernel $q$ is called stable if $\bar{q}(x) \triangleq \sup_{a \in A(x)} q(x,a) < \infty$, where $q(x,a) \triangleq -q(\{x\}|x,a)$, $x \in X$. Set $q(\cdot|x,a) \triangleq 0$ for all $(x,a) \notin Gr(A)$.

Everywhere in this chapter the following assumption holds.

**Assumption 5.2.1** *The transition kernel $q$ is stable.*

Assumption 5.2.1 is needed to define and analyze the following two models considered in this chapter : the relaxed CTMDP and the MDP corresponding to the relaxed CTMDPs. It is also needed for continuity assumption in Sect. 5.5. Assumption 5.2.1 is not needed for Theorems 5.4.3 and 5.5.4, and for the statements of Theorems 5.4.1 and 5.4.2, if the two models mentioned above are excluded from their formulations.

Adjoin the isolated points $x_\infty$, $a_\infty$ to $X$ and $A$, respectively, and define $X_\infty \triangleq X \cup \{x_\infty\}, A_\infty \triangleq A \cup \{a_\infty\}$ as well as the $\sigma$-fields $\mathscr{X}_\infty = \sigma(\mathscr{X}, \{x_\infty\})$ and $\mathscr{A}_\infty = \sigma(\mathscr{A}, \{a_\infty\})$. Set $A(x_\infty) \triangleq a_\infty$ and $q(x_\infty, a_\infty) \triangleq 0$. Let $\mathbb{R}_+ \triangleq (0, \infty)$, $\mathbb{R}_+^0 \triangleq [0, \infty)$ and $\bar{\mathbb{R}}_+ \triangleq (0, \infty]$. Let $\mathscr{R}_+$ and $\bar{\mathscr{R}}_+$ be the Borel $\sigma$-fields of $\mathbb{R}_+$ and $\bar{\mathbb{R}}_+$ respectively. Define the basic measurable space $(\Omega, \mathscr{F})$, where $\Omega = (X \times \bar{\mathbb{R}}_+)^\infty$ and $\mathscr{F} = (\mathscr{X} \times \bar{\mathscr{R}}_+)^\infty$.

Following the construction of the jump Markov model in Kitaev [23], we briefly describe the stochastic process under study. For $\omega = \{x_0, \theta_1, x_1, \ldots\} \in \Omega$, define the random variables $x_n(\omega) = x_n$, $\theta_{n+1}(\omega) = \theta_{n+1}$, $n \ge 0$, $t_0 = 0$, $t_n = \theta_1 + \theta_2 + \cdots + \theta_n$, $n \ge 1$, $t_\infty = \lim_{n \to \infty} t_n$. Let $\omega_n = \{x_0, \theta_1, \ldots, \theta_n, x_n\}$, where $n \ge 0$ (for n = 0, omit $\theta_0$) denote the history up to and including the $n$th jump. The jump process of interest is denoted by $\xi_t(\omega)$:

$$\xi_t(\omega) \triangleq \sum_{n \ge 0} I\{t_n \le t < t_{n+1}\} x_n + I\{t_\infty \le t\} x_\infty.$$

The function $\xi_t(\omega)$ is piecewise continuous. Thus, the values $\xi_{t-}$ are well-defined when $t < t_\infty$.

A policy $\phi$ is defined by a sequence $\{\phi_n, n = 0, 1, \ldots\}$ of measurable mappings of $((X \times \mathbb{R}_+)^{n+1}, (\mathscr{X} \times \mathscr{R}_+)^{n+1}) \to (A, \mathscr{A})$ such that $\phi_n(x_0, \theta_1, x_1, \ldots, \theta_n, x_n, s) \in A(x_n)$, where $x_0 \in X$, $x_i \in X$, and $\theta_i \in \mathbb{R}_+$, for all $i = 1, \ldots, n$, and $s \in \mathbb{R}_+$. A policy $\phi$ selects actions

$$a_t \triangleq \sum_{n \geq 0} I\{t_n < t \leq t_{n+1}\} \phi_n(x_0, \theta_1, \ldots, \theta_n, x_n, t - t_n) + I\{t_\infty \leq t\} a_\infty, \quad t > 0.$$

We say that a policy $\phi$ is *simple* if $\phi_n(x_0, \theta_1, \ldots, \theta_n, x_n, s) = \phi_n(x_n)$. Here, following the tradition, we slightly abuse notations by using the same notation $\phi_n$ on both sides of the last equality. Our use of the term "simple" is consistent with its use in Dynkin and Yushkevich [5], where it means a nonrandomized Markov policy for discrete time. We use this term both for discrete and continuous time. A policy is *called switching simple* if $\phi_n(x_0, \theta_1, \ldots, x_n, s) = \phi_n(x_n, s)$. A policy $\phi$ is called deterministic if there exists a mapping $\phi : X \to A$ such that $\phi_n(x_0, \theta_1, \ldots, \theta_n, x_n, s) = \phi(x_n)$, $n = 0, 1, \ldots$, for all $x_i \in X$, where $i = 0, \ldots, n$, for all $\theta_i \in \mathbb{R}_+$, where $i = 1, \ldots, n$, and for all $s \in \mathbb{R}_+$. Of course, $\phi$ is a measurable mapping and $\phi(x) \in A(x)$ for all $x \in X$. The same is true for mappings $\phi_n$ for a simple policy $\phi$.

Any policy $\phi$ defines the predictable random measure $\nu^\phi$ on $(\mathbb{R}_+ \times X, \mathscr{R}_+ \times \mathscr{X})$ defined by

$$\nu^\phi(\omega, (0, t] \times Y) = \int_0^{t \wedge t_\infty} q(Y \setminus \{\xi_{s-}\} | \xi_{s-}, a_s) ds, \quad t \in \mathbb{R}_+, Y \in \mathscr{X}.$$

According to Jacod [19, Theorem 3.6], this random measure and the initial probability distribution $\gamma$ of the initial state $x_0 \in X$ uniquely define a marked point process $x_0, \theta_1, x_1, \theta_2 \ldots$ such that $\nu^\phi$ is the compensator of its random measure. We denote $P_\gamma^\phi$ and $E_\gamma^\phi$ the probabilities and expectations associated with this process. If $\gamma(\{x\}) = 1$ for some $x \in X$, we shall write $P_x^\phi$ and $E_x^\phi$, respectively.

Let $c^+ = \max\{c, 0\}$ and $c^- = \min\{c, 0\}$ for any number $c$. For a positive discount rate $\alpha$, consider the expected total discounted rewards:

$$V_+(x, \phi) = E_x^\phi \left[ \int_0^{t_\infty} e^{-\alpha t} r^+(\xi_t, a_t) dt + \sum_{n=1}^\infty e^{-\alpha t_n} R^+(\xi_{t_{n-1}}, a_{t_n}, \xi_{t_n}) \right]$$

and

$$V_-(x, \phi) = E_x^\phi \left[ \int_0^{t_\infty} e^{-\alpha t} r^-(\xi_t, a_t) dt + \sum_{n=1}^\infty e^{-\alpha t_n} R^-(\xi_{t_{n-1}}, a_{t_n}, \xi_{t_n}) \right].$$

We follow the convention that $(+\infty) + (-\infty) = -\infty$ throughout this chapter. For almost all problems considered in the literature, the following assumption (the so-called general convergence condition) holds:

$$V_+(x,\phi) < \infty \qquad \text{for any policy } \phi \text{ and for any state } x \in X. \tag{5.3}$$

Everywhere in this chapter, we assume that the following weaker assumption holds.

**Assumption 5.2.2** *For any policy $\phi$ and for any initial state $x \in X$, either $V_+(x,\phi) < \infty$ or $V_-(x,\phi) > -\infty$.*

The total expected discounted rewards are defined as

$$V(x,\phi) = V_+(x,\phi) + V_-(x,\phi) = E_x^\phi \left[ \int_0^{t_\infty} e^{-\alpha t} r(\xi_t, a_t) dt + \sum_{n=1}^{\infty} e^{-\alpha t_n} R(\xi_{t_{n-1}}, a_{t_n}, \xi_{t_n}) \right]$$

$$= E_x^\phi \left[ \sum_{n=1}^{\infty} \left\{ \int_{t_{n-1}}^{t_n} e^{-\alpha t} r(\xi_t, a_t) dt + e^{-\alpha t_n} R(\xi_{t_{n-1}}, a_{t_n}, \xi_{t_n}) \right\} \right],$$

where the second and third equalities hold because of Assumption 5.2.2.

Let $\Pi$ be the set of all policies and $V(x) = \sup_{\phi \in \Pi} V(x,\phi)$. A policy $\phi$ is called optimal if $V(x,\phi) = V(x)$ for all $x \in X$.

Define

$$r^*(x,a) = \frac{r(x,a) + \int_{X \setminus \{x\}} R(x,a,y)q(dy|x,a)}{\alpha + q(x,a)}, \qquad x \in X,\ a \in A(x). \tag{5.4}$$

Following Feinberg [10], consider the occupancy measures

$$\tilde{M}_{x,n}^\phi(Y,B) = E_x^\phi \int_{t_n}^{t_{n+1}} e^{-\alpha t} I\{\xi_t \in Y, a_t \in B\}(\alpha + q(\xi_t, a_t)) dt, \tag{5.5}$$

where $n = 0,1,\ldots$, $Y \in \mathscr{X}$, $B \in \mathscr{A}$. Then, according to [10, Corollary 4.4 and (4.20)],

$$V(x,\phi) = \sum_{n=0}^{\infty} \int_X \int_A r^*(z,a) \tilde{M}_{x,n}^\phi(dz, da). \tag{5.6}$$

In particular, (5.6) means that the introduced model is equivalent to the model without instant rewards at jump epochs and with the reward rate function

$$\hat{r}(x,a) = r(x,a) + \int_{X \setminus \{x\}} R(x,a,y)q(dy|x,a), \qquad x \in X,\ a \in A(x). \tag{5.7}$$

For a standard Borel space $(E, \mathscr{E})$, denote by $\mathscr{P}(E)$ the set of probability measures on $(E, \mathscr{E})$. Let $\mathscr{M}(E)$ be the minimal $\sigma$-field on $\mathscr{P}(E)$ such that for any $C \in \mathscr{E}$ the function $\mu(C)$ defined on $\mathscr{P}(E)$ is measurable. Then $(\mathscr{P}(E), \mathscr{M}(E))$ is a standard Borel space. In particular, if $(E, \mathscr{E})$ is a Polish space, then $(\mathscr{P}(E), \mathscr{M}(E))$ is the Borel $\sigma$-field in the topology of weak convergence.

**Relaxed CTMDP and relaxed policies.** Consider a CTMDP with a stable kernel $q$. We relax the model by replacing the action set $A$ with $\mathscr{P}(A)$, the action sets $A(x)$ with $\mathscr{P}(A(x))$, $x \in X$; the transition kernels $q(Y|x,a)$ with

$$\bar{q}(Y|x,\mu) \triangleq \int_{A(x)} q(Y|x,a)\mu(da),$$

where $\mu \in \mathscr{P}(A(x))$; and the reward functions $r$ and $R$ with the functions $\bar{r}$ and $\bar{R}(x,\mu) \equiv 0$, respectively, where

$$\bar{r}(x,\mu) \triangleq \int_{A(x)} \hat{r}(x,a)\mu(da).$$

Let again $\bar{q}(x,\mu) \triangleq -\bar{q}(X \setminus \{x\}|x,\mu)$. Since $q$ is stable, $\bar{q}(x,\mu) < \infty$ for all $x \in X$ and $\mu \in \mathscr{P}(A(x))$.

As shown in Feinberg [10], the relaxed CTMDP satisfies the same basic assumptions (i)–(vi) as the original model. So, we can consider a policy $\pi$ for this CTMDP. A policy for the relaxed CTMDP is called a relaxed policy for the original CTMDP. If a policy is simple in the relaxed CTMDP, it is called a relaxed simple policy in the original CTMDP. Relaxed policies are usually called randomized policies in the literature.

For a fixed initial state or initial distribution, a relaxed policy also defines a stochastic process up to time $t_\infty$, and we denote by $P_x^\pi$ and $E_x^\pi$ the probabilities and expectations for a relaxed policy $\pi$ and initial state $x$. Let $\Pi^R$ be the set of all relaxed policies. If a relaxed policy $\pi$ selects at any time instance and along any trajectory a measure $\delta_{a_t}$ concentrated at one point, then it can be interpreted as a usual policy. Following this interpretation, we write $\Pi \subseteq \Pi^R$. Let $\pi_t(da|\omega)$ be the probability distribution of actions that a relaxed policy $\pi$ selects at time $t < t_\infty$ along the trajectory $\omega$. The corresponding measure is defined as $\pi_t(\omega)$. Here two different notations are used for the same objects because $\pi_t(B|\omega) = \pi_t(\omega)(B)$ for any $B \in \mathscr{A}$. As above with the random variables $\xi_t(\omega)$, we will not write $\omega$ explicitly whenever possible. Thus, we shall write $\pi_t(da)$ and $\pi_t$ instead of $\pi_t(da|\omega)$ and $\pi_t(\omega)$, respectively.

The expected total discounted reward for a relaxed policy $\pi$ and an initial state $x$ is the expected total discounted reward for the relaxed CTMDP. It can be written as

$$V(x,\pi) = V_+(x,\pi) + V_-(x,\pi).$$

Here we use for the relaxed CTMDP the notations introduced for the original CTMDP. If either $V_+(x,\pi) < \infty$ or $V_-(x,\pi) > -\infty$, then

$$V(x,\pi) = E_x^\pi \left[ \int_0^{t_\infty} e^{-\alpha t} \bar{r}(\xi_t, \pi_t) dt \right] = E_x^\pi \left[ \sum_{n=1}^{\infty} \int_{t_{n-1}}^{t_n} e^{-\alpha t} \bar{r}(\xi_t, \pi_t) dt \right],$$

or, in a more explicit form,

$$V(x,\pi) = E_x^{\pi}\left[\int_0^{t_{\infty}} e^{-\alpha t}\int_A \hat{r}(\xi_t,a)\pi_t(da)dt\right] = E_x^{\pi}\left[\sum_{n=1}^{\infty}\int_{t_{n-1}}^{t_n} e^{-\alpha t}\int_A \hat{r}(\xi_t,a)\pi_t(da)dt\right].$$

We also recall that, following the convention in this chapter, $V(x,\pi) = -\infty$ when $V_+(x,\pi) = \infty$ and $V_-(x,\pi) = -\infty$.

We also define $V^R(x) = \sup_{\pi\in\Pi^R} V(x,\pi)$. A relaxed policy $\pi$ is called optimal if $V(x,\pi) = V^R(x)$ for all $x \in X$.

## 5.3  Definition of a Discrete-Time Total-Reward MDP

A discrete-time MDP is defined by a multiplet $\{X,A,A(x),p(Y|x,a),\tilde{r}(x,a)\}$, where the state set $X$, action set $A$, and sets of available actions $A(x)$, $x \in X$, have the same properties as in CTMDPs; $p(\cdot|x,a)$ is the transition probability from $Gr(A)$ to $X$, that is, $p(\cdot|x,a)$ is a probability distribution on $(X,\mathscr{X})$ for any $(x,a) \in Gr(A)$ and $p(Y|x,a)$ is a measurable function on $Gr(A)$ for any $Y \in \mathscr{X}$; $\tilde{r}(x,a)$ is the one-step reward function measurable on $Gr(A)$.

A trajectory is a sequence $x_0,a_0,x_1,a_1,\ldots$ from $(X \times A)^{\infty}$. A policy $\sigma$ is a sequence of transition probabilities $\sigma_n$, $n = 0,1,\ldots$, from $(X \times A)^n \times X$ to $A$ such that $\sigma_n(A(x_n)|x_0,a_0,\ldots,x_{n-1},a_{n-1},x_n) = 1$. A policy $\sigma$ is called randomized Markov if $\sigma_n(B|x_0,a_0,\ldots,x_{n-1},a_{n-1},x_n) = \sigma_n(B|x_n)$, $x_n \in X$, $n = 0,1,\ldots$, for all $B \in \mathscr{A}$. A simple policy $\phi$ is defined as a sequence of measurable mappings $\phi_n : X \to A$, $n = 0,1,\ldots$ such that $\phi_n(x) \in A(x)$, $x \in X$. A deterministic (stationary deterministic) policy is defined as a measurable mapping from $X$ to $A$ satisfying $\phi(x) \in A(x)$ for all $x \in X$. This means that in state $x$, the action $\phi(x)$ is selected. Let $\Delta$ be the set of all policies.

As usual, in view of the Ionesu Tulcea theorem [16], any initial state distribution $\gamma$ and policy $\sigma$ define a probability measure $P_{\gamma}^{\sigma}$ on the sequence of trajectories. We denote by $E_{\gamma}^{\sigma}$ the expectation with respect to $P_{\gamma}^{\sigma}$. We write $P_x^{\sigma}$ and $E_x^{\sigma}$, $x \in X$, instead of $P_{\gamma}^{\sigma}$ and $E_{\gamma}^{\sigma}$, respectively, when $\gamma(\{x\}) = 1$.

Let

$$W_+(x,\sigma) = E_x^{\sigma}\sum_{n=0}^{\infty}\tilde{r}^+(x_n,a_n), \quad W_-(x,\sigma) = E_x^{\sigma}\sum_{n=0}^{\infty}\tilde{r}^-(x_n,a_n),$$

and

$$W(x,\sigma) = W_+(x,\sigma) + W_-(x,\sigma).$$

If either $W_+(x,\sigma) < \infty$ or $W_-(x,\sigma) > -\infty$, then

$$W(x,\sigma) = E_x^{\sigma}\sum_{n=0}^{\infty}\tilde{r}(x_n,a_n),$$

and $W(x,\sigma) = -\infty$, if $W_+(x,\sigma) = \infty$ and $W_-(x,\sigma) = -\infty$.

Similar to the continuous-time case, consider the value function $W(x) = \sup_{\sigma \in \Delta}$ $W(x, \sigma)$. A policy $\sigma$ is called optimal if $W(x, \sigma) = W(x)$ for all $x \in X$.

We also consider occupancy measures for the MDP:

$$M_{x,n}^{\sigma}(Y,B) \triangleq E_x^{\sigma} \mathbf{I}\{x_n \in Y, a_n \in B\}, \quad n = 0, 1, \ldots, \ Y \in \mathscr{X}, \ B \in \mathscr{A}. \tag{5.8}$$

Then, for $x \in X$ and for $\sigma \in \Delta$,

$$W(x, \sigma) = \sum_{n=0}^{\infty} \int_X \int_A \tilde{r}(z, a) M_{x,n}^{\sigma}(\mathrm{d}z, \mathrm{d}a). \tag{5.9}$$

For any policy $\sigma$ and for a fixed $x \in X$, consider a randomized Markov policy $\sigma^*$ such that

$$\sigma_n^*(B|x_n) = \frac{P_x^{\sigma}\{dx_n, a_n \in B)}{P_x^{\sigma}\{dx_n\}}, \qquad P_x^{\sigma} - \text{a.s.} \tag{5.10}$$

It is well known [34] that

$$P_x^{\sigma^*}\{x_n \in Y, a_n \in B\} = P_x^{\sigma}\{x_n \in Y, a_n \in B\}, \quad Y \in \mathscr{X}, \ B \in \mathscr{A}.$$

This implies

$$M_{x,n}^{\sigma^*} = M_{x,n}^{\sigma}, \qquad n = 0, 1, \ldots, \tag{5.11}$$

and, in view of (5.9), for any reward function $\tilde{r}$

$$W(x, \sigma^*) = W(x, \sigma). \tag{5.12}$$

## 5.4   Main Results

For a total-reward discounted CTMDP defined by a multiplet $\{X, A, A(x), q(\cdot|x,a),$ $r(x,a), R(x,a,y)\}$, consider a discrete-time MDP defined by a multiplet $\{X_\infty, A_\infty,$ $A(x), p(\cdot|x,a), \tilde{r}(x,a)\}$, where $A(x_\infty) = \{a_\infty\}$,

$$p(Y|x,a) = \begin{cases} \frac{q(Y\setminus\{x\}|x,a)}{\alpha + q(x,a)}, & \text{if } Y \in \mathscr{X}, x \in X, a \in A(x), \\ \frac{\alpha}{\alpha + q(x,a)}, & \text{if } Y = \{x_\infty\}, x \in X, a \in A(x), \\ 1, & \text{if } Y = \{x_\infty\}, x = x_\infty, a = a_\infty, \end{cases} \tag{5.13}$$

and $\tilde{r}(x,a) = r^*(x,a)$ (see (5.4)) and $\tilde{r}(x_\infty, a_\infty) = 0$.

Thus, the original CTMDP $\{X, A, A(x), q(\cdot|x,a), r(x,a), R(x,a,y)\}$ defines two other models: the relaxed CTMDP $\{X, \mathscr{P}(A), \mathscr{P}(A(x)), \bar{q}(\cdot|x,\mu), \bar{r}(x,\mu), \bar{R}(x,\mu,y)\}$, where $\mu \in \mathscr{P}(A(x))$, and the discrete-time MDP $\{X_\infty, A_\infty, A(x), p(\cdot|x,a), \tilde{r}(x,a)\}$ with the transition probabilities and rewards defined in (5.4,5.13) and in the line following (5.13). These three models have the sets of policies $\Pi$, $\Pi^R$, and $\Delta$, respectively, where $\Pi \subseteq \Pi^R$. The objective criteria for the two CTMDPs are the expected total discounted rewards, and the objective criterion for the discrete-time MDP is the expected total reward.

In addition, we can consider the fourth model: the MDP corresponding to the relaxed CTMDP. Let $\Delta^R$ be the set of policies for this MDP; $\Delta \subseteq \Delta^R$. Observe that for $x \in X$ and $\mu \in \mathscr{P}(A(x))$, the one-step reward for this MDP is

$$\tilde{r}(x,\mu) = \frac{\hat{r}(x,\mu)}{\alpha + \bar{q}(x,\mu)}. \tag{5.14}$$

Observe that the sets of states, actions, and transition rates define the structure of the models and the sets of policies $\Pi$, $\Pi^R$, $\Delta$, and $\Delta^R$, while the reward functions $r$ and $R$ define the objective functions, which are functionals on these sets.

**Theorem 5.4.1** *Consider a CTMDP and its three associated models: the relaxed CTMDP, the corresponding MDP, and the MDP corresponding to the relaxed CTMDP. For any policy $\phi$ in the CTMDP and any state $x \in X$, in each of three associated models, there exists a policy $\sigma$ such that $V(x,\phi) = G(x,\sigma)$ and this equality hold for all reward functions $r$ and $R$, where $G$ denotes the expected total reward in the two associated MDPs and the expected total discounted reward in the relaxed CTMDP.*

*Proof.* (i) Consider the relaxed CTMDP. Since $\phi \in \Pi \subseteq \Pi^R$, the policy $\phi$ itself is the required policy in the relaxed CTMDP. Thus, we can set $\sigma = \phi$.

(ii) Fix $x \in X$. Recall that $\tilde{M}_{x,n}^\phi$, $n = 0, 1, \ldots$, are occupancy measures for the CTMDP defined in (5.5), when the initial state is $x$. Consider on $(X, \mathscr{X})$ finite nonnegative measures $\tilde{m}_{x,n}^\phi$, $n = 0, 1, \ldots$,

$$\tilde{m}_{x,n}^\phi(Y) \triangleq \tilde{M}_{x,n}^\phi(Y, A), \qquad Y \in \mathscr{X}. \tag{5.15}$$

For the corresponding MDP, consider a randomized Markov policy $\sigma$ satisfying for each $B \in \mathscr{A}$

$$\sigma_n(B|z) = \frac{\tilde{M}_{x,n}^\phi(dz, B)}{\tilde{m}_{x,n}^\phi(dz)}, \qquad \tilde{m}_{\mu,n}^\phi - \text{a. e.}, \ n = 0, 1, \ldots . \tag{5.16}$$

According to Feinberg [10, Theorem 4.5 and Lemma A.2], such a randomized Markov policy $\sigma$ exists and

$$M_{x,n}^\sigma = \tilde{M}_{x,n}^\phi, \qquad n = 0, 1, \ldots, \tag{5.17}$$

and therefore $V(x, \phi) = W(x, \sigma)$ for any reward functions $r$ and $R$. We remark that [10, Lemma A.2] deals with a SMDP. For a randomized Markov policy, decisions depend on the current state and jump number. They do not depend on the current or past values of the time parameter $t$. Therefore, these policies are also randomized Markov policies for the MDP corresponding to the CTMDP; see [10, Corollary A.4] with $\bar{\beta} = 1$, where the parameter $\bar{\beta}$ is considered in that corollary. We remark that $\bar{\beta} < 1$ is chosen in [10, Corollary A.4] to ensure that the discount factor is less than 1. As shown in [10], the discount factor less than 1 can be chosen if the jump rates are uniformly bounded. Since the jump rates may be unbounded in the current chapter, it may be impossible to set the discount factor $\bar{\beta} < 1$, but it is possible to set the discount factor $\bar{\beta} = 1$.

(iii) As shown in (i), a policy $\pi$ belongs to the relaxed CTMDP. By applying (ii) to the relaxed CTMDP, we obtain a Markov policy with the required properties in the MDP corresponding to the relaxed CTMDP.

$\square$

**Definition 5.4.1** *Consider a CTMDP and its three associated models: the relaxed CTMDP, the corresponding MDP, and the MDP corresponding to the relaxed MDP. Two of these four models are called equivalent, if for each state $x \in X$, for any policy in one of the models, there exists a policy in another model such that the values of the total-reward criteria for these policies are equal in the corresponding models, when the initial state $x$ is fixed. If this is true for any reward functions $r$ and $R$, then the models are called strongly equivalent.*

**Theorem 5.4.2** *Consider a CTMDP and its three associated models: the relaxed CTMDP, the corresponding MDP, and the MDP corresponding to the relaxed CTMDP. Then*

(a) *The following three models are strongly equivalent: the relaxed CTMDP, the corresponding MDP, and the MDP corresponding to the relaxed CTMDP.*
(b) *If the set A is countable, then all four models are strongly equivalent.*

We consider two lemmas needed for the proof of Theorem 5.4.2,

**Lemma 5.4.1** *Theorem 5.4.2 holds under the addition condition that the reward functions $r$ and $R$ are nonnegative.*

*Proof.* (a) Fix the initial state $x \in X$. Observe that $\Delta \subseteq \Delta^R$ and that, according to Theorem 5.4.1, for any policy in the relaxed CTMDP there exists a policy with the required property in the MDP corresponding to the relaxed CTMDP. Thus, to show the strong equivalence of the MDP corresponding to the CTMDP and the MDP corresponding to the relaxed CTMDP, it is sufficient to show that for any policy in the MDP corresponding to the relaxed CTMDP, there exists a policy in the MDP corresponding to the CTMDP such that these two policies have equal objective functions and this is true for all nonnegative reward functions $r$ and $R$.

Observe that for an arbitrary state $x \in X$ and for an arbitrary action $\mu \in \mathscr{P}(A(x))$ in the MDP corresponding to the relaxed CTMDP there exists a randomized action $\pi \in \mathscr{P}(A(x))$ in the MDP corresponding to the CTMDP such that these two actions yield the same transition probabilities and expected one-step rewards. More precisely, for $\mu \in \mathscr{P}(A(x))$, we define $\pi \in \mathscr{P}(A(x))$ by

$$\pi(da) = \frac{\alpha + q(x,a)}{\alpha + \int_{A(x)} q(x,a)\mu(da)} \mu(da). \tag{5.18}$$

Indeed, if an action is selected according to the distribution $\pi$ in the MDP corresponding to the CTMDP, then the transition probability $p$ and expected one-step reward $\tilde{r}$ are

$$p(Y|x,\pi) = \int_{A(x)} p(Y|x,a)\pi(da) = \int_{A(x)} \frac{q(Y \setminus \{x\}|x,a)\pi(da)}{\alpha + q(x,a)}, \quad Y \in \mathscr{X},$$

$$\tilde{r}(x,\pi) = \int_{A(x)} \frac{\hat{r}(x,a)\pi(da)}{\alpha + q(x,a)}.$$

If an action $\mu$ is selected in the MDP corresponding to the relaxed CTMDP, then the transition probability $\bar{p}$ and expected one-step reward $\bar{r}$ are

$$\bar{p}(Y|x,\mu) = \frac{\int_{A(x)} q(Y \setminus \{x\}|x,a)\mu(da)}{\alpha + \int_{A(x)} q(x,a)\mu(da)}, \quad Y \in \mathscr{X},$$

$$\bar{r}(x,\mu) = \frac{\int_{A(x)} \hat{r}(x,a)\mu(da)}{\alpha + \int_{A(x)} q(x,a)\mu(da)}.$$

In view of (5.18), $p(Y|x,\pi) = \bar{p}(Y|x,\mu)$ and $\tilde{r}(x,\pi) = \bar{r}(x,\mu)$. This implies that, if the initial state is fixed, for any randomized Markov policy in the MDP corresponding to the relaxed CTMDP, there exists a randomized Markov policy in the MDP corresponding to the CTMDP such that they objective functions coincide. In addition, this is true for any nonnegative reward functions $r$ and $R$. Thus, the MDP corresponding to the CTMDP and the MDP corresponding to the relaxed CTMDP are strongly equivalent.

According to cases (i) and (ii) in the proof of Theorem 5.4.1, for any policy $\pi$ in the relaxed CTMDP, there exists a policy with the same performance for all reward functions in the MDP corresponding to the relaxed CTMDP. Now consider a policy $\sigma$ for the MDP corresponding to the CTMDP. Without loss of generality [34], let $\sigma$ be randomized Markov. The policy $\sigma$ can be considered as a policy that selects actions only at jump epochs and decisions depend only on the state to which the process jumps in and the jump number. In other words, $\sigma$ is a policy for a SMDP defined by the underlying CTMDP model. Define the relaxed switching simple policy $\pi$ by

$$\pi_n(da|z,s) = \frac{e^{-q(z,a)s}\sigma_n(da|z)}{\int_A e^{-q(z,a)s}\sigma_n(da|z)}, \quad n = 0,1,\ldots, \ z \in X, \ s \geq 0. \tag{5.19}$$

The marked point processes defined by the policies $\sigma$ and $\pi$ have the same sojourn-times distributions because for every $n = 0, 1, \ldots$ they have the same hazard rates after $n$th jump:

$$\frac{\int_A q(x_n, a) e^{-q(x_n, a)s} \sigma_n(da|x_n)}{\int_A e^{-q(x_n, a)s} \sigma_n(da|x_n)}.$$

In addition, they have the same transition probabilities,

$$\frac{\int_A q(Y \setminus \{x\}|x_n, a) e^{-q(x_n, a)\theta_{n+1}} \sigma_n(da|x_n)}{\int_A e^{-q(x_n, a)\theta_{n+1}} \sigma_n(da|x_n)},$$

at jump epochs. The initial state $x$ is fixed, and it is the same for marked point processes defined by the policies $\sigma$ and $\pi$. Thus, the policies $\sigma$ and $\pi$ define the same marked point processes.

In addition, the reward rates are equal:

$$\bar{r}(x_n, \pi_n(x_n, s)) = \frac{\int_A r(x_n, a) e^{-q(x_n, a)s} \sigma_n(da|x_n)}{\int_A e^{-q(x_n, a)s} \sigma_n(da|x_n)}.$$

Thus, for any discount rate $\alpha \in [0, 1]$, for any policy $\sigma$ in the MDP corresponding to the original CTMDP, there exist a relaxed policy $\pi$ in the CTMDP such that the expected total discounted rewards are equal for these two policies. We remark that this correspondence is proved in Feinberg [10, (8.5), (8.6)] by using compensators.

(b) Without loss of generality, let $A = \{1, 2, \ldots\}$. Fix $x \in X$ and consider a randomized Markov policy $\sigma$ for the MDP corresponding to a CTMDP. For $n = 0, 1, \ldots$, we set

$$\theta_i(z, n) = -(\alpha + q(z, i))^{-1} \ln \left(1 - \frac{\sigma_n(i|z)}{\sum_{k=i}^{\infty} \sigma_n(k|z)}\right), \quad i = 1, 2, \ldots, z \in X, a \in A,$$

$$(5.20)$$

where $\frac{0}{0} = 0$ and $\ln(0) = -\infty$. We also set $\theta_0(z, n) = 0$, $\theta_i(z, n) = \sum_{k=1}^{i} \theta_k(z, n)$. Consider a switching simple policy $\phi$:

$$\phi_n(z, s) = i \quad \text{for } z \in X, n = 0, 1, \ldots, \theta_{i-1}(z, n) \le s < \theta_i(z, n), i = 1, 2, \ldots .$$

$$(5.21)$$

According to [10, Theorem 5.2], $\sigma$ satisfies (5.16) written for $\pi = \phi$. In addition, $V(x, \phi) = W(x, \sigma)$ for any nonnegative reward functions $r$ and $R$. In conclusion, we notice that technical conditions in Feinberg [10] and here are different. Instead of Condition 5.1 from [10], we use the assumption that $A$ is countable. These conditions achieve the same goal: they guarantee that the functions $\phi_n(z, s)$ are well defined in (5.21) and measurable in $(z, s)$.

$$\square$$

**Lemma 5.4.2** *Assumption 5.2.2 implies that the following statements hold for all* $x \in X$:

(a) *Either* $\sup_{\phi \in \Pi} V_+(x, \phi) < \infty$ *or* $\inf_{\phi \in \Pi} V_-(x, \phi) > -\infty$.
(b) *Either* $\sup_{\pi \in \Pi^R} V_+(x, \pi) < \infty$ *or* $\inf_{\pi \in \Pi^R} V_-(x, \pi) > -\infty$.
(c) *Either* $\sup_{\sigma \in \Delta} W_+(x, \sigma) < \infty$ *or* $\inf_{\sigma \in \Delta} W_-(x, \sigma) > -\infty$.
(d) *Either* $\sup_{\sigma^* \in \Delta^R} W_+(x, \sigma^*) < \infty$ *or* $\inf_{\sigma^* \in \Delta^R} W_-(x, \sigma^*) > -\infty$.

*Proof.* (c) Fix $x \in X$. We shall prove that if statement (c) does not hold, then Assumption 5.2.2 does not hold either. Let statement (c) does not hold. Define the function $\hat{r}_+$ by (5.7) with the functions $r$ and $R$ replaced with the functions $r^+$ and $R^+$, respectively. Similarly, define the function $\hat{r}_-$ by (5.7) with the functions $r$ and $R$ replaced with the functions $r^-$ and $R^-$, respectively. Then define the functions $\tilde{r}_+$ and $\tilde{r}_-$ by (5.14) with the function $\hat{r}$ replaced with the functions $\hat{r}_+$ and $\hat{r}_-$, respectively. For $\sigma \in \Delta$, let $W_{++}(x, \sigma)$ and $W_{--}(x, \sigma)$ be the discrete-time expected total rewards with the reward functions $\tilde{r}_+$ and $\tilde{r}_-$, respectively. Then $W_{++}(x, \sigma) \geq W_+(x, \sigma)$ and $W_{--}(x, \sigma) \leq W_-(x, \sigma)$ for all $\sigma \in \Delta$. Thus, since (c) does not hold,

$$\sup_{\sigma \in \Delta} W_{++}(x, \sigma) = \infty v \tag{5.22}$$

and

$$\inf_{\sigma \in \Delta} W_{--}(x, \sigma) = -\infty. \tag{5.23}$$

The rest of the proof shows that (5.22) and (5.23) imply that Assumption 5.2.2 does not hold. Equality (5.22) means that the value for a positive dynamic programming problem is infinite. Thus, for every $n = 1, 2, \ldots$, there exists a deterministic policy $\phi_n$ satisfying $W_{++}(x, \phi_n) \geq 2^n$. Similarly, for every $n = 1, 2, \ldots$, there exists a deterministic policy $\psi_n$ satisfying $W_{--}(x, \phi_n) \leq -2^n$. Consider a policy $\sigma \in \Delta$ that is defined in the following way: this policy independently selects one of the policies $\{\phi_1, \psi_1, \phi_2, \psi_2, \ldots\}$, and policy $\phi_n$ and each policy $\psi_n$ are selected with the probability $2^{-n+1}$, $n = 1, 2, \ldots$ (here we apply the same method as in [5, p. 108]). We have $W_{++}(x, \sigma) = \infty$ and $W_{--}(x, \sigma) = -\infty$.

We also have that $W_{++}(x, \sigma^*) = \infty$ and $W_{--}(x, \sigma^*) = -\infty$ for a randomized Markov policy $\sigma^*$ satisfying (5.10). Since the policy $\sigma^*$ uses a countable number of actions at any state $z$, for the policy $\phi$ for the CTMDP, defined by (5.21) with $\sigma = \sigma^*$ in (5.20), we have $V_+(x, \phi) = \infty$ and $V_-(x, \phi) = -\infty$. Thus, Assumption 5.2.2 does not hold.

(a) Theorem 5.4.1 implies that if statement (a) does not hold, then both (5.22) and (5.23) hold. Therefore, Assumption 5.2.2 does not hold.
(b) Lemma 5.4.1 implies that if statement (b) does not hold, then both (5.22) and (5.23) hold. Therefore, Assumption 5.2.2 does not hold.
(d) The proof is the same as for (b).    $\square$

*Proof of Theorem 5.4.2.*    Lemma 5.4.1 implies the theorem under additional conditions that either both $r$ and $R$ are nonnegative or they both are nonpositive. Lemma 5.4.2 implies that this is also true for general functions $r$ and $R$.    □

Let $f : X \rightarrow \mathbb{R}^0_+$. A policy $\phi$ is called $f$-optimal for a CTMDP if $V(x, \phi) \geq V(x) - f(x)$ for all $x \in X$. A policy $\phi$ is called $f$-optimal for an MDP if $W(x, \phi) \geq W(x) - f(x)$ for all $x \in X$.

**Theorem 5.4.3** (a) *The function $V$ is universally measurable and $V(x) = V^R(x) = W(x) = W^R(x)$ for all $x \in X$.*
(b) *If for a function $f : X \rightarrow \mathbb{R}^0_+$, a simple policy $\varphi$ is $f$-optimal for the corresponding discrete-time MDP, it is also $f$-optimal for the original CTMDP. In particular, a deterministic $f$-optimal policy for the discrete-time MDP is $f$-optimal for the CTMDP.*

*Proof.* (a) Fix arbitrary $x \in X$. According to Theorem 5.4.2, $V^R(x) = W(x) = W^R(x)$. Theorem 5.4.1 implies that $W(x) \geq V(x)$. Let $\sigma$ be a policy in the MDP corresponding to the CTMDP. According to Feinberg [6, Theorem 3], for any $K > 0$, there exists a simple policy $\phi$ such that

$$W(x, \phi) \geq \begin{cases} W(x, \sigma), & \text{if } W(x, \sigma) < \infty, \\ K, & \text{if } W(x, \sigma) = \infty. \end{cases}$$

In addition, $V(x) \geq V(x, \phi) = W(x, \phi)$ for any simple policy $\phi$. Thus, for any policy $\sigma$ in the CTMDP and for any $K > 0$,

$$V(x) \geq \begin{cases} W(x, \sigma), & \text{if } W(x, \sigma) < \infty, \\ K, & \text{if } W(x, \sigma) = \infty. \end{cases}$$

This implies $V(x) \geq W(x)$. Thus, $V(x) = W(x)$. Universal measurability of the function $W$ follows from [7, Theorem 3.1.B]. Statement (b) follows from statement (a).    □

## 5.5  Applications: Value Iterations and Optimality of Deterministic Policies

Theorems 5.4.1–5.4.3 can be used to prove the existence of optimal policies for CTMDPs, describe the structure of optimal policies, and compute them. For example, an optimal deterministic policy for the corresponding MDP is optimal for the original CTMDP. These theorems also demonstrate that there is no need to consider relaxed policies for CTMDPs, because they do not overperform standard policies. We conjecture that the assumption that $A$ is countable is not necessary for the validity of Theorem 5.4.2.

The classic results on the convergence of value iterations and on other properties of positive dynamic programs [3,4,34] imply the following fact, which is a stronger version of Piunovskiy and Zhang [29, Theorem 1], where the case $R(x,a,y) = R(x)$ was considered.

**Theorem 5.5.4** *Let the function $r^*$, defined in (5.4), be nonnegative. Then the function $V$ is the minimal nonnegative solution to the following Bellman equation*

$$V(x) = \sup_{a \in A(x)} \left\{ r^*(x,a) + \int_{X \setminus \{x\}} \frac{q(dy|x,a)}{\alpha + q(x,a)} V(y) \right\}, \quad x \in X, \tag{5.24}$$

*and can be computed by the value iteration procedure: $W_k(x) \nearrow V(x)$ when $k \to \infty$, where $W_0(x) \triangleq 0$ and*

$$W_{k+1}(x) \triangleq \sup_{a \in A(x)} \left\{ r^*(x,a) + \int_{X \setminus \{x\}} \frac{q(dy|x,a)}{\alpha + q(x,a)} W_k(y) \right\}, \quad x \in X. \tag{5.25}$$

*Proof.* Since the function $r^*$ is nonnegative, the corresponding MDP has nonnegative one-step rewards and, according to [3, Proposition 9.14], its $k$-horizon values $W_k$ form a nondecreasing sequence converging to $W$. According to Theorem 5.4.3, $V = W$.                                                                                                    ☐

As stated in Theorem 5.4.3, the function $V$ is universally measurable and, more precisely, it is upper semianalytic; see [5] or [3] for details. The same is true for the $k$-horizon values $W_k$, $k = 1,2,\ldots$, of the MDP corresponding to the CTMDP. As the following theorem states, the function $V$ satisfies the optimality equation under broad conditions. However, optimal policies may not exist under conditions of Theorem 5.5.4 even when the action set $A$ is finite; see [4] or [9, Example 6.8].

**Theorem 5.5.5** *If $V(x) < \infty$ for all $x \in X$, then the function $V$ satisfies the Bellman equation (5.24). If, in addition, $r^*(x,a) \leq 0$ for all $(x,a) \in Gr(A)$, then $V$ is the maximal nonpositive function satisfying (5.24).*

*Proof.* According to Theorem 5.4.3, $V = W$. The value function for an MDP satisfies the optimality equation [5, Sect. 6.2], and $W$ is its maximal solution when all one-step rewards are nonpositive, [34] or [3, Proposition 9.10].                                    ☐

Value iterations may not converge to the optimal value function, if the reward function is not nonnegative [34, Example 6.1]. However, for MDPs with nonpositive rewards, value iterations converge to the optimal value function under compactness and continuity conditions introduced by Schäl [31, 32] for MDPs with compact action sets. Extensions of these conditions to MDPs with noncompact action sets are currently well-understood; see [16, Lemma 4.2.8] for MDPs with setwise continuous transition probabilities and [11, Proposition 3.1] for MDPs with weakly continuous transition probabilities, where discounting was considered. The same is true for MDPs with nonpositive one-step rewards—so-called negative MDPs.

These conditions also imply the existence of deterministic optimal policies for discounted and for negative MDPs. We present below such conditions for CTMDPs.

Let $q^+(Y|x,a) \triangleq q(Y \setminus \{x\}|x,a)$, $Y \in \mathcal{X}$. Observe that $q^+(X|x,a) = q(X \setminus \{x\}|x,a) < \infty$. The definition (5.4) can be rewritten as

$$r^*(x,a) = \frac{r(x,a) + \int_X R(x,a,y)q^+(dy|x,a)}{\alpha + q(x,a)}, \qquad x \in X, \ a \in A(x). \qquad (5.26)$$

**Condition (Sc).**

(i) For each $x \in X$, the kernel $q^+(\cdot|x,a)$ is setwise continuous in $a \in A(x)$; that is, the function $\int_X f(y)q^+(dy|x,a)$ is continuous in $(x,a) \in Gr(A)$ for any bounded measurable function $f$ on $X$.

(ii) The reward function $r^*(x,a)$ is bounded above on $Gr(A)$, and for each $x \in X$, it is sup-compact on $A(x)$, that is, the set $\{a \in A(x)| r^*(x,a) \geq c\}$ is compact for each finite constant $c$.

**Condition (Wc).**

(i) The kernel $q^+(\cdot|x,a)$ is weakly continuous in $(x,a) \in Gr(A)$; that is, the function $\int_X f(y)q^+(dy|x,a)$ is continuous in $(x,a) \in Gr(A)$ for any bounded continuous function $f$ on $X$.

(ii) The reward function $r^*(x,a)$ is sup-compact on $Gr(A)$, that is, the set $\{(x,a) \in Gr(A)| r^*(x,a) \geq c\}$ is compact for each finite constant $c$.

Observe that Condition (Sc)(i) implies that $q(x,a)$ is continuous in $a \in A(x)$ for each $x \in X$. This and the inequality $\alpha > 0$ imply that the transition probability $p(\cdot|x,a)$ is setwise continuous in $a \in A(x)$ for each $x \in X$. Similarly, Condition (Wc)(i) implies that $q(x,a)$ is continuous in $(x,a) \in Gr(A)$. This and the inequality $\alpha > 0$ imply that the transition probability $p(\cdot|x,a)$ is weakly continuous in $(x,a) \in Gr(A)$.

Thus, Conditions (Sc) and (Wc) imply respectively the validity of the following conditions for the corresponding MDP. These conditions are the versions of Schäl's [31, 32] conditions (S) and (W) originally introduced for compact action sets.

**Condition (Su).**

(i) For each $x \in X$, the transition probability kernel $p(\cdot|x,a)$ is setwise continuous in $a \in A(x)$.

(ii) Condition (Sc) (ii) holds.

**Condition (Wu).**

(i) The transition probability $p(\cdot|x,a)$ is weakly continuous in $(x,a) \in A(x)$.

(ii) Condition (Wc) (ii) holds.

**Theorem 5.5.6** Let $r^*(x,a) \leq 0$ for all $(x,a) \in Gr(A)$. If Condition (Sc) holds, then the value function $V$ is measurable. If the Condition (Wc) holds, then the value function $V$ is sup-compact. In either case, the following statements hold:

*(a) There exists a deterministic optimal policy,*
*(b) A deterministic policy is optimal if and only if for all $x \in X$*

$$V(x) = r^*(x, \phi(x)) + \int_{X \setminus \{x\}} \frac{q(dy|x, \phi(x))}{\alpha + q(x, \phi(x))} V(y) \qquad (5.27)$$

$$= \max_{a \in A(x)} \left\{ r^*(x, a) + \int_{X \setminus \{x\}} \frac{q(dy|x, a)}{\alpha + q(x, a)} V(y) \right\}. \qquad (5.28)$$

*(c) The value function $V$ can be computed by the value iteration procedure: $W_k(x) \searrow V(x)$ when $k \to \infty$, where $W_0(x) \triangleq 0$ and $W_{k+1}$, $k = 1, 2, \ldots$, are defined in (5.25).*

*Proof.* If Conditions (Sc) or (Wc) hold for a CTMDP, then Conditions (Su) or (Wu) hold for the corresponding MDP, respectively. For MDPs, the statements of the theorem are standard facts. For example, similar facts are presented in [16, Lemma 4.2.8] for MDPs satisfying Condition (Su) and in [11, Proposition 3.1] for MDPs satisfying Condition (Wu). Though in [16] and in [11] a discount factor is considered, the arguments from [16] and [11] are applicable when $r$ is nonpositive and conditions (Su) or (Wu) hold for the corresponding MDP. □

If $t_\infty = \infty$ ($P_x^\pi$-a.s.) for all $\pi$ and $x$, then the problem studied in this chapter becomes an infinite-horizon problem in continuous time. Conditions for $t_\infty = \infty$ ($P_x^\pi$-a.s.) are described in the literature; see [12,14]. In addition, under certain conditions, a policy $\pi$, for which the probability of $t_\infty < \infty$ is positive, cannot be optimal. For example, such a condition is $R(x, a, y) < c < 0$ for some number $c$. In this case, $V(x, \pi) = -\infty$ for any policy $\pi$ with $P_x^\pi \{t_\infty < \infty\} > 0$. In conclusion, we formulate the following statement.

**Theorem 5.5.7** *Let either Condition (Sc) or Condition (Wc) be satisfied. If the following two assumptions hold:*

*(i) The function $\hat{r}$, defined in (5.7), is bounded above,*
*(ii) For any initial state $x \in X$ and for any policy $\pi \in \Pi$, the inequality $V(x, \pi) > -\infty$ implies $t_\infty = \infty$ ($P_x^\pi$-a.s.),*

*then there exists a deterministic optimal policy.*

*Proof.* Let $\hat{r}(x, a) \leq K < \infty$ for some finite constant $K > 0$. Let us subtract $K$ from the reward rate $r$. Then the corresponding MDP will have nonpositive one-step rewards. Therefore, Theorem 5.5.6 implies the existence of a deterministic optimal policy for the CTMDP with the reduced reward rate. Assumption (ii) implies that the expected total rewards $V(x, \pi)$ will be reduced for any policy $\pi$ and any for $x \in X$ by the constant $K/\alpha = K \int_0^\infty e^{-\alpha t} dt$. Thus, an optimal policy for the CTMDP with the reduced reward rates is also optimal for the original CTMDP. □

**Acknowledgements** This research was partially supported by NSF grants CMMI-0900206 and CMMI-0928490.

# References

1. Bather, J.: Optimal stationary policies for denumerable Markov chains in continuous time. Adv. Appl. Prob. **8**, 144–158 (1976)
2. Bellman, R.: Dynamic Programming. Princeton University Press, Princeton, N.J. (1957)
3. Bertsekas, D.P. and Shreve, S.E.: Stochastic Optimal Control: the Discrete Time Case. Academic Press, New York (1978)
4. Blackwell, D.: Positive dynamic programming. In Proceedings of the 5th Berkeley Symposium on Mathematical Statistics and Probability **1** 415–418, University of California Press, Berkeley (1967).
5. Dynkin, E.B. and Yushkevich, A.A.: Controlled Markov Processes. Springer, Berlin (1979)
6. Feinberg, E.A.: Non-randomized Markov and semi-Markov strategies in dynamic programming. Theory Probability Appl., **27**, 116–126 (1982)
7. Feinberg, E.A.: Controlled Markov processes with arbitrary numerical criteria. Theory Probability Appl., **27**, 486–503 (1982)
8. Feinberg, E.A.: A generalization of "expectation equals reciprocal of intensity" to nonstationary distributions. J. Appl. Probability, **31**, 262–267 (1994)
9. Feinberg, E.A.: Total reward criteria. In Feinberg, E.A., Shwartz, A. (eds.) Handbook of Markov Decision Processes, pp. 173–207, Kluwer, Boston (2002)
10. Feinberg, E.A.: Continuous time discounted jump Markov Decision Processes: a discrete-event approach. Math. Oper. Res. **29**, 492–524 (2004)
11. Feinberg, E.A., Lewis, M.E.: Optimality inequalities for average cost Markov decision processes and the stochastic cash balance problem. Mathematics of Operations Research **32** 769–783 (2007)
12. Guo, X. Hernández-Lerma, O.: Continuous-Time Markov Decision Processes: Theory and Applications. Springer, Berlin (2009)
13. Guo, X. Hernández-Lerma, O., Prieto-Rumeau, T: A survey of recent results on continuous-time Markov Decision Processes. Top **14**, 177–261 (2006)
14. Guo, X., Piunovskiy, A. Discounted continuous-time Markov decision processes with constraints: unbounded transition and loss rates. Mathematics of Operations Research **36** 105–132 (2011)
15. Hernández-Lerma O.: Lectures on Continuous-Time Markov Control Processes. Aportaciones Mathemáticas. **3**. Sociedad Mathemática Mexicana, México (1994)
16. Hernández-Lerma, O., Lasserre J.: Discrete-Time Markov Control Processes: Basic Optimality Criteria. Springer, New York (1996)
17. Hordijk, A. and F. A. van der Duyn Schouten: Discretization procedures for continuous time decision processes. In Transactions of the Eighth Prague Conference on Information Theory, Statistical Decision Functions, Random Processes, Volume **C**, pp. 143–154, Academia, Prague (1979)
18. Howard, R.: Dynamic Programming and Markov Processes. Wiley, New York (1960)
19. Jacod, J.: Multivariate point processes: predictable projections, Radon-Nikodym derivatives, representation of martingales. Z. Wahr. verw. Geb. **31** 235–253 (1975)
20. Kakumanu, P.: Continuously discounted Markov decision models with countable state and action space. Ann. Math. Stat. **42** 919–926 (1971).
21. Kakumanu, P.: Relation between continuous and discrete time Markovian decision problems. Naval. Res. Log. Quart. **24** 431–439 (1977)
22. Kallianpur, G.:Stochastic Filtering Theory. Springer, New York (1980)
23. Kitaev, Yu. M.: Semi-Markov and jump Markov controlled models: average cost criterion. Theory Probab. Appl. **30** 272–288 (1985).
24. Kitaev, Yu. M. and V. V. Rykov.: Controlled Queueing Systems. CRC Press, New York (1995)
25. Lippman, S.: Applying a new device in the optimization of exponential queueing systems. Oper. Res. **23**, 687–710 (1975)

26. Martin-Löf, A.: Optimal control of a continuous-time Markov chain with periodic transition probabilities. Oper. Res. **15**, 872–881 (1966)
27. Miller, B.L.: 1968. Finite state continuous time Markov decision processes with a finite planning horizon. SIAM J. Control **6** 266–280 (1968)
28. Miller, B.L.: Finite state continuous time Markov decision processes with an infinite planning horizon. J. Math. Anal. Appl. **22** 552–569 (1968)
29. Piunovskiy, A., Zhang, Y.: The transformation method for continuous-time Markov decision processes, Preprint, University of Liverpool, 2011
30. Rykov, V.V.: Markov Decision Processes with finite spaces of states and decisions. Theory Prob. Appl. **11**, 302–311 (1966)
31. Schäl, M.: Conditions for optimality and for the limit of $n$-stage optimal policies to be optimal. Z. Wahrs. verw. Gebr. **32** 179–196 (1975)
32. Schäl, M.: On dynamic programming: compactness of the space of policies. Stoch. Processes Appl. **3** 345–364, 1975.
33. Serfozo, R.F.: An equivalence between continuous and discrete time Markov decision processes. Oper. Res. **27**, 616–620 (1979)
34. Strauch, R.E.: Negative dynamic programming, Ann. Math. Stat, **37**, 871–890 (1966)
35. Yushkevich, A. A.: Controlled Markov models with countable state space and continuous time. Theory Probab. Appl. **22** 215–235 (1977)
36. Zachrisson, L.E.: Markov games. In: Dresher, M., Shapley, L., Tucker A. (eds.) Advances in Game Theory, pp. 211–253. Princeton University Press, Princeton, N.J. (1964)

# Chapter 6
# Continuous-Time Controlled Jump Markov Processes on the Finite Horizon

Mrinal K. Ghosh and Subhamay Saha

## 6.1  Introduction

This chapter studies continuous-time Markov decision processes and continuous-time zero-sum stochastic dynamic games. In the continuous-time setup, although the infinite horizon cases have been well studied, the corresponding literature on finite horizon case is few and far between. Infinite horizon continuous-time Markov decision processes have been studied by many authors (e.g. see [5] and the references therein). In the finite horizon case, Pliska [7] has used a semi-group approach to characterise the value function and the optimal control. But his approach yields only existential results. In this chapter, we show that the value function is a smooth solution of an appropriate dynamic programming equation. Our method of proof gives algorithms for computing the value function and an optimal control.

The situation is analogous for continuous-time stochastic dynamic Markov games. In this problem as well, the infinite horizon case has been studied in the literature [6]. To our knowledge, the finite horizon case has not been studied. In this chapter, we prove that the value of the game on the finite horizon exists and is the solution of an appropriate Isaacs equation. This leads to the existence of saddle point equilibrium.

The rest of our chapter is structured as follows. In Sect. 6.2 we analyse the finite horizon continuous-time MDP. Section 6.3 deals with zero-sum stochastic dynamic games. We conclude our chapter in Sect. 6.4 with a few remarks.

M.K. Ghosh (✉) • S. Saha
Department of Mathematics, Indian Institute of Science, Bangalore 560012, India
e-mail: mkg@math.iisc.ernet.in; subhamay@math.iisc.ernet.in

D. Hernández-Hernández and A. Minjárez-Sosa (eds.), *Optimization, Control, and Applications of Stochastic Systems*, Systems & Control: Foundations & Applications, DOI 10.1007/978-0-8176-8337-5_6, © Springer Science+Business Media, LLC 2012

## 6.2   Finite Horizon Continuous-Time MDP

Throughout this chapter the time horizon is $T$. The control model we consider is given by

$$\{X, U, (\lambda(t,x,u), t \in [0,T], x \in X, u \in U), Q(t,x,u,\mathrm{d}z), c(t,x,u)\}$$

where each element is described below.

**The state space** $X$. The state space $X$ is the set of states of the process under observation which is assumed to be a Polish space.

**The action space** $U$. The decision-maker dynamically takes his action from the action space $U$. We assume that $U$ is a compact metric space.

**The instantaneous transition rate** $\lambda$. $\lambda : [0,T] \times X \times U \to [0,\infty)$ is a given function satisfying the following assumption:

(A1)  $\lambda$ is continuous and there exists a constant $M$ such that

$$\sup_{t,x,u} \lambda(t,x,u) \le M.$$

**The transition probability kernel** $Q$. For a fixed $t \in [0,T], x \in X, u \in U$, $Q(t,x,u,.)$ is a probability measure on $X$ with $Q(t,x,u,\{x\}) = 0$. $Q$ satisfies the following:

(A2)  $Q$ is weakly continuous, i.e. if $x_n \to x$, $t_n \to t$, $u_n \to u$, then for any $f \in C_b(X)$

$$\int_X f(z)Q(t_n,x_n,u_n,\mathrm{d}z) \to \int_X f(z)Q(t,x,u,\mathrm{d}z).$$

**The cost rate** $c$. $c : [0,T] \times X \times U \to [0,\infty)$ is a given function satisfying the following assumption:

(A3)  $c$ is continuous and there exists a finite constant $\tilde{C}$ such that

$$\sup_{t,x,u} c(t,x,u) \le \tilde{C}.$$

Next we give an informal description of the evolution of the controlled system. Suppose that the system is in state $x$ at time $t \ge 0$ and the controller or the decision-maker takes an action $u \in U$. Then the following happens on the time interval $[t, t+\mathrm{d}t]$:

1. The decision maker has to pay an infinitesimal cost $c(t,x,u)\mathrm{d}t$, and
2. A transition from state $x$ to a set $A$ (not containing $x$) occurs with probability

$$\lambda(t,x,u)\mathrm{d}t \int_A Q(t,x,u,\mathrm{d}z) + o(\mathrm{d}t);$$

or the system remains in state $x$ with probability

$$1 - \lambda(t,x,u)dt + o(dt).$$

Now we describe the optimal control problem. To this end we first describe the set of admissible controls. Let

$$\mathbf{u} : [0,T] \times X \to U$$

be a measurable function. Let $\mathcal{U}$ denote the set of all such measurable functions which is the set of admissible controls. Such controls are called Markov controls. For each $\mathbf{u} \in \mathcal{U}$, it can be shown that there exists is a strong Markov process $\{X_t\}$ (see [1, 3]) having the generator

$$\mathcal{A}_t^{\mathbf{u}} f(x) = -\lambda(t,x,\mathbf{u}(t,x))f(x) + \int_X f(z)Q(t,x,\mathbf{u}(t,x),dz)$$

where $f$ is a bounded measurable function.

For each $\mathbf{u} \in \mathcal{U}$, define

$$V^{\mathbf{u}}(t,x) = \mathbb{E}_{t,x}^{\mathbf{u}} \left[ \int_t^T c(s,X_s,\mathbf{u}(s,X_s))ds + g(X_T) \right] \tag{6.1}$$

where $g : X \to \mathbb{R}_+$ is the terminal cost function which is assumed to be bounded, continuous and $\mathbb{E}_{t,x}^{\mathbf{u}}$ is the expectation operator under the control $\mathbf{u}$ with initial condition $X_t = x$. The aim of the controller is to minimise $V^{\mathbf{u}}$ over all $\mathbf{u} \in \mathcal{U}$. Define

$$V(t,x) = \inf_{\mathbf{u} \in \mathcal{U}} \mathbb{E}_{t,x}^{\mathbf{u}} \left[ \int_t^T c(s,X_s,\mathbf{u}(s,X_s))ds + g(X_T) \right]. \tag{6.2}$$

The function $V$ is called the value function. If $\mathbf{u}^* \in \mathcal{U}$ satisfies

$$V^{\mathbf{u}^*}(t,x) = V(t,x) \quad \forall(t,x),$$

then $\mathbf{u}^*$ is called an optimal control.
The associated dynamic programming equation is

$$\begin{cases} \frac{d\varphi}{dt}(t,x) + \inf_{u \in U}\left[ c(t,x,u) - \lambda(t,x,u)\varphi(t,x) \right. \\ \left. + \lambda(t,x,u)\int_X \varphi(t,z)Q(t,x,u,dz) \right] = 0 \\ \text{on } X \times [0,T) \quad \text{and} \\ \varphi(T,x) = g(x). \end{cases} \tag{6.3}$$

The importance of (6.3) is illustrated by the following verification theorem.

**Theorem 6.2.1** *If* (6.3) *has a solution* $\varphi$ *in* $C_b^{1,0}([0,T] \times X)$, *then* $\varphi = V$, *the value function. Moreover, if* $\boldsymbol{u}^*$ *is such that*

$$\left[ c(t,x,\boldsymbol{u}^*(t,x)) - \lambda(t,x,\boldsymbol{u}^*(t,x))\varphi(t,x) + \lambda(t,x,\boldsymbol{u}^*(t,x)) \int_X \varphi(t,z)Q(t,x,\boldsymbol{u}^*(t,x),\mathrm{d}z) \right]$$

$$= \inf_{u \in U} \left[ c(t,x,u) - \lambda(t,x,u)\varphi(t,x) + \lambda(t,x,u) \int_X \varphi(t,z)Q(t,x,u,\mathrm{d}z) \right], \qquad (6.4)$$

*then* $\boldsymbol{u}^*$ *is an optimal control.*

*Proof.* Using Ito-Dynkin formula to the solution $\varphi$ of (6.3), we obtain

$$\varphi(t,x) \leq \inf_{\boldsymbol{u} \in \mathscr{U}} \mathbb{E}_{t,x}^{\boldsymbol{u}} \left[ \int_t^T c(s,X_s,\boldsymbol{u}(s,X_s))\mathrm{d}s + g(X_T) \right].$$

For $\boldsymbol{u} = \boldsymbol{u}^*$ as in the statement of the theorem, we get the equality

$$\varphi(t,x) = \mathbb{E}_{t,x}^{\boldsymbol{u}^*} \left[ \int_t^T c(s,X_s,\boldsymbol{u}^*(s,X_s))\mathrm{d}s + g(X_T) \right].$$

The existence of such a $\boldsymbol{u}^*$ follows by a standard measurable selection theorem [2]. □

In view of the above theorem, it suffices to show that (6.3) has a solution in $C_b^{1,0}([0,T] \times X)$.

**Theorem 6.2.2** *Under* (A1)–(A3), *the dynamic programming equation* (6.3) *has a unique solution in* $C_b^{1,0}([0,T] \times X)$.

*Proof.* Let $\varphi(t,x) = e^{-\gamma t}\psi(t,x)$ for some $\gamma < \infty$. Then from (6.3) we get,

$$\begin{cases} e^{-\gamma t}\frac{\mathrm{d}\psi}{\mathrm{d}t}(t,x) - \gamma e^{-\gamma t}\psi(t,x) + \inf_{u \in U}\left[ c(t,x,u) - \lambda(t,x,u)e^{-\gamma t}\psi(t,x) \right. \\ \left. +\lambda(t,x,u)\int_X e^{-\gamma t}\psi(t,z)Q(t,x,u,\mathrm{d}z) \right] = 0 \\ \text{on} \quad X \times [0,T) \quad \text{and} \\ \psi(T,x) = e^{\gamma T}g(x). \end{cases}$$

Thus (6.3) has a solution if and only if

$$\begin{cases} \frac{\mathrm{d}\psi}{\mathrm{d}t}(t,x) - \gamma\psi(t,x) + \inf_{u \in U}\left[ e^{\gamma t}c(t,x,u) - \lambda(t,x,u)\psi(t,x) \right. \\ \left. +\lambda(t,x,u)\int_X \psi(t,z)Q(t,x,u,\mathrm{d}z) \right] = 0 \\ \text{on} \quad X \times [0,T) \quad \text{and} \\ \psi(T,x) = e^{\gamma T}g(x) \end{cases}$$

has a solution. The above differential equation is equivalent to the following integral equation:

$$\psi(t,x) = e^{\gamma t} g(x) + e^{\gamma t} \int_t^T e^{-\gamma s} \inf_{u \in U} \left[ e^{\gamma s} c(s,x,u) - \lambda(s,x,u) \psi(s,x) \right.$$

$$\left. + \lambda(s,x,u) \int_X \psi(s,z) Q(s,x,u,dz) \right] ds.$$

Let $C_b^{\text{unif}}([0,T] \times X)$ be the space of bounded continuous functions $\varphi$ on $[0,T] \times X$ with the additional property that given $\varepsilon > 0$ there exists $\delta > 0$ such that

$$\sup_x |\varphi(t+h,x) - \varphi(t,x)| < \varepsilon \quad \text{whenever} \quad |h| < \delta.$$

Suppose $\varphi_n \in C_b^{\text{unif}}([0,T] \times X)$ and $\varphi_n \to \varphi$ uniformly. Then

$$|\varphi(t+h,x) - \varphi(t,x)| \le |\varphi(t+h,x) - \varphi_n(t+h,x)| + |\varphi_n(t+h,x) - \varphi_n(t,x)|$$
$$|\varphi_n(t,x) - \varphi(t,x)|.$$

Given $\varepsilon > 0$, there exists $n_0$ such that $\sup_{t,x} |\varphi_{n_0}(t,x) - \varphi(t,x)| < \frac{\varepsilon}{3}$, and for this $n_0$, there exists $\delta > 0$ such that $\sup_x |\varphi_{n_0}(t+h,x) - \varphi_{n_0}(t,x)| < \frac{\varepsilon}{3}$ whenever $|h| < \delta$. Putting $n = n_0$, we get from the above inequality

$$\sup_x |\varphi(t+h,x) - \varphi(t,x)| < \varepsilon \quad \text{whenever} \quad |h| < \delta.$$

Thus $C_b^{\text{unif}}([0,T] \times X)$ is a closed subspace of $C_b([0,T] \times X)$, and hence it is a Banach space.

Now for $\varphi \in C_b^{\text{unif}}([0,T] \times X)$, it follows from the assumption on $Q$ that $\int_X \varphi(t,z) Q(t,x,u,dz)$ is continuous in $t,x$ and $u$. Define

$$\mathscr{T} : C_b^{\text{unif}}([0,T] \times X) \to C_b^{\text{unif}}([0,T] \times X) \quad \text{by}$$

$$\mathscr{T}\psi(t,x) = e^{\gamma t} g(x) + e^{\gamma t} \int_t^T e^{-\gamma s} \inf_{u \in U} \left[ e^{\gamma s} c(s,x,u) - \lambda(s,x,u) \psi(s,x) \right.$$

$$\left. + \lambda(s,x,u) \int_X \psi(s,z) Q(s,x,u,dz) \right] ds.$$

For $\psi_1, \psi_2 \in C_b^{\text{unif}}([0,T] \times X)$, we have

$$|\mathscr{T}\psi_1(t,x) - \mathscr{T}\psi_2(t,x)| \le e^{\gamma t} \int_t^T e^{-\gamma s} 2M \|\psi_1 - \psi_2\| ds$$

$$= \frac{2M}{\gamma} e^{\gamma t} [e^{-\gamma t} - e^{-\gamma T}] \|\psi_1 - \psi_2\|$$

$$= \frac{2M}{\gamma} [1 - e^{-\gamma(T-t)}] \|\psi_1 - \psi_2\|$$

$$\le \frac{2M}{\gamma} \|\psi_1 - \psi_2\|.$$

Thus if we choose $\gamma = 2M + 1$, then $\mathscr{T}$ is a contraction and hence has a fixed point. Let $\varphi$ be the fixed point. Then $e^{-(2M+1)t}\varphi$ is the unique solution of (6.3).  $\square$

## 6.3  Zero-Sum Stochastic Game

In this section, we consider a zero-sum stochastic game. The control model we consider here is given by

$$\{X, U, V, (\lambda(t,x,u,v), t \in [0,T], x \in X, u \in U, v \in V, Q(t,x,u,v,dz), r(t,x,u,v)\}$$

where $X$ is the state space as before; $U$ and $V$ are the action spaces for player I and player II, respectively; $\lambda$ and $Q$ denote the rate and transition kernel, respectively, which now depend on the additional parameter $v$; and $r$ is the reward rate. The dynamics of the game is similar to that of MDP with appropriate modifications. Here player I receives a payoff from player II. The aim of player I is to maximise his payoff, and player II seeks to minimise the payoff to player I.

Now we describe the strategies of the players. In order to solve the problem, we will need to consider Markov relaxed strategies. We denote the space of strategies of player I by $\mathscr{U}$ and that of player *II* by $\mathscr{V}$ where

$$\mathscr{U} = \{\mathbf{u} \,|\, \mathbf{u} : [0,T] \times X \to \mathscr{P}(U) \text{ measurable}\},$$

$$\mathscr{V} = \{\mathbf{v} \,|\, \mathbf{v} : [0,T] \times X \to \mathscr{P}(V) \text{ measurable}\}.$$

Now corresponding to $\lambda$, $Q$ and $r$, define

$$\tilde{\lambda}(t,x,\mu,v) = \int_V \int_U \lambda(t,x,u,v)\mu(du)v(dv),$$

$$\tilde{Q}(t,x,\mu,v) = \int_V \int_U Q(t,x,u,v)\mu(du)v(dv),$$

$$\tilde{r}(t,x,\mu,v) = \int_V \int_U r(t,x,u,v)\mu(du)v(dv),$$

where $\mu \in \mathscr{P}(U)$ and $v \in \mathscr{P}(V)$. As in the previous section, we make the following assumptions:

(A1′) $\lambda$ is continuous and there exists a finite constant $M$ such that

$$\sup_{t,x,u,v} \lambda(t,x,u,v) \leq M.$$

(A2′) $Q$ is weakly continuous, i.e. if $x_n \to x$, $t_n \to t$, $u_n \to u$ and $v_n \to v$, then for any $f \in C_b(X)$

$$\int_X f(z)Q(t_n,x_n,u_n,v_n,\mathrm{d}z) \to \int_X f(z)Q(t,x,u,v,\mathrm{d}z).$$

(A3′) $r$ is continuous and there exists a finite constant $\tilde{C}$ such that

$$\sup_{t,x,u,v} r(t,x,u,v) \leq \tilde{C}.$$

If the players use strategies $(\mathbf{u},\mathbf{v}) \in \mathscr{U} \times \mathscr{V}$, then the expected payoff to player I is given by

$$\mathbb{E}_{t,x}^{\mathbf{u},\mathbf{v}} \left[ \int_t^T \tilde{r}(s,X_s,\mathbf{u}(s,X_s),\mathbf{v}(s,X_s))\mathrm{d}s + g(X_T) \right]$$

where $g$ is the terminal reward function which is assumed to be bounded and continuous. Now we define the upper and lower values for our game. Define

$$\overline{V}(t,x) = \inf_{\mathbf{v}\in\mathscr{V}} \sup_{\mathbf{u}\in\mathscr{U}} \mathbb{E}_{t,x}^{\mathbf{u},\mathbf{v}} \left[ \int_t^T \tilde{r}(s,X_s,\mathbf{u}(s,X_s),\mathbf{v}(s,X_s))\mathrm{d}s + g(X_T) \right].$$

Also define

$$\underline{V}(t,x) = \sup_{\mathbf{u}\in\mathscr{U}} \inf_{\mathbf{v}\in\mathscr{V}} \mathbb{E}_{t,x}^{\mathbf{u},\mathbf{v}} \left[ \int_t^T \tilde{r}(s,X_s,\mathbf{u}(s,X_s),\mathbf{v}(s,X_s))\mathrm{d}s + g(X_T) \right].$$

The function $\overline{V}$ is called the upper value function of the game, and $\underline{V}$ is called the lower value function of the game. In the game, player I is trying to maximise his payoff and player II is trying to minimise the payoff of player I. Thus $\underline{V}$ is the minimum payoff that player I is guaranteed to receive and $\overline{V}$ is the guaranteed greatest amount that player II can lose to player I. In general $\underline{V} \leq \overline{V}$. If $\overline{V}(t,x) = \underline{V}(t,x)$, then the game is said to have a value. A strategy $\mathbf{u}^*$ is said to be an optimal strategy for player I if

$$\mathbb{E}_{t,x}^{\mathbf{u}^*,\mathbf{v}} \left[ \int_t^T \tilde{r}(s,X_s,\mathbf{u}^*(s,X_s),\mathbf{v}(s,X_s))\mathrm{d}s + g(X_T) \right] \geq \overline{V}(t,x)$$

for any $t,x,\mathbf{v}$.

Similarly, $\mathbf{v}^*$ is called an optimal policy for player II if

$$\mathbb{E}_{t,x}^{\mathbf{u},\mathbf{v}^*}\left[\int_t^T \tilde{r}(s,X_s,\mathbf{u}(s,X_s),\mathbf{v}^*(s,X_s))\mathrm{d}s + g(X_T)\right] \leq \underline{V}(t,x)$$

for any $t,x,\mathbf{u}$. Such a pair $(\mathbf{u}^*,\mathbf{v}^*)$, if it exists, is called a saddle point equilibrium. Our aim is to find the value of the game and to find optimal strategies for both the players. To this end, consider the following pair of Isaacs equations:

$$\begin{cases} \frac{\mathrm{d}\varphi}{\mathrm{d}t}(t,x) + \inf_{\nu\in\mathscr{P}(V)} \sup_{\mu\in\mathscr{P}(U)} \left[\tilde{r}(t,x,\mu,\nu) - \tilde{\lambda}(t,x,\mu,\nu)\varphi(t,x)\right. \\ \left. + \tilde{\lambda}(t,x,\mu,\nu)\int_X \varphi(t,z)\tilde{Q}(t,x,\mu,\nu,\mathrm{d}z)\right] = 0 \\ \text{on } X\times[0,T) \quad \text{and} \\ \varphi(T,x) = g(x). \end{cases} \tag{6.5}$$

$$\begin{cases} \frac{\mathrm{d}\psi}{\mathrm{d}t}(t,x) + \sup_{\mu\in\mathscr{P}(U)} \inf_{\nu\in\mathscr{P}(V)} \left[\tilde{r}(t,x,\mu,\nu) - \tilde{\lambda}(t,x,\mu,\nu)\psi(t,x)\right. \\ \left. + \tilde{\lambda}(t,x,\mu,\nu)\int_X \psi(t,z)\tilde{Q}(t,x,\mu,\nu,\mathrm{d}z)\right] = 0 \\ \text{on } X\times[0,T) \quad \text{and} \\ \varphi(T,x) = g(x). \end{cases} \tag{6.6}$$

By Fan's minimax theorem [4], we have that if $\varphi \in C_b^{1,0}([0,T]\times X)$ is a solution of (6.5), then it is also a solution of (6.6) and vice versa. The importance of Isaacs equations is illustrated by the following theorem.

**Theorem 6.3.1** *Let $\varphi^* \in C_b^{1,0}([0,T]\times X)$ be a solution of (6.5) and (6.6). Then*

*(i) $\varphi^*$ is the value of the game.*
*(ii) Let $(\mathbf{u}^*,\mathbf{v}^*) \in \mathscr{U}\times\mathscr{V}$ be such that*

$$\inf_{\nu\in\mathscr{P}(V)}\left[\tilde{r}(t,x,\mathbf{u}^*(t,x),\nu) - \tilde{\lambda}(t,x,\mathbf{u}^*(t,x),\nu)\varphi^*(t,x) + \tilde{\lambda}(t,x,\mathbf{u}^*(t,x),\nu)\right.$$

$$\left.\int_X \varphi^*(t,z)\tilde{Q}(t,x,\mathbf{u}^*(t,x),\nu,\mathrm{d}z)\right]$$

$$= \sup_{\mu\in\mathscr{P}(U)} \inf_{\nu\in\mathscr{P}(V)}\left[\tilde{r}(t,x,\mu,\nu) - \tilde{\lambda}(t,x,\mu,\nu)\psi(t,x) + \tilde{\lambda}(t,x,\mu,\nu)\right.$$

$$\left.\int_X \psi(t,z)\tilde{Q}(t,x,\mu,\nu,\mathrm{d}z)\right] \tag{6.7}$$

*and*

$$\sup_{\mu \in \mathscr{P}(U)} \left[ \tilde{r}(t,x,\mu,v^*(t,x)) - \tilde{\lambda}(t,x,\mu,v^*(t,x))\varphi^*(t,x) + \tilde{\lambda}(t,x,\mu,v^*(t,x)) \right.$$

$$\left. \int_X \varphi^*(t,z)\tilde{Q}(t,x,\mu,v^*(t,x),\mathrm{d}z) \right]$$

$$= \inf_{v \in \mathscr{P}(V)} \sup_{\mu \in \mathscr{P}(U)} \left[ \tilde{r}(t,x,\mu,v) - \tilde{\lambda}(t,x,\mu,v)\varphi(t,x) + \tilde{\lambda}(t,x,\mu,v) \right.$$

$$\left. \int_X \varphi(t,z)\tilde{Q}(t,x,\mu,v,\mathrm{d}z) \right]. \tag{6.8}$$

*Then $u^*$ is an optimal policy for player I and $v^*$ is an optimal policy for player II.*

*Proof.* Let $\mathbf{u}^*$ be as in (6.7) and $\mathbf{v}$ be any arbitrary strategy of player II. Then by Ito-Dynkin formula applied to the solution $\varphi$, we obtain

$$\varphi^*(t,x) \leq \mathbb{E}_{t,x}^{\mathbf{u}^*,\mathbf{v}} \left[ \int_t^T \tilde{r}(s,X_s,\mathbf{u}^*(s,X_s),\mathbf{v}(s,X_s))\mathrm{d}s + g(X_T) \right]$$

$$\leq \inf_{\mathbf{v} \in \mathscr{V}} \mathbb{E}_{t,x}^{\mathbf{u}^*,\mathbf{v}} \left[ \int_t^T \tilde{r}(s,X_s,\mathbf{u}^*(s,X_s),\mathbf{v}(s,X_s))\mathrm{d}s + g(X_T) \right]$$

$$\leq \underline{V}(t,x).$$

Now let $\mathbf{v}^*$ be as in (6.8) and let $\mathbf{u}$ be any arbitrary strategy of player I. Then again by Ito's formula we obtain

$$\varphi^*(t,x) \geq \mathbb{E}_{t,x}^{\mathbf{u},\mathbf{v}^*} \left[ \int_t^T \tilde{r}(s,X_s,\mathbf{u}(s,X_s),\mathbf{v}^*(s,X_s))\mathrm{d}s + g(X_T) \right]$$

$$\geq \inf_{\mathbf{v} \in \mathscr{V}} \mathbb{E}_{t,x}^{\mathbf{u},\mathbf{v}^*} \left[ \int_t^T \tilde{r}(s,X_s,\mathbf{u}(s,X_s),\mathbf{v}^*(s,X_s))\mathrm{d}s + g(X_T) \right]$$

$$\geq \overline{V}(t,x).$$

From the above two inequalities, it follows that

$$\varphi^*(t,x) = \overline{V}(t,x) = \underline{V}(t,x).$$

Hence $\varphi^*$ is the value of the game. Moreover it follows that $(\mathbf{u}^*,\mathbf{v}^*)$ is a saddle point equilibrium. $\square$

Now our aim is to find a solution of (6.5) (and hence of (6.6)) in $C_b^{1,0}([0,T] \times X)$. Our next theorem asserts the existence of such a solution.

**Theorem 6.3.2** *Under* $(A1')$–$(A3')$, *equation* (6.5) *has a unique solution in* $C_b^{1,0}([0,T] \times X)$.

*Proof.* Let $\varphi(t,x) = e^{-\gamma t} \psi(t,x)$ for some $\gamma < \infty$. Substituting in (6.5), we get

$$
\begin{cases}
e^{-\gamma t} \frac{d\psi}{dt}(t,x) - \gamma e^{-\gamma t} \psi(t,x) + \inf_{v \in \mathscr{P}(V)} \sup_{\mu \in \mathscr{P}(U)} \big[ \tilde{r}(t,x,\mu,v) - \tilde{\lambda}(t,x,\mu,v) e^{-\gamma t} \psi(t,x) \\
\quad + \tilde{\lambda}(t,x,\mu,v) \int_X e^{-\gamma t} \psi(t,z) \tilde{Q}(t,x,\mu,v,dz) \big] = 0 \\
\text{on} \quad X \times [0,T) \quad \text{and} \\
\psi(T,x) = e^{\gamma T} g(x).
\end{cases}
$$

Thus (6.5) has a solution if and only if

$$
\begin{cases}
\frac{d\psi}{dt}(t,x) - \gamma \psi(t,x) + \inf_{v \in \mathscr{P}(V)} \sup_{\mu \in \mathscr{P}(U)} \big[ e^{\gamma t} \tilde{r}(t,x,\mu,v) - \tilde{\lambda}(t,x,\mu,v) \psi(t,x) \\
\quad + \tilde{\lambda}(t,x,\mu,v) \int_X \psi(t,z) \tilde{Q}(t,x,\mu,v,dz) \big] = 0 \\
\text{on} \quad X \times [0,T) \quad \text{and} \\
\psi(T,x) = e^{\gamma T} g(x)
\end{cases}
$$

has a solution. The above differential equation is equivalent to the following integral equation:

$$
\psi(t,x) = e^{\gamma t} g(x) + e^{\gamma t} \int_t^T e^{-\gamma s} \inf_{v \in \mathscr{P}(V)} \sup_{\mu \in \mathscr{P}(U)} \Big[ e^{\gamma s} \tilde{r}(s,x,\mu,v) - \tilde{\lambda}(s,x,\mu,v) \psi(s,x)
$$
$$
+ \tilde{\lambda}(s,x,\mu,v) \int_X \psi(s,z) \tilde{Q}(s,x,\mu,,v,dz) \Big] ds.
$$

Let $C_b^{\mathrm{unif}}([0,T] \times X)$ be the same space as defined in the previous section. Define

$$
\mathscr{T} : C_b^{\mathrm{unif}}([0,T] \times X) \to C_b^{\mathrm{unif}}([0,T] \times X) \quad \text{by}
$$

$$
\mathscr{T}\psi(t,x) = e^{\gamma t} g(x) + e^{\gamma t} \int_t^T e^{-\gamma s} \inf_{v \in \mathscr{P}(V)} \sup_{\mu \in \mathscr{P}(U)} \Big[ e^{\gamma s} \tilde{r}(s,x,\mu,v)
$$
$$
- \tilde{\lambda}(s,x,\mu,v) \psi(s,x) + \tilde{\lambda}(s,x,\mu,v) \int_X \psi(s,z) \tilde{Q}(s,x,\mu,v,dz) \Big] ds.
$$

For $\psi_1, \psi_2 \in C_b^{\mathrm{unif}}([0,T] \times X)$, we have

$$|\mathcal{T}\psi_1(t,x) - \mathcal{T}\psi_2(t,x)| \leq e^{\gamma t}\int_t^T e^{-\gamma s}2M\|\psi_1 - \psi_2\|ds$$

$$= \frac{2M}{\gamma}e^{\gamma t}[e^{-\gamma t} - e^{-\gamma T}]\|\psi_1 - \psi_2\|$$

$$= \frac{2M}{\gamma}[1 - e^{-\gamma(T-t)}]\|\psi_1 - \psi_2\|$$

$$\leq \frac{2M}{\gamma}\|\psi_1 - \psi_2\|.$$

Thus if we choose $\gamma = 2M + 1$, then $\mathcal{T}$ is a contraction and hence has a fixed point. Let $\varphi$ be the fixed point. Then $e^{-(2M+1)t}\varphi$ is the unique solution of (6.5). $\square$

## 6.4 Conclusion

In this chapter we have established smooth solutions of dynamic programming equations for continuous-time controlled Markov chains on the finite horizon. This has led to the existence of an optimal Markov strategy for continuous-time MDP and saddle point equilibrium in Markov strategies for zero-sum games. We have used the boundedness condition on the cost function $c$ for simplicity. For continuous-time MDP, if $c$ is unbounded above, then we can show that $V(t,x)$ is the minimal non-negative solution of (6.3) by approximating the cost function $c$ by $c \wedge n$ for a positive integer $n$ and then letting $n \to \infty$. If $c$ is unbounded on both sides and it satisfies a suitable growth condition, then again we can prove the existence of unique solutions of dynamic programming equations in $C^{1,0}([0,T] \times X)$ with appropriate weighted norm; see [5] and [6] for analogous results.

## References

1. A. Arapostathis, V. S. Borkar and M. K. Ghosh, *Ergodic Control of Diffusion Processes*, Cambridge University Press, 2011.
2. V. E. Benes, *Existence of optimal strategies based on specified information for a class of stochastic decision problems*, SIAM J. Control 8 (1970), 179–188.
3. M. H. A. Davis, *Markov Models and Optimization*, Chapman and Hall, 1993.
4. K. Fan, *Fixed-point and minimax theorems in locally convex topological linear spaces*, Proc. of the Natl. Academy of Sciences of the United States of America 38 (1952), 121–126.
5. X. Guo and O. Hernández-Lerma, *Continuous-Time Markov Decision Processes. Theory and Applications*, Springer-Verlag, 2009.
6. X. Guo and O. Hernández-Lerma, *Zero-sum games for continuous-time jump Markov processes in Polish spaces: discounted payoffs*, Adv. in Appl. Probab. 39 (2007), 645–668.
7. S. R. Pliska, *Controlled jump processes*, Stochastic Processes Appl. 3 (1975), 259–282.

# Chapter 7
# Existence and Uniqueness of Solutions of SPDEs in Infinite Dimensions

T.E. Govindan

## 7.1 Introduction

In this chapter, we study a neutral stochastic partial differential equation in a real Hilbert space of the form

$$d[x(t) + f(t, x(\rho(t)))] = [Ax(t) + a(t, x(\rho(t)))]dt$$
$$+ b(t, x(\rho(t)))dw(t), \quad t > 0; \tag{7.1}$$

$$x(t) = \varphi(t), \quad t \in [-r, 0] \quad (0 \le r < \infty); \tag{7.2}$$

where $a : R^+ \times C \to X$ $(R^+ = [0, \infty))$, $b : R^+ \times C \to L(Y, X)$, and $f : R^+ \times C \to D((-A)^{-\alpha})$, $0 < \alpha \le 1$, are Borel measurable, and $-A : D(-A) \subset X$ is the infinitesimal generator of a strongly continuous semigroup $\{S(t), t \ge 0\}$ defined on $X$. Here $w(t)$ is a $Y$-valued $Q$-Wiener process, and the past stochastic process $\{\varphi(t), t \in [-r, 0]\}$ has almost sure (a.s) continuous paths with $E\|\varphi\|_C^p < \infty$, $p \ge 2$. The delay function $\rho : [0, \infty) \to [-r, \infty)$ is measurable, satisfying $-r \le \rho(t) \le t, t \ge 0$.

To motivate this kind of equations, consider a semilinear equation with a finite delay in the deterministic case; see Hernandez and Henriquez [7]:

$$\frac{dx(t)}{dt} = Ax(t) + f(x_t) + Bu(t), \quad t > 0, \tag{7.3}$$

where $x(t) \in X$ represents the state, $u(t) \in R^m$ denotes the control, $x_t(s) = x(t+s)$, $-r \le s \le 0, A : X \to X$, and $B : R^m \to X$. The feedback control $u(t)$ will be defined by

$$u(t) = K_0 x(t) - \frac{d}{dt} \int_{-r}^{t} K_1(t-s)x(s)ds, \tag{7.4}$$

T.E. Govindan (✉)
ESFM, Instituto Politécnico Nacional, México D.F. 07738, México
e-mail: tegovindan@yahoo.com

D. Hernández-Hernández and A. Minjárez-Sosa (eds.), *Optimization, Control, and Applications of Stochastic Systems*, Systems & Control: Foundations & Applications, DOI 10.1007/978-0-8176-8337-5_7, © Springer Science+Business Media, LLC 2012

where $K_0 : X \to R^m$ is a bounded linear operator, and $K_1 : [0, \infty) \to L(X, R^m)$ is an operator-valued map. The closed system corresponding to the control (7.4) takes the form:

$$\frac{\mathrm{d}}{\mathrm{d}t}\left[x(t) + \int_{-r}^{t} K_1(t-s)x(s)\mathrm{d}s\right] = (A + BK_0)x(t) + f(x_t), \quad t > 0.$$

A special case of (7.1), see (7.6) below, was studied in Govindan [4, 5] and Luo [8] by using Lipschitz conditions on all the nonlinear terms and in Govindan [6] using local Lipschitz conditions on $a(t, u)$ and $b(t, u)$. In this chapter, our goal is to study the existence and uniqueness problem for (7.1) and (7.2) using Lipschitz conditions.

The rest of the chapter is organized as follows: In Sect. 7.2, we give the preliminaries. For a general theory of stochastic differential equations in infinite dimensions, we refer to the excellent texts of Ahmed [1] and Da Prato and Zabczyk [2]. Since we work in the same framework as in Govindan [4] and Taniguchi [10], we shall be quite brief here. In Sect. 7.3, we present our main result. Some examples are given in Sect. 7.4.

## 7.2 Preliminaries

Let $X, Y$ be real separable Hilbert spaces and $L(Y, X)$ be the space of bounded linear operators mapping $Y$ into $X$. We shall use the same notation $|\cdot|$ to denote norms in $X, Y$ and $L(Y, X)$. Let $(\Omega, B, P, \{B_t\}_{t \geq 0})$ be a complete probability space with an increasing right continuous family $\{B_t\}_{t \geq 0}$ of complete sub-$\sigma$-algebras of $B$. Let $\beta_n(t)(n = 1, 2, 3, \ldots)$ be a sequence of real-valued standard Brownian motions mutually independent defined on this probability space. Set $w(t) = \sum_{n=1}^{\infty} \sqrt{\lambda_n}\beta_n(t)e_n$, $t \geq 0$, where $\lambda_n \geq 0$ $(n = 1, 2, 3, \ldots)$ are nonnegative real numbers and $\{e_n\}$ $(n = 1, 2, 3, \ldots)$ is a complete orthonormal basis in $Y$. Let $Q \in L(Y, Y)$ be an operator defined by $Qe_n = \lambda_n e_n$. The above $Y$-valued stochastic processes $w(t)$ is called a $Q$-Wiener process. Let $h(t)$ be an $L(Y, X)$-valued function and let $\lambda$ be a sequence $\{\sqrt{\lambda_1}, \sqrt{\lambda_2}, \ldots\}$. Then we define $|h(t)|_\lambda = \left\{\sum_{n=1}^{\infty} |\sqrt{\lambda_n}h(t)e_n|^2\right\}^{1/2}$. If $|h(t)|_\lambda^2 < \infty$, then $h(t)$ is called $\lambda$-Hilbert-Schmidt operator.

Next, we define the $X$-valued stochastic integral with respect to the $Y$-valued $Q$-Wiener process $w(t)$; see Taniguchi [10].

**Definition 7.2.1.** Let $\Phi : [0, \infty) \to \sigma(\lambda)(Y, X)$ be a $B_t$-adapted process. Then, for any $\Phi$ satisfying $\int_0^t E|\Phi(s)|_\lambda^2 \mathrm{d}s < \infty$, we define the $X$-valued stochastic integral $\int_0^t \Phi(s)\mathrm{d}w(s) \in X$ with respect to $w(t)$ by

$$\left(\int_0^t \Phi(s)\mathrm{d}w(s), h\right) = \int_0^t < \Phi^*(s)h, \mathrm{d}w(s) >, \qquad h \in X,$$

where $\Phi^*$ is the adjoint operator of $\Phi$.

A semigroup $\{S(t), t \geq 0\}$ is said to be exponentially stable if there exist positive constants $M$ and $a$ such that $||S(t)|| \leq M\exp(-at)$, $t \geq 0$, where $||\cdot||$ denotes the operator norm in $X$. If $M = 1$, the semigroup is said to be a contraction.

Let $C := C([-r,0];X)$ denote the space of continuous functions $\varphi : [-r,0] \to X$ endowed with the norm $||\varphi||_C = \sup_{-r \leq s \leq 0} |\varphi(s)|$. Let $B_T = B_T(\varphi)$ be the space of measurable random processes $\phi(t,\omega)$ with a.s. continuous paths; for each $t \in [0,T]$, $\phi(t,\omega)$ is measurable with respect to $B_t$ and $\phi(s,\omega) = \varphi(s,\omega)$ for $-r \leq s \leq 0$ with the norm $||\phi||_{B_T} = (E||\phi(.,\omega)||_C^p)^{1/p}$, $1 \leq p < \infty$. $B_T$ is a Banach space, see Govindan [4] and the references therein.

If $\{S(t), t \geq 0\}$ is an analytic semigroup with an infinitesimal generator $-A$ such that $0 \in \rho(-A)$ (the resolvent set of $-A$), then it is possible to define the fractional power $(-A)^\alpha$, for $0 < \alpha \leq 1$, as a closed linear operator on its domain $D((-A)^\alpha)$. Furthermore, the subspace $D((-A)^\alpha)$ is dense in $X$ and the expression

$$||x||_\alpha = |(-A)^\alpha x|, \qquad x \in D((-A)^\alpha),$$

defines a norm on $D((-A)^\alpha)$.

**Definition 7.2.2.**  A stochastic process $\{x(t), t \in [0,T]\}(0 < T < \infty)$ is called a mild solution of equation (7.1) if

(i)  $x(t)$ is $B_t$-adapted with $\int_0^T |x(t)|^2 dt < \infty$, a.s., and
(ii)  $x(t)$ satisfies the integral equation

$$x(t) = S(t)[\varphi(0) + f(0,\varphi)] - f(t,x(\rho(t)))$$

$$- \int_0^t AS(t-s)f(s,x(\rho(s)))ds$$

$$+ \int_0^t S(t-s)a(s,x(\rho(s)))ds$$

$$+ \int_0^t S(t-s)b(s,x(\rho(s)))dw(s), \quad \text{a.s.,} \quad t \in [0,T]. \qquad (7.5)$$

For convenience of the reader, we will state below some results that will be needed in the sequel.

**Lemma 7.2.1.**  *[3] Let $W_A^\Phi(t) = \int_0^t S(t-s)\Phi(s)dw(s), t \in [0,T]$. For any arbitrary $p > 2$, there exists a constant $c(p,T) > 0$ such that for any $T \geq 0$ and a proper modification of the stochastic convolution $W_A^\Phi$, one has*

$$E\sup_{t \leq T} |W_A^\Phi(t)|^p \leq c(p,T)\sup_{t \leq T}||S(t)||^p E \int_0^T |\Phi(s)|_\lambda^p ds.$$

*Moreover if $E \int_0^T |\Phi(s)|_\lambda^p ds < \infty$, then there exists a continuous version of the process $\{W_A^\Phi, t \geq 0\}$.*

**Lemma 7.2.2.** *[2, Theorem 6.10] Suppose $-A$ generates a contraction semigroup. Then the process $W_A^\Phi(.)$ has a continuous modification and there exists a constant $\kappa > 0$ such that*

$$E \sup_{s \in [0,t]} |W_A^\Phi(s)|^2 \leq \kappa E \int_0^t |\Phi(s)|_\lambda^2 ds, \qquad t \in [0,T].$$

**Theorem 7.2.1.** *[1, 9] Let $-A$ be the infinitesimal generator of an analytic semigroup $\{S(t), t \geq 0\}$. If $0 \in \rho(A)$, then*

(a) *$S(t) : X \to X_\alpha$ for every $t > 0$ and $\alpha \geq 0$.*
(b) *For every $x \in X_\alpha$, we have*

$$S(t)A^\alpha x = A^\alpha S(t)x.$$

(c) *For every $t > 0$, the operator $A^\alpha S(t)$ is bounded and*

$$\|A^\alpha S(t)\| \leq M_\alpha t^{-\alpha} e^{-at}, \quad a > 0.$$

(d) *Let $0 < \alpha \leq 1$ and $x \in D(A^\alpha)$; then*

$$\|S(t)x - x\| \leq C_\alpha t^\alpha \|A^\alpha x\|.$$

## 7.3  Existence and Uniqueness of a Mild Solution

In this section we study the existence and uniqueness of a mild solution of the equation (7.1). But, before that, it shall be interesting to recall some recent existence and uniqueness results for a mild solution of a simpler equation given by

$$d[x(t) + f(t, x_t)] = [Ax(t) + a(t, x_t)]dt + b(t, x_t)dw(t), \quad t > 0; \qquad (7.6)$$

$$x(t) = \varphi(t), \quad t \in [-r, 0] \quad (0 \leq r < \infty); \qquad (7.7)$$

where $x_t(s) = x(t + s)$, $-r \leq s \leq 0$, and all the other terms are as defined before.

To begin with, we present our first result for (7.6) from Luo [8].

**Hypothesis (A):**

Let the following assumptions hold a.s.:

(A1) *$-A$ is the infinitesimal generator of an analytic semigroup of bounded linear operators $\{S(t), t \geq 0\}$ in $X$, and the semigroup is exponentially stable.*
(A2) *The mappings $a(t, u)$ and $b(t, u)$ satisfy the following Lipschitz and linear growth conditions:*

$$|a(t,u) - a(t,v)| \leq L_1 ||u - v||_C, \quad L_1 > 0,$$
$$|b(t,u) - b(t,v)|_\lambda \leq L_2 ||u - v||_C, \quad L_2 > 0,$$
$$|a(t,u)|^2 + |b(t,u)|_\lambda^2 \leq L_3^2 (1 + ||u||_C^2), \quad L_3 > 0,$$

for all $u, v \in C$.

(A3)  $f(t, u)$ is a function continuous in $t$ and satisfies

$$|(-A)^\alpha f(t,u) - (-A)^\alpha f(t,v)| \leq L_4 ||u - v||_C, \quad L_4 > 0,$$
$$|(-A)^\alpha f(t,u)| \leq L_5 (1 + ||u||_C), \quad L_5 > 0,$$

for all $u, v \in C$.

**Theorem 7.3.2.** *[8] Let Assumptions (A1)–(A3) hold. Then there exists a unique mild solution $\{x(t), 0 \leq t \leq T\}$ of the problem (7.6) and (7.7) provided*

$$L||(-A)^{-\alpha}|| + LM_{1-\alpha} \Gamma(\alpha)/a^\alpha < 1,$$

*where $L = \max\{L_4, L_5\}$ and $\Gamma(\cdot)$ is the Gamma function.*

In Govindan [5], a new successive approximation procedure was introduced to study the existence and uniqueness of a mild solution of equation (7.6) under the following hypothesis.

**Hypothesis (B):**

Let the following assumptions hold a.s.:

(B1)  Same as (A1).

(B2)  For $p \geq 2$, the functions $a(t,u)$ and $b(t,u)$ satisfy the Lipschitz and linear growth conditions:

$$|a(t,u) - a(t,v)|^p \leq L_6 ||u - v||_C^p, \quad L_6 > 0,$$
$$|b(t,u) - b(t,v)|_\lambda^p \leq L_7 ||u - v||_C^p, \quad L_7 > 0,$$
$$|a(t,u)|^p + |b(t,u)|_\lambda^p \leq L_8 (1 + ||u||_C^p), \quad L_8 > 0,$$

for all $u, v \in C$.

(B3)  Same as (A3).

**Theorem 7.3.3.** *[5] Let Assumptions (B1)–(B3) hold. For $p = 2$, $\{S(t), t \geq 0\}$ is a contraction. Then there exists a unique mild solution $x(t)$ of the problem (7.6) and (7.7) provided $1/p < \alpha < 1$, and*

$$L||(-A)^{-\alpha}|| < 1,$$

*where $L = \max\{L_4, L_5\}$.*

Recently, Govindan [6] studied the existence and uniqueness problem for (7.6) by using only a local Lipschitz condition. To consider this result, we need the following

**Hypothesis (C):**
Let the following assumptions hold a.s.:

(C1) Same as (A1) but now the semigroup is a contraction.
(C2) The functions $a(t,u)$ and $b(t,u)$ are continuous and that there exist positive constants $M_i = M_i(T)$, $i = 1,2$ such that

$$|a(t,u) - a(t,v)| \leq M_1 \|u - v\|_C,$$
$$|b(t,u) - b(t,v)|_\lambda \leq M_2 \|u - v\|_C;$$

for all $t \in [0,T]$ and $u, v \in C$.
(C3) The function $f(t,u)$ is continuous and that there exists a positive constant $M_3 = M_3(T)$ such that

$$\|f(t,u) - f(t,v)\|_\alpha \leq M_3 \|u - v\|_C,$$

for all $t \in [0,T]$ and $u, v \in C$.
(C4) $f(t,u)$ is continuous in the quadratic mean sense:

$$\lim_{(t,u)\to(s,v)} E\|f(t,u) - f(s,v)\|_\alpha^2 \to 0.$$

**Theorem 7.3.4.** *[6] Suppose that Assumptions (C1)–(C4) are satisfied. Then, there exists a time $0 < t_m = t_{\max} \leq \infty$ such that (7.6) has a unique mild solution. Further, if $t_m < \infty$, then $\lim_{t\uparrow t_m} E|x(t)|^2 = \infty$.*

Finally, we state the following hypothesis to consider the main result of the chapter.

**Hypothesis (D):**
Let the following assumptions hold a.s.:

(D1) $-A$ is the infinitesimal generator of an analytic semigroup of bounded linear operators $\{S(t), t \geq 0\}$ in $X$, and the semigroup is exponentially stable.
(D2) For $p \geq 2$, the functions $a(t,u)$ and $b(t,u)$ satisfy the Lipschitz and linear growth conditions:

$$|a(t,u) - a(t,v)|^p \leq C_1 \|u - v\|_C^p, \qquad C_1 > 0,$$
$$|b(t,u) - b(t,v)|_\lambda^p \leq C_2 \|u - v\|_C^p, \qquad C_2 > 0,$$

for all $u, v \in C$.
There exist positive constants $C_3, P$ and $\delta > a$ and a continuous function $\psi : R^+ \to R^+$ satisfying $|\psi(t)|^2 < P\exp(-\delta t)$, $t \geq 0$ such that

$$|a(t,u)|^p + |b(t,u)|_\lambda^p \leq C_3 \|u\|_C^p + \psi(t), \qquad t \geq 0,$$

for all $u \in C$.

(D3)  $f(t,u)$ is a function continuous in $t$ and satisfies

$$|(-A)^\alpha f(t,u) - (-A)^\alpha f(t,v)| \leq L_4 ||u-v||_C, \quad L_4 > 0,$$
$$|(-A)^\alpha f(t,u)| \leq L_5(1 + ||u||_C), \quad L_5 > 0,$$

for all $u,v \in C$.

**Theorem 7.3.5.** *Let Assumptions (D1)–(D3) hold. Suppose that for the case $p = 2$, the semigroup $\{S(t), t \geq 0\}$ is a contraction. Then there exists a unique mild solution $x(t)$ of (7.1) provided*

$$L||(-A)^{-\alpha}|| + LM_{1-\alpha}\Gamma(\alpha)/a^\alpha < 1,$$

*where $L = \max\{L_4, L_5\}$.*

To prove this theorem, let us introduce the following iteration procedure: Define for each integer $n = 1, 2, 3, \ldots$,

$$
\begin{aligned}
x^{(n)}(t) = {} & S(t)[\varphi(0) + f(0,\varphi)] - f(t, x^{(n)}(\rho(t))) \\
& - \int_0^t AS(t-s)f(s, x^{(n)}(\rho(s)))ds \\
& + \int_0^t S(t-s)a(s, x^{(n-1)}(\rho(s)))ds \\
& + \int_0^t S(t-s)b(s, x^{(n-1)}(\rho(s)))dw(s), \qquad t \in [0,T], \qquad (7.8)
\end{aligned}
$$

and for $n = 0$,

$$x^{(0)}(t) = S(t)\varphi(0), \qquad t \in [0,T], \qquad (7.9)$$

while for $n = 0, 1, 2, \ldots$,

$$x^{(n)}(t) = \varphi(t), \qquad t \in [-r, 0]. \qquad (7.10)$$

*Proof of Theorem 7.3.5.* Let $T$ be any fixed time with $0 < T < \infty$. Rewriting (7.8) as

$$
\begin{aligned}
x^{(n)}(s) = {} & S(s)\varphi(0) + S(s)(-A)^{-\alpha}(-A)^\alpha f(0,\varphi) \\
& - (-A)^{-\alpha}(-A)^\alpha f(x, x^{(n)}(\rho(t))) \\
& + \int_0^s (-A)^{1-\alpha}S(s-\tau)(-A)^\alpha f(\tau, x^{(n)}(\rho(\tau)))d\tau \\
& + \int_0^s S(s-\tau)a(\tau, x^{(n-1)}(\rho(\tau)))d\tau \\
& + \int_0^s S(s-\tau)b(\tau, x^{(n-1)}(\rho(\tau)))dw(\tau), \quad \text{a.s.} \quad s \in [0,T].
\end{aligned}
$$

By Theorem 7.2.1, Assumptions (D1) and (D3) and Luo [8], we get

$$|x^{(n)}(s)| \leq Me^{-as}|\varphi(0)| + Me^{-as}||(-A)^{-\alpha}||L_5(1+||\varphi||_C)$$

$$+L_5||(-A)^{-\alpha}||(1+||x_s^{(n)}||_C)$$

$$+\int_0^s L_5 \frac{M_{1-\alpha}e^{-a(s-\tau)}}{(s-\tau)^{1-\alpha}}(1+||x_\tau^{(n)}||_C)d\tau$$

$$+\left|\int_0^s S(s-\tau)a(\tau,x^{(n-1)}(\rho(\tau)))d\tau\right|$$

$$+\left|\int_0^s S(s-\tau)b(\tau,x^{(n-1)}(\rho(\tau)))dw(\tau)\right| \quad \text{a.s..}$$

Note that $(-A)^{-\alpha}$, for $0 < \alpha \leq 1$ is a bounded operator, see Pazy [9, Lemma 6.3 p. 71]. An application of Lemma 7.2.1 (or Lemma 7.2.2 for $p = 2$) then yields

$$\left[1 - L_5||(-A)^{-\alpha}|| - L_5M_{1-\alpha}\Gamma(\alpha)/a^\alpha\right]^p E \sup_{0\leq s\leq t} |x^{(n)}(s)|^p$$

$$\leq 3^{p-1}\left\{E\left[M|\varphi(0)| + L_5(1+||\varphi||_C)((M+1)||(-A)^{-\alpha}|| + M_{1-\alpha}\Gamma(\alpha)/a^\alpha)\right]^p\right.$$

$$+M^pT^{p-1}\int_0^t E|a(s,x^{(n-1)}(\rho(s)))|^p ds$$

$$\left. +M^pc(p,T)\int_0^t E|b(s,x^{(n-1)}(\rho(s)))|_\lambda^p ds\right\}.$$

Next, by Assumption (D2), we have

$$E \sup_{0\leq s\leq t} |x^{(n)}(s)|^p \leq \frac{3^{p-1}}{[1 - L_5||(-A)^{-\alpha}|| - L_5M_{1-\alpha}\Gamma(\alpha)/a^\alpha]^p}$$

$$\times \left\{E\left[M|\varphi(0)| + L_5(1+||\varphi||_C)((M+1)||(-A)^{-\alpha}|| + M_{1-\alpha}\Gamma(\alpha)/a^\alpha\right]^p\right.$$

$$\left. +M^p(T^{p-1}+c(p,T))\int_0^t (C_3E||x_s^{(n-1)}||_C^p + \psi(s))ds\right\}, \quad n=1,2,3,\ldots.$$

Since $E||\varphi||_C^p < \infty$, it follows from the last inequality that

$$E \sup_{0\leq s\leq t} |x^{(n)}(s)|^p < \infty, \quad \text{for all} \quad n=1,2,3,\ldots \quad \text{and} \quad t \in [0,T],$$

proving the boundedness of $\{x^{(n)}(t)\}_{n\geq 1}$.

Let us next show that $\{x^{(n)}\}$ is Cauchy in $B_T$. For this, consider

$$x^{(1)}(s) - x^{(0)}(s) = S(s)f(0,\varphi) - f(s,x^{(1)}(\rho(s)))$$

$$+ \int_0^s (-A)S(s-\tau)f(\tau,x^{(1)}(\rho(\tau)))d\tau$$

$$+ \int_0^s S(s-\tau)a(\tau,x^{(0)}(\rho(\tau)))d\tau$$

$$+ \int_0^s S(s-\tau)b(\tau,x^{(0)}(\rho(\tau)))dw(\tau).$$

By Assumptions (D1) and (D3), we have

$$|x^{(1)}(s) - x^{(0)}(s)| \le M||(-A)^{-\alpha}||L_5(1 + ||\varphi||_C) + ||(-A)^{-\alpha}||L_4||x_s^{(1)} - x_s^{(0)}||_C$$

$$+ ||(-A)^{-\alpha}||L_5(1 + ||x_s^{(0)}||_C)$$

$$+ C_3 M_{1-\alpha} \frac{\Gamma(\alpha)}{a^\alpha} \left(1 + \sup_{0 \le s \le t} ||x_s^{(0)}||_C\right)$$

$$+ \left| \int_0^s (-A)^{1-\alpha} S(s-\tau)(-A)^\alpha [f(\tau,x^{(1)}(\rho(\tau))) \right.$$

$$\left. - f(\tau,x^{(0)}(\rho(\tau)))]d\tau \right|$$

$$+ \left| \int_0^s S(s-\tau)a(\tau,x^{(0)}(\rho(\tau)))d\tau \right|$$

$$+ \left| \int_0^s S(s-\tau)b(\tau,x^{(0)}(\rho(\tau)))w(\tau) \right| \quad \text{a.s..}$$

Next, using Lemma 7.2.1 (or Lemma 7.2.2 for $p = 2$) and Assumptions (D2) and (D3), we have

$$E||x_t^{(1)} - x_t^{(0)}||_C^p \le 3^{p-1} \left[1 - L_4||(-A)^{-\alpha}|| - L_4 M_{1-\alpha}\Gamma(\alpha)/a^\alpha\right]^{-p}$$

$$\times \left\{ E\left[ M||(-A)^{-\alpha}||L_5(1 + ||\varphi||_C) \right.\right.$$

$$+ ||(-A)^{-\alpha}||L_5(1 + \sup_{0 \le s \le t} ||x_s^{(0)}||_C)$$

$$\left. + L_5 M_{1-\alpha}\frac{\Gamma(\alpha)}{a^\alpha}(1 + \sup_{0 \le s \le t} ||x_s^{(0)}||_C)\right]^p$$

$$+ M^p(T^{p-1} + c(p,T)) \int_0^t (C_3 E||x_s^{(0)}||_C^p + \psi(s))ds \right\}.$$

Next, consider

$$x^{(n)}(s) - x^{(n-1)}(s) = f(s, x^{(n-1)}(\rho(s))) - f(s, x^{(n)}(\rho(s)))$$
$$- \int_0^s AS(s-\tau)[f(\tau, x^{(n)}(\rho(\tau))) - f(\tau, x^{(n-1)}(\rho(\tau)))]d\tau$$
$$+ \int_0^s S(s-\tau)[a(\tau, x^{(n-1)}(\rho(\tau))) - a(\tau, x^{(n-2)}(\rho(\tau)))]d\tau$$
$$+ \int_0^s S(s-\tau)[b(\tau, x^{(n-1)}(\rho(\tau))) - b(\tau, x^{(n-2)}(\rho(\tau)))]dw(\tau).$$

Estimating as before, we have

$$[1 - L_4||(-A)^{-\alpha}|| - L_4 M_{1-\alpha}\Gamma(\alpha)/a^\alpha||]^p E||x_t^{(n)} - x_t^{(n-1)}||_C^p$$
$$\leq 2^{p-1} M^p T^{p-1} C_1 \int_0^t E||x_s^{(n-1)} - x_s^{(n-2)}||_C^p ds$$
$$+ 2^{p-1} M^p c(p,T) C_2 \int_0^t E||x_s^{(n-1)} - x_s^{(n-2)}||_C^p ds.$$

Thus,

$$E||x_t^{(n)} - x_t^{(n-1)}||_C^p$$
$$\leq \frac{2^{p-1} M^p (C_1 T^{p-1} + C_2 c(p,T))}{[1 - L_4||(-A)^{-\alpha}|| - L_4 M_{1-\alpha}\Gamma(\alpha)/a^\alpha||]^p} \int_0^t E||x_s^{(n-1)} - x_s^{(n-2)}||_C^p ds.$$

Using the familiar Cauchy formula,

$$E||x_t^{(n)} - x_t^{(n-1)}||_C^p \leq \frac{[2^{p-1} M^p (C_1 T^{p-1} + C_2 c(p,T))]^{n-1}}{[1 - L_4||(-A)^{-\alpha}|| - L_4 M_{1-\alpha}\Gamma(\alpha)/a^\alpha||]^{(n-1)p}}$$
$$\times \int_0^t \frac{(t-s)^{n-2}}{(n-2)!} E||x_s^{(1)} - x_s^{(0)}||_C^p ds$$
$$\leq \frac{[2^{p-1} M^p (C_1 T^{p-1} + C_2 c(p,T))]^{n-1}}{[1 - L_4||(-A)^{-\alpha}|| - L_4 M_{1-\alpha}\Gamma(\alpha)/a^\alpha||]^{(n-1)p}}$$
$$\times E||x_s^{(1)} - x_s^{(0)}||_C^p \frac{T^{n-1}}{(n-1)!}.$$

This shows that $\{x_n\}$ is Cauchy in $B_T$. Then the standard Borel-Cantelli lemma argument can be used to show that $x_n(t) \to x(t)$ as $n \to \infty$ uniformly in $t$ on $[0,T]$ and $x(t)$ is indeed a mild solution of equation (7.1).

Finally, to show the uniqueness, let $x(t)$ and $y(t)$ be two mild solutions of equation (7.1). Consider

$$x(s) - y(s) = f(s, y(\rho(s))) - f(s, x(\rho(s)))$$
$$- \int_0^s AS(s-\tau)[f(\tau, x(\rho(\tau))) - f(\tau, y(\rho(\tau)))]d\tau$$
$$+ \int_0^s S(s-\tau)[a(\tau, x(\rho(\tau))) - a(\tau, y(\rho(\tau)))]d\tau$$
$$+ \int_0^s S(s-\tau)[b(\tau, x(\rho(\tau))) - b(\tau, y(\rho(\tau)))]dw(\tau).$$

Proceeding as before, we have

$$E||x_t - y_t||_C^p \leq \frac{2^{p-1}M^p(C_1 T^{p-1} + C_2 c(p,T))}{[1 - L_4||(-A)^{-\alpha}|| - L_4 M_{1-\alpha}\Gamma(\alpha)/a^\alpha|||^p} \int_0^t E||x_s - y_s||_C^p ds.$$

Applying Gronwall's lemma,

$$E||x_t - y_t||_C^p = 0, \qquad t \in [0,T],$$

and the uniqueness follows. This completes the proof.                $\square$

## 7.4   Applications

In this section, we consider two examples as applications of the results considered in the previous section.

*Example 7.4.1.*  Consider the neutral stochastic partial functional differential equation with finite delays $r_1, r_2$, and $r_3$ ($r > r_i \geq 0, i = 1,2,3$):

$$d\left[z(t,x) + \frac{\ell_3(t)}{||(-A)^{3/4}||} \int_{-r_3}^0 z(t+u,x)du\right] = \left[\frac{\partial^2}{\partial x^2} z(t,x) + \ell_1(t) \int_{-r_1}^0 z(t+u,x)du\right]dt$$
$$+ \ell_2(t)z(t-r_2,x)d\beta(t), \quad t > 0,$$

$$(7.11)$$

$$\ell_i : R^+ \to R^+, \quad i = 1,2,3; \quad z(t,0) = z(t,\pi) = 0, \quad t > 0,$$
$$z(s,x) = \varphi(s,x), \quad \varphi(\cdot,x) \in C \quad a.s.,$$
$$\varphi(s,\cdot) \in L^2[0,\pi], \quad -r \leq s \leq 0, \quad 0 \leq x \leq \pi;$$

where $\beta(t)$ is a standard one-dimensional Wiener process; $\ell_i(t), i = 1,2,3$ is a continuous function; and $E||\varphi||_C^2 < \infty$.

Take $X = L^2[0, \pi]$, $Y = R^1$. Define $-A : X \to X$ by $-A = \partial^2/\partial x^2$ with domain $D(-A) = \{w \in X : w, \partial w/\partial x$ are absolutely continuous, $\partial^2 w/\partial x^2 \in X$, $w(0) = w(\pi) = 0\}$. Then

$$-Aw = \sum_{n=1}^{\infty} n^2(w, w_n)w_n, \quad w \in D(-A),$$

where $w_n(x) = \sqrt{2/\pi} \sin nx$, $n = 1, 2, 3, \ldots$, is the orthonormal set of eigenvectors of $-A$.

It is well-known that $-A$ is the infinitesimal generator of an analytic semigroup $\{S(t), t \geq 0\}$ in $X$ and is given by

$$S(t)w = \sum_{n=1}^{\infty} e^{-n^2 t}(w, w_n)w_n, \quad w \in X,$$

that satisfies $||S(t)|| \leq \exp(-\pi^2 t), t \geq 0$, and hence is a contraction semigroup.

Define now

$$f(t, z_t) = \ell_3(t) \int_{-r_3}^{0} z(t + u, x)\mathrm{d}u,$$

$$a(t, z_t) = \ell_1(t) \int_{-r_1}^{0} z(t + u, x)\mathrm{d}u,$$

$$b(t, z_t) = \ell_2(t)z(t - r_2, x).$$

Next,

$$||f(t, z_t)||_{3/4} = \frac{\ell_3(t)}{||(-A)^{3/4}||} \left|(-A)^{3/4} \int_{-r_3}^{0} z(t + u, x)\mathrm{d}u\right|$$

$$\leq \ell_3(T)r_3||z_t||_C, \quad \text{a.s..}$$

This shows that $f : R^+ \times C \to D((-A)^{3/4})$ with $C_4(T) = \ell_3(T)r_3$. Similarly, $a : R^+ \times C \to X$ and $b : R^+ \times C \to L(R, X)$. Thus, (7.11) can be expressed as (7.6) with $-A, f, a$, and $b$ as defined above. Hence, there exists a unique mild solution by Theorem 7.3.4.

The existence results Theorems 7.3.2 and 7.3.3 are not applicable to (7.11).

*Example 7.4.2.* Consider the stochastic partial differential equation with a variable delay:

$$\mathrm{d}z(t, x) = \left[\frac{\partial^2}{\partial x^2}z(t, x) + F(t, z(\rho(t), x))\right]\mathrm{d}t$$

$$+ G(t, z(\rho(t), x))\mathrm{d}\beta(t), \quad t > 0, \tag{7.12}$$

$$z(t,0) = z(t,\pi) = 0, \quad t > 0,$$

$$z(s,x) = \varphi(s,x), \quad \varphi(\cdot,x) \in C,$$

$$\varphi(s,\cdot) \in L^2[0,\pi], \quad -r \le s \le 0, \quad 0 \le x \le \pi;$$

where $\beta(t)$ is a standard one-dimensional Wiener process and $E\|\varphi\|_C^2 < \infty$.
Let us assume that $F : R^+ \times R \to R$ and $G : R^+ \times R \to R$ satisfy the conditions:
For $t \ge 0$, there exists a constant $k > 0$ such that

$$|F(t,z_1) - F(t,z_2)|^2 + |G(t,z_1) - G(t,z_2)|^2 \le k|z_1 - z_2|^2, \quad z_1, z_2 \in R,$$

and there exist positive constants $L, P, \delta$ and a continuous function $\psi : R^+ \to R^+$
satisfying $|\psi(t)|^2 \le P\exp(-\delta t)$, for all $t \ge 0$ such that

$$|F(t,z)|^2 + |G(t,z)|^2 \le L|z|^2 + \psi(t), \quad z \in R.$$

Take $X = L^2[0,\pi], Y = R$. Define $-A : X \to X$ by $-A = \partial^2/\partial x^2$ with domain $D(-A)$
given as before in Example 7.4.1. So, $-A$ is the infinitesimal generator of an analytic
semigroup $\{S(t), t \ge 0\}$ in $X$ that satisfies $\|S(t)\| \le \exp(-\pi^2 t), t \ge 0$.
    Define now

$$a(t,z(\rho(t))) = F(t,z(\rho(t),x)),$$

and

$$b(t,z(\rho(t))) = G(t,z(\rho(t),x)).$$

Next,

$$|a(t,z(\rho(t)))|^2 + |b(t,z(\rho(t)))|_\lambda^2 = \int_0^\pi [|F(t,z(\rho(t),x))|^2 + |G(t,z(\rho(t),x))|^2]dx$$

$$\le \pi[L\|z\|_C^2 + \psi(t)].$$

This shows that $a : R^+ \times C \to X$ and $b : R^+ \times C \to L(R,X)$. Thus, (7.12) can be
expressed as (7.1) with $-A, a$, and $b$ as defined above and with $f = 0$. It can be
verified as above that

$$|a(t,z_1(\rho(t))) - a(t,z_2(\rho(t)))|^2 + |b(t,z_1(\rho(t))) - b(t,z_2(\rho(t)))|_\lambda^2 \le \pi k\|z_1 - z_2\|_C^2.$$

Hence, (7.12) has a unique mild solution by Theorem 7.3.5.

**Acknowledgements** The author wishes to thank SIP and COFAA both from IPN, Mexico for
financial support.

# References

1. Ahmed, N.U.: Semigroup Theory with Applications to Systems and Control, Pitman Research Notes in Math., Vol. 246, (1991).
2. Da Prato, G., Zabczyk, J.: Stochastic Equations in Infinite Dimensions, Cambridge Univ. Press, Cambridge, (1992).
3. Da Prato, G., Zabczyk, J.: A note on stochastic convolution, Stochastic Anal. Appl. 10, 143–153 (1992).
4. Govindan, T.E.: Almost sure exponential stability for stochastic neutral partial functional differential equations, Stochastics 77, 139–154 (2005).
5. Govindan, T.E.: A new iteration procedure for stochastic neutral partial functional differential equations, Internat. J. Pure Applied Math. 56, 285–298 (2009).
6. Govindan, T.E.: Mild solutions of neutral stochastic partial functional differential equations, Internat. J. Stochastic Anal. vol. 2011, Article ID 186206, 13 pages, (2011). doi:10.1155/2011/186206.
7. Hernandez, E., Henriquez, H.R.: Existence results for partial neutral functional differential equations with unbounded delay, J. Math. Anal. Appl. 221, 452–475 (1998).
8. Luo, L.: Exponential stability for stochastic neutral partial functional differential equations, J. Math. Anal. Appl. 355, 414–425 (2009).
9. Pazy, A.: Semigroups of Linear operators and Applications to Partial Differential Equations, Springer, New York, (1983).
10. Taniguchi, T.: Asymptotic stability theorems of semilinear stochastic evolution equations in Hilbert spaces, Stochastics Stochastics Reports 53, 41–52 (1995).

# Chapter 8
# A Constrained Optimization Problem with Applications to Constrained MDPs

Xianping Guo, Qingda Wei, and Junyu Zhang

## 8.1 Introduction

Constrained optimization problems form an important aspect in control theory, for instance, constrained Markov decision processes (MDPs) [2, 3, 10–13, 15–21, 24, 25, 27, 29–33, 35, 36], and constrained diffusion processes [4–9]. In this chapter, we are concerned with a constrained optimization problem, in which the objective function is defined on the product space of a linear space and a convex set. The constrained optimization problem is to maximize the values of the function with any fixed variable in the linear space, over a constrained subset of the convex set which is given by the function with *another* fixed variable from the linear space and with a given constraint. The basic idea for the constrained optimization problem comes from the studies on the discounted and average optimality for discrete- and continuous-time MDPs with a constraint. We aim to develop a *unified* approach to dealing with such constrained MDPs. More precisely, for discrete- and continuous-time MDPs with a constraint, the linear space can be taken as a set of some real-valued functions such as reward and cost functions in such MDPs, and the convex set can be chosen as the set of all randomized Markov policies, the set of all randomized stationary policies, or the set of all the occupation measures according to a specified case of MDPs with different criteria. The objective function can be taken as one of the expected discounted (average) criteria, while the first variable in the objective function can be taken as the reward/cost functions in MDPs and the second one as a policy in a class of policies. Thus, MDPs with a constraint can be reduced to our constrained optimization problem.

Research supported by NSFC, GDUPS, RFDP, and FRFCU.

X. Guo (✉) • Q. Wei • J. Zhang
Sun Yat-Sen University, Guangzhou 510275, China
e-mail: mcsgxp@mail.sysu.edu.cn; wwqingda@sina.com; mcszhjy@mail.sysu.edu.cn

D. Hernández-Hernández and A. Minjárez-Sosa (eds.), *Optimization, Control, and Applications of Stochastic Systems*, Systems & Control: Foundations & Applications, DOI 10.1007/978-0-8176-8337-5_8, © Springer Science+Business Media, LLC 2012

A fundamental question on the constrained optimization problem is whether there exists a constrained-optimal solution. The Lagrange multiplier technique is a classical approach to proving the existence of a constrained-optimal solution for such an optimization problem. There are several authors using the Lagrange multiplier technique to study MDPs with a constraint; see, for instance, discrete-time constrained MDPs with the discounted and average criteria [3, 32, 33] and continuous-time constrained MDPs with the discounted and average criteria [18, 20, 35]. All the aforementioned works [3, 18, 20, 32, 33, 35] require the nonnegativity assumption on the costs. We also apply this approach to analyze the constrained optimization problem. Following the arguments in [3, 18, 20, 32, 33, 35], we give conditions under which we prove the existence of a constrained-optimal solution to the constrained optimization problem, and also give a characterization of a constrained-optimal solution for a particular case.

Then, we apply our main results to discrete- and continuous-time constrained MDPs with discounted and average criteria. More precisely, in Sect. 8.4.1, we use the results to show the existence of a constrained-optimal policy for the discounted discrete-time MDPs with a constraint in which the state space is a Polish space and the rewards/costs may be unbounded from above and from below. To the best of our knowledge, there are no any existing works dealing with constrained discounted discrete-time MDPs in Borel spaces and with unbounded rewards/costs. In Sect. 8.4.2, we investigate an application of the main results to constrained discrete-time MDPs with state-dependent discount factors and extend the results in [32] to the case in which discount factors can depend on states and rewards/costs can be unbounded from above and from below. In Sects. 8.4.3 and 8.4.4, we consider the average and discounted continuous-time MDPs with a constraint, respectively. Removing the nonnegativity assumption on the cost function as in [18, 20, 35], we prove that the results in [18, 20, 35] still hold using the results in this chapter.

The rest of this chapter is organized as follows. In Sect. 8.2, we introduce the constrained optimization problem under consideration and give some preliminary facts needed to prove the existence of a constrained-optimal solution to the optimization problem. In Sect. 8.3, we state and prove our main results on the existence of a constrained-optimal solution. In Sect. 8.4, we provide some applications of our main results to constrained MDPs with different optimality criteria.

## 8.2 A Constrained Optimization Problem

In this section, we state the constrained optimization problem under consideration and give some preliminary results needed to prove the existence of a constrained-optimal solution. To do so, we introduce some notation below:

1. Let $C$ be a linear space and $D$ a convex set.

2. Suppose that $G$ is a real-valued function on the product space $C \times D$ and satisfies the following property:

$$G(k_1 c_1 + k_2 c_2, d) = k_1 G(c_1, d) + k_2 G(c_2, d) \qquad (8.1)$$

for any $c_1, c_2 \in C$, $d \in D$ and any constants $k_1, k_2 \in \mathbb{R} := (-\infty, +\infty)$.

For any *fixed* $c \in C$, let

$$U := \{ d \in D : G(c, d) \leq \rho \},$$

which depends on the given $c$ and a so-called constraint constant $\rho$.

Then, for another given $r \in C$, we consider a constrained optimization problem below:

$$\text{Maximize } G(r, \cdot) \text{ over } U. \qquad (8.2)$$

**Definition 8.2.1.** $d^* \in U$ is said to be a constrained-optimal solution to the problem (8.2) if $d^*$ maximizes $G(r, d)$ over $d \in U$, that is,

$$G(r, d^*) = \sup_{d \in U} G(r, d).$$

*Remark 8.2.1.* When $D$ is a compact and convex metric space, and $G(r, d)$ and $G(c, d)$ are continuous in $d \in D$, it follows from the Weierstrass theorem [1, p.40] that there exists a constrained-optimal solution. In general, however, $D$ may be unmetrizable in some cases, such as the set of all randomized Markov policies in continuous-time MDPs [20, p.10]; see continuous-time constrained MDPs with the discounted criteria in Sect. 8.4.4. In order to solve (8.2), we *assume that there exists a subset $D' \subseteq D$, which is assumed to be a compact metric space throughout this chapter.*

To analyze problem (8.2), we define the following unconstrained optimization problem by introducing a Lagrange multiplier $\lambda \geq 0$,

$$b^\lambda := r - \lambda c, \quad G^*(b^\lambda) := \sup_{d \in D} G(b^\lambda, d), \qquad (8.3)$$

and then give the conditions below.

**Assumption 8.2.1**

(i) *For each fixed $\lambda \geq 0$, $D_\lambda^* := \{ d^\lambda \in D' \mid G(b^\lambda, d^\lambda) = G^*(b^\lambda) \} \neq \emptyset$.*

(ii) *There exists a constant $M > 0$, such that $\max \{ |G(r, d)|, |G(c, d)| \} \leq M$ for all $d \in D$.*

(iii) *$G(r, d)$ and $G(c, d)$ are continuous in $d \in D'$.*

Assumption 8.2.1(i) implies that there exists at least an element $d^\lambda \in D'$ such that $G(b^\lambda, \cdot)$ attains its maximum. In addition, the boundedness and continuity hypotheses in Assumptions 8.2.1(ii) and (iii) are commonly used in optimization control theory.

**Assumption 8.2.2** *For the given* $c \in C$, *there exists an element* $d' \in D$ *(depending on* $c$*) such that* $G(c, d') < \rho$, *which means that* $\{d \in D \mid G(c, d) < \rho\} \neq \emptyset$.

*Remark 8.2.2.* Assumption 8.2.2 is a Slater-like hypothesis, typical for the constrained optimization problems; see, for instance, [3, 17, 18, 20, 32, 33, 35].

In order to prove the existence of a constrained-optimal solution, we need the following preliminary lemmas.

**Lemma 8.2.1.** *Suppose that Assumption 8.2.1(i) holds. Then,* $G(c, d^\lambda)$ *is nonincreasing in* $\lambda \in [0, \infty)$, *where* $d^\lambda \in D_\lambda^*$ *is arbitrary but fixed for each* $\lambda \geq 0$.

*Proof.* For each $d \in D$, by (8.1) and (8.3), we have

$$G(b^\lambda, d) = G(r, d) - \lambda G(c, d) \text{ for all } \lambda \geq 0.$$

Moreover, since $G(b^\lambda, d^\lambda) = G^*(b^\lambda)$ for all $\lambda \geq 0$ and $d^\lambda \in D_\lambda^*$, we have, for any $h > 0$,

$$
\begin{aligned}
-hG(c, d^\lambda) &= G(b^{\lambda+h}, d^\lambda) - G(b^\lambda, d^\lambda) \\
&\leq G(b^{\lambda+h}, d^{\lambda+h}) - G(b^\lambda, d^\lambda) \\
&\leq G(b^{\lambda+h}, d^{\lambda+h}) - G(b^\lambda, d^{\lambda+h}) \\
&= -hG(c, d^{\lambda+h}),
\end{aligned}
$$

which implies that

$$G(c, d^\lambda) \geq G(c, d^{\lambda+h}).$$

Hence, $G(c, d^\lambda)$ is nonincreasing in $\lambda \in [0, \infty)$.                                           $\square$

*Remark 8.2.3.* Under Assumption 8.2.1(i), it follows from Lemma 8.2.1 that the following nonnegative constant

$$\widetilde{\lambda} := \inf\{\lambda \geq 0 : G(c, d^\lambda) \leq \rho, d^\lambda \in D_\lambda^*\} \tag{8.4}$$

is well defined.

**Lemma 8.2.2.** *Suppose that Assumptions 8.2.1(i), (ii) and 8.2.2 hold. Then, the constant* $\widetilde{\lambda}$ *in (8.4) is finite; that is,* $\widetilde{\lambda}$ *is in* $[0, \infty)$.

*Proof.* Let $\kappa := \rho - G(c, d') > 0$, with $d'$ as in Assumption 8.2.2. Since $\lim_{\lambda \to \infty} \frac{2M}{\lambda} = 0$ for the constant $M$ as in Assumption 8.2.1(ii), there exists $\delta > 0$ such that

$$\frac{2M}{\lambda} - \kappa < 0 \ \text{ for all } \lambda \geq \delta. \tag{8.5}$$

Thus, for any $d^\lambda \in D^*_\lambda$ with $\lambda \geq \delta$, we have

$$G(r,d^\lambda) - \lambda G(c,d^\lambda) = G(b^\lambda,d^\lambda) \geq G(b^\lambda,d') = G(r,d') - \lambda G(c,d').$$

That is,

$$\frac{G(r,d^\lambda) - G(r,d')}{\lambda} + G(c,d') - \rho \geq G(c,d^\lambda) - \rho,$$

which, together with Assumption 8.2.1(ii) and (8.5), yields

$$G(c,d^\lambda) - \rho \leq \frac{|G(r,d^\lambda)| + |G(r,d')|}{\lambda} - \kappa \leq \frac{2M}{\lambda} - \kappa < 0 \ \text{ for all } \lambda \geq \delta. \tag{8.6}$$

Hence, it follows from (8.6) that $\widetilde{\lambda} \leq \delta < \infty$. $\qquad\square$

**Lemma 8.2.3.** *Suppose that Assumptions 8.2.1(i) and (iii) hold. If* $\lim_{k\to\infty} \lambda_k = \lambda$, *and* $d^{\lambda_k} \in D^*_{\lambda_k}$ *(for each $k \geq 1$) is such that* $\lim_{k\to\infty} d^{\lambda_k} = \overline{d} \in D'$, *then* $\overline{d} \in D^*_\lambda$.

*Proof.* As $d^{\lambda_k} \in D^*_{\lambda_k}$ for all $k \geq 1$, by (8.1) and (8.3), we have

$$G(r,d^{\lambda_k}) - \lambda_k G(c,d^{\lambda_k}) = G(b^{\lambda_k},d^{\lambda_k}) \geq G(b^{\lambda_k},d) = G(r,d) - \lambda_k G(c,d) \tag{8.7}$$

for all $d \in D$. Letting $k \to \infty$ in (8.7) and using Assumption 8.2.1(iii), we get

$$G(b^\lambda,\overline{d}) = G(r,\overline{d}) - \lambda G(c,\overline{d}) \geq G(r,d) - \lambda G(c,d) = G(b^\lambda,d) \ \text{ for all } d \in D.$$

Thus, $\overline{d} \in D^*_\lambda$. $\qquad\square$

**Lemma 8.2.4.** *If there exist $\lambda_0 \geq 0$ and $d^* \in U$ such that*

$$G(c,d^*) = \rho \ \text{ and } \ G(b^{\lambda_0},d^*) = G^*(b^{\lambda_0}),$$

*then $d^*$ is a constrained-optimal solution to the problem (8.2).*

*Proof.* For any $d \in U$, since $G(b^{\lambda_0},d^*) = G^*(b^{\lambda_0}) \geq G(b^{\lambda_0},d)$, we have

$$G(r,d^*) - \lambda_0 G(c,d^*) \geq G(r,d) - \lambda_0 G(c,d). \tag{8.8}$$

As $G(c,d^*) = \rho$ and $G(c,d) \leq \rho$ (because $d \in U$), from (8.8) we get

$$G(r,d^*) \geq G(r,d^*) + \lambda_0(G(c,d) - \rho) \geq G(r,d) \ \text{ for all } d \in U,$$

which implies the desired result. $\qquad\square$

## 8.3  Main Results

In this section, we focus on the existence of a constrained-optimal solution. To do so, in addition to Assumptions 8.2.1 and 8.2.2, we also impose the following condition.

**Assumption 8.3.1**

(i) For each $\theta \in [0,1]$, $d_1, d_2 \in D^*_{\widetilde{\lambda}}$, $d_\theta := \theta d_1 + (1-\theta)d_2$ satisfies $G(b^{\widetilde{\lambda}}, d_\theta) = G^*(b^{\widetilde{\lambda}})$.

(ii) $G(c, d_\theta)$ is continuous in $\theta \in [0,1]$.

*Remark 8.3.4.* For each fixed $c_1 \in C$, if $G(c_1, \cdot)$ satisfies the following property

$$G(c_1, d_\theta) = \theta G(c_1, d_1) + (1-\theta)G(c_1, d_2)$$

for all $d_1, d_2 \in D$ and $\theta \in [0,1]$, then Assumption 8.3.1 is obviously true.

Now we give our first main result on the problem (8.2).

**Theorem 8.3.1.** *Under Assumptions 8.2.1, 8.2.2, and 8.3.1, the following statements hold:*

(a) *If $\widetilde{\lambda} = 0$, then there exists a constrained-optimal solution $\widetilde{d} \in D'$.*

(b) *If $\widetilde{\lambda} > 0$, then a constrained-optimal solution $d^* \in D$ exists, and moreover, there exist a number $\theta^* \in [0,1]$ and $d^1, d^2 \in D^*_{\widetilde{\lambda}}$ such that*

$$G(c, d^1) \geq \rho, \;\; G(c, d^2) \leq \rho, \;\; and \;\; d^* = \theta^* d^1 + (1-\theta^*)d^2.$$

*Proof.* (a) The case $\widetilde{\lambda} = 0$: By the definition of $\widetilde{\lambda}$, there exists a sequence $d^{\lambda_k} \in D^*_{\lambda_k} \subset D'$ such that $\lambda_k \downarrow 0$ as $k \to \infty$. Because $D'$ is compact, without loss of generality, we may assume that $d^{\lambda_k} \to \widetilde{d} \in D'$. Thus, by Lemma 8.2.1, we have $G(c, d^{\lambda_k}) \leq \rho$ for all $k \geq 1$, and then it follows from Assumption 8.2.1(iii) that $\widetilde{d} \in U$. Moreover, for each $d \in U$, we have $G^*(b^{\lambda_k}) = G(b^{\lambda_k}, d^{\lambda_k}) \geq G(b^{\lambda_k}, d)$, which, together with Assumption 8.2.1(ii), implies

$$G(r, d^{\lambda_k}) - G(r, d) \geq \lambda_k(G(c, d^{\lambda_k}) - G(c, d)) \geq -2\lambda_k M. \qquad (8.9)$$

Letting $k \to \infty$ in (8.9), by Assumption 8.2.1(iii), we have

$$G(r, \widetilde{d}) \geq G(r, d) \;\; \text{for all } d \in U,$$

which means that $\widetilde{d}$ is a constrained-optimal solution.

(b) The case $\widetilde{\lambda} \in (0, \infty)$: Since $\widetilde{\lambda}$ is in $(0, \infty)$, there exist two sequences of positive numbers $\{\lambda_k\}$ and $\{\delta_k\}$ such that $d^{\lambda_k} \in D^*_{\lambda_k}$, $d^{\delta_k} \in D^*_{\delta_k}$, $\lambda_k \uparrow \widetilde{\lambda}$, and $\delta_k \downarrow \widetilde{\lambda}$

as $k \to \infty$. By the compactness of $D'$, we may suppose that $d^{\lambda_k} \to d^1 \in D'$ and $d^{\delta_k} \to d^2 \in D'$. By Lemma 8.2.3, we have $d^1, d^2 \in D_{\tilde{\lambda}}^*$. By Assumption 8.2.1(iii) and Lemma 8.2.1, we have

$$G(c,d^1) \geq \rho \text{ and } G(c,d^2) \leq \rho. \tag{8.10}$$

Define the following map:

$$\theta \mapsto G(c, \theta d^1 + (1-\theta)d^2) \text{ for each } \theta \in [0,1].$$

Thus, it follows from Assumption 8.3.1(ii) and (8.10) that there exists $\theta^* \in [0,1]$ such that

$$G(c, \theta^* d^1 + (1-\theta^*)d^2) = \rho. \tag{8.11}$$

Let $d^* := \theta^* d^1 + (1-\theta^*)d^2$. Then, by Assumption 8.3.1(i), we have $G(b^{\tilde{\lambda}}, d^*) = G^*(b^{\tilde{\lambda}})$, which together with (8.11) and Lemma 8.2.4 yields that $d^* \in D$ is a constrained-optimal solution.                    □

To further characterize a constrained-optimal solution, we next consider a particular case of the problem (8.2).

**A special case:** Let $X := \{1,2,\ldots\}$, $Y$ be a metric space, and $\mathscr{P}(Y)$ the set of all probability measures on $Y$. For each $i \in X$, $Y(i) \subset Y$ is assumed to be a compact metric space. Let $D := \{\psi|\ \psi : X \to \mathscr{P}(Y) \text{ such that } \psi(\cdot|i) \in \mathscr{P}(Y(i))\ \forall i \in X\}$, and $D' := \{d|\ d : X \to Y \text{ such that } d(i) \in Y(i)\ \forall i \in X\}$.

*Remark 8.3.5.* (a) The set $D$ is convex. That is, if $\psi_k (k = 1,2)$ are in $D$, and $\psi^p(\cdot|i) := p\psi_1(\cdot|i) + (1-p)\psi_2(\cdot|i)$ for any $p \in [0,1]$ and $i \in X$, then $\psi^p \in D$.
(b) A function $d \in D'$ may be identified with the element $\psi \in D$, for which $\psi(i)$ is the Dirac measure at the point $d(i)$ for all $i \in X$. Hence, we have $D' \subset D$.
(c) Note that $D'$ can be written as the product space $D' = \prod_{i \in X} Y(i)$. Hence, by the compactness of $Y(i)$ and the Tychonoff theorem, $D'$ is a compact metric space.

In order to obtain the characterization of a constrained-optimal solution for this particular case, we also need the following condition.

**Assumption 8.3.2** *For each $\lambda \geq 0$, if $d^1, d^2 \in D_\lambda^*$, then $d \in D_\lambda^*$ for each $d \in \{d \in D' : d(i) \in \{d^1(i), d^2(i)\}\ \forall i \in X\}$.*

Then, we have the second main result on the problem (8.2) as follows.

**Theorem 8.3.2.** *(For the special case.) Suppose that Assumptions 8.2.1, 8.2.2, 8.3.1, and 8.3.2 hold for the special case. Then there exists a constrained-optimal solution $d^*$, which is of one of the following two forms (i) and (ii): (i) $d^* \in D'$ and (ii) there exist $g^1, g^2 \in D_{\tilde{\lambda}}^*$, a point $i^* \in X$, and a number $\theta_0 \in [0,1]$ such that $g^1(i) = g^2(i)$ for all $i \neq i^*$, and, in addition,*

$$d^*(y|i) = \begin{cases} \theta_0 & for\ y = g^1(i)\ when\ i = i^*, \\ 1 - \theta_0 & for\ y = g^2(i)\ when\ i = i^*, \\ 1 & for\ y = g^1(i)\ when\ i \neq i^*. \end{cases}$$

*Proof.* Let $\widetilde{\lambda}$ be as in (8.4). If $\widetilde{\lambda} = 0$, by Theorem 8.3.1 we have $d^* \in D'$. Thus, we only need to consider the other case $\widetilde{\lambda} > 0$. By Theorem 8.3.1(b), there exist $d^1, d^2 \in D^*_{\widetilde{\lambda}}$ such that $G(c, d^1) \geq \rho$ and $G(c, d^2) \leq \rho$. If $G(c, d^1)$ (or $G(c, d^2)) = \rho$, it follows from Lemma 8.2.4 that $d^1$ (or $d^2$) is a constrained-optimal solution. Hence, to complete the proof, we shall consider the following case:

$$G(c, d^1) > \rho \quad and \quad G(c, d^2) < \rho. \tag{8.12}$$

Using $d^1$ and $d^2$, we construct a sequence $\{d_n\}$ as follows. For all $n \geq 1$ and $i \in X$, let

$$d_n(i) = \begin{cases} d^1(i) & i < n, \\ d^2(i) & i \geq n. \end{cases}$$

Obviously, $d_1 = d^2$ and $\lim_{n \to \infty} d_n = d^1$. Since $d^1, d^2 \in D^*_{\widetilde{\lambda}}$, by Assumption 8.3.2, we see that $d_n \in D^*_{\widetilde{\lambda}}$ for all $n \geq 1$. As $d_1 = d^2$, by (8.12) we have $G(c, d_1) < \rho$. If there exists $n^*$ such that $G(c, d_{n^*}) = \rho$, then $d_{n^*}$ is a constrained-optimal solution (by Lemma 8.2.4). Thus, in the remainder of the proof, we may assume that $G(c, d_n) \neq \rho$ for all $n \geq 1$. If $G(c, d_n) < \rho$ for all $n \geq 1$, then by Assumption 8.2.1(iii), we have

$$\lim_{n \to \infty} G(c, d_n) = G(c, d^1) \leq \rho,$$

which is a contradiction to (8.12). Hence, there exists some $n > 1$ such that $G(c, d_n) > \rho$, which, together with $G(c, d_1) < \rho$, gives the existence of some $\widetilde{n}$ such that

$$G(c, d_{\widetilde{n}}) < \rho \quad and \quad G(c, d_{\widetilde{n}+1}) > \rho. \tag{8.13}$$

Obviously, $d_{\widetilde{n}}$ and $d_{\widetilde{n}+1}$ differ in at most the point $\widetilde{n}$.

Let $g^1 := d_{\widetilde{n}}$, $g^2 := d_{\widetilde{n}+1}$ and $i^* := \widetilde{n}$. For any $\theta \in [0,1]$, using $g^1$ and $g^2$, we construct $d_\theta \in D$ as follows. For each $i \in X$,

$$d_\theta(y|i) = \begin{cases} \theta & for\ y = g^1(i)\ when\ i = i^*, \\ 1 - \theta & for\ y = g^2(i)\ when\ i = i^*, \\ 1 & for\ y = g^1(i)\ when\ i \neq i^*. \end{cases}$$

Then, we have

$$d_\theta(\cdot|i) = \theta \delta_{g^1(i)}(\cdot) + (1 - \theta) \delta_{g^2(i)}(\cdot) \tag{8.14}$$

for all $i \in X$ and $\theta \in [0,1]$, where $\delta_y(\cdot)$ denotes the Dirac measure at any point $y$. Hence, by (8.13), (8.14), and Assumption 8.3.1, there exists $\theta_0 \in (0,1)$ such that

$$G(c,d_{\theta_0}) = \rho \ \text{ and } \ G(b^{\widetilde{\lambda}},d_{\theta_0}) = G^*(b^{\widetilde{\lambda}}),$$

which, together with Lemma 8.2.4, yield that $d_{\theta_0}$ is a constrained-optimal solution. Obviously, $d_{\theta_0}$ randomizes between $g^1$ and $g^2$, which differ in at most the point $i^*$, and so the theorem follows. $\qquad\square$

## 8.4  Applications to MDPs with a Constraint

In this section, we show applications of the constrained optimization problem to MDPs with a constraint. In Sect. 8.4.1, we use Theorem 8.3.1 to show the existence of a constrained-optimal policy for the constrained discounted discrete-time MDPs in a Polish space and with unbounded rewards/costs. In Sect. 8.4.2, we investigate an application of Theorem 8.3.2 to discrete-time constrained MDPs with state-dependent discount factors. In Sects. 8.4.3 and 8.4.4, we will improve the corresponding results in [18, 20, 35] using Theorem 8.3.2 above.

### 8.4.1  Discrete-Time Constrained MDPs with Discounted Criteria

The constrained discounted discrete-time MDPs with a constant discount factor have been studied; see, for instance, [2, 11, 32] for the case of a countable state space and [15, 16, 24, 27, 29] for the case of a Borel state space. Except [2] dealing with the case in which the rewards may be unbounded from above and from below, all the aforementioned works investigate the case in which rewards are assumed to be bounded from above. To the best of our knowledge, in this subsection we first deal with the case in which the state space is a Polish space and the rewards may be unbounded from above and from below.

The model of discrete-time constrained MDPs under consideration is as follows [22, 23]:

$$\left\{X, A, (A(x), x \in X), Q(\cdot|x,a), r(x,a), c(x,a), \rho\right\}, \tag{8.15}$$

where $X$ and $A$ are state and action spaces, which are assumed to be Polish spaces with Borel $\sigma$-algebras $\mathscr{B}(X)$ and $\mathscr{B}(A)$, respectively. We denote by $A(x) \in \mathscr{B}(A)$ the set of admissible actions at state $x \in X$. Let $K := \{(x,a)|x \in X, a \in A(x)\}$, which is assumed to be a closed subset of $X \times A$. Furthermore, the transition law $Q(\cdot|x,a)$ with $(x,a) \in K$ is a stochastic kernel on $X$ given $K$. Finally, the function $r(x,a)$ on

$K$ denotes rewards, while the function $c(x,a)$ on $K$ and the number $\rho$ denote costs and a constraint, respectively. We assume that $r(x,a)$ and $c(x,a)$ are real-valued Borel-measurable on $K$.

We denote by $\Pi$, $\Phi$, and $F$ the classes of all randomized history-dependent policies, randomized stationary policies, and stationary policies, respectively; see [22, 23] for details.

Let $\Omega := (X \times A)^{\infty}$ and $\mathscr{F}$ the corresponding product $\sigma$-algebra. Then, for an arbitrary policy $\pi \in \Pi$ and an arbitrary initial distribution $\nu$ on $X$, the well-known Tulcea theorem [22, p.178] gives the existence of a unique probability measure $P_{\nu}^{\pi}$ on $(\Omega, \mathscr{F})$ and a stochastic process $\{(x_k, a_k), k \geq 0\}$. The expectation operator with respect to $P_{\nu}^{\pi}$ is denoted by $E_{\nu}^{\pi}$, and we write $E_{\nu}^{\pi}$ as $E_x^{\pi}$ when $\nu(\{x\}) = 1$.

Fix a discount factor $\alpha \in (0,1)$ and an initial distribution $\nu$ on $X$. We define the expected discounted reward $V(r, \pi)$ and the expected discounted cost $V(c, \pi)$ as follows:

$$V(r,\pi) := E_{\nu}^{\pi}\left[\sum_{k=0}^{\infty} \alpha^k r(x_k, a_k)\right] \text{ and } V(c,\pi) := E_{\nu}^{\pi}\left[\sum_{k=0}^{\infty} \alpha^k c(x_k, a_k)\right] \text{ for all } \pi \in \Pi.$$

Then, the constrained optimization problem for the model (8.15) is as follows:

$$\text{Maximize } V(r,\cdot) \text{ over } U_1 := \{\pi \in \Pi \mid V(c,\pi) \leq \rho\}. \tag{8.16}$$

To solve (8.16), we consider the following conditions:

(B1)    There exist a continuous function $\omega_1 \geq 1$ on $X$ and positive constants $L_1$, $m$, and $\beta_1 < 1$ such that, for each $(x,a) \in K$,

$$|r(x,a)| \leq L_1 \omega_1(x), \ |c(x,a)| \leq L_1 \omega_1(x), \text{ and } \int_X \omega_1^2(y)Q(dy|x,a) \leq \beta_1 \omega_1^2(x) + m.$$

(B2)    The function $\omega_1$ is a moment function on $K$, that is, there exists a nonde-creasing sequence of compact sets $K_n \uparrow K$ such that $\liminf_{n \to \infty} \{\omega_1(x) : (x,a) \notin K_n\} = \infty$.

(B3)    $\nu(\omega_1^2) := \int_X \omega_1^2(x)\nu(dx) < \infty$.

(B4)    $Q(\cdot|x,a)$ is weakly continuous on $K$, that is, the function $\int_X u(y)Q(dy|x,a)$ is continuous in $(x,a) \in K$ for each bounded continuous function $u$ on $X$.

(B5)    The functions $r(x,a)$ and $c(x,a)$ are continuous on $K$.

(B6)    There exists $\pi' \in \Pi$ such that $V(c, \pi') < \rho$.

*Remark 8.4.6.* (a) Condition (B1) is known as the statement of the Lyapunov-like inequality and the growth condition on the rewards/costs. Conditions (B4) and (B5) are the usual continuity conditions. Condition (B6) is the Slater-like condition.

(b) Conditions (B1) and (B3) are used to guarantee the finiteness of the expected discounted rewards/costs. The role of condition (B2) is to prove the compactness of the set of all the discount occupation measures in the $\omega_1$-weak topology (see Lemma 8.4.5).

To state our main results of Sect. 8.4.1, we need to introduce some notation. Let $\omega_1$ be as in condition (B1). We denote by $B_{\omega_1}(X)$ the Banach space of real-valued measurable functions $u$ on $X$ with the finite norm $\|u\|_{\omega_1} := \sup_{x \in X} \frac{|u(x)|}{\omega_1(x)}$, that is, $B_{\omega_1}(X) := \{u| \|u\|_{\omega_1} < \infty\}$. Moreover, we say that a function $v$ on $K$ belongs to $B_{\omega_1}(K)$ if $x \mapsto \sup_{a \in A(x)} |v(x,a)|$ is in $B_{\omega_1}(X)$. We denote by $C_{\omega_1}(K)$ the set of all continuous functions on $K$ which also belong to $B_{\omega_1}(K)$, and $\mathcal{M}_{\omega_1}(K)$ stands for the set of all measures $\mu$ on $\mathcal{B}(K)$ such that $\int_K \omega_1(x)\mu(dx,da) < \infty$. Moreover, $\mathcal{M}_{\omega_1}(K)$ is endowed with $\omega_1$-weak topology. Recall that the $\omega_1$-weak topology on $\mathcal{M}_{\omega_1}(K)$ is the coarsest topology for which all mappings $\mu \mapsto \int_K v(x,a)\mu(dx,da)$ are continuous for each $v \in C_{\omega_1}(K)$. Since $X$ and $A$ are both Polish spaces, by Corollary A.44 in [14, p.423] we see that $\mathcal{M}_{\omega_1}(K)$ is metrizable with respect to the $\omega_1$-weak topology.

By Lemma 24 in [29, p.141], it suffices to consider the discount occupation measures induced by randomized stationary policies in $\Phi$. For each $\varphi \in \Phi$, we define the discount occupation measure by

$$\eta^\varphi(B \times E) := \sum_{k=0}^\infty \alpha^k P_v^\varphi(x_k \in B, a_k \in E) \quad \text{for all } B \in \mathcal{B}(X) \text{ and } E \in \mathcal{B}(A).$$

The set of all the discount occupation measures is denoted by $\mathcal{N}$, i.e., $\mathcal{N} := \{\eta^\varphi : \varphi \in \Phi\}$. From the conditions (B1) and (B3), we have

$$\int_K \omega_1(x)\eta^\varphi(dx,da) = \sum_{k=0}^\infty \alpha^k E_v^\varphi[\omega_1(x_k)] \leq \frac{v(\omega_1^2)}{1-\alpha} + \frac{m}{(1-\beta_1)(1-\alpha)} < \infty \tag{8.17}$$

for all $\varphi \in \Phi$, which yields $\mathcal{N} \subset \mathcal{M}_{\omega_1}(K)$.

Then, the constrained optimization problem (8.16) is equivalent to the following form:

$$\text{Maximize } \int_K r(x,a)\eta(dx,da) \text{ over } \left\{\eta \in \mathcal{N}\left| \int_K c(x,a)\eta(dx,da) \leq \rho\right.\right\} =: U_o. \tag{8.18}$$

Now we provide a characterization of the discount occupation measures below.

**Lemma 8.4.5.** *Under conditions (B1)–(B4), the following statements hold:*

*(a) If $\eta \in \mathcal{M}_{\omega_1}(K)$, then $\eta$ is in $\mathcal{N}$ if and only if*

$$\int_K u(x)\eta(dx,da) = \int_X u(x)v(dx) + \alpha \int_K \int_X u(y)Q(dy|x,a)\eta(dx,da)$$

*for each bounded continuous function $u$ on $X$.*
*(b) $\mathcal{N}$ is convex and compact in the $\omega_1$-weak topology.*

*Proof.* (a) See Lemma 25 in [29, p.141] for the proof of part (a).

(b) The convexity property follows directly from part (a). To prove that $\mathcal{N}$ is compact, we will first show that $\mathcal{N}$ is closed in the $\omega_1$-weak topology. Let $\{\eta^{\varphi_n}\} \subset \mathcal{N}$ be a sequence converging to some measure $\eta$ on $X \times A$ in the $\omega_1$-weak topology. Thus, there exists a positive integer $N_1$ such that for each $n \geq N_1$, we have

$$\left| \int_K \omega_1(x)\eta^{\varphi_n}(dx,da) - \int_{X \times A} \omega_1(x)\eta(dx,da) \right| \leq 1,$$

which together with (8.17) yields

$$\int_{X \times A} \omega_1(x)\eta(dx,da) \leq \frac{v(\omega_1^2)}{1-\alpha} + \frac{m}{(1-\beta_1)(1-\alpha)} + 1 < \infty,$$

and so $\eta \in \mathcal{M}_{\omega_1}(X \times A)$. Moreover, since $K$ is assumed to be closed and $\eta^{\varphi_n}$ weakly converges to $\eta$, by Theorem A.38 in [14, p.420] we have

$$0 = \liminf_{n \to \infty} \eta^{\varphi_n}(K^c) \geq \eta(K^c) \geq 0,$$

which implies that $\eta$ concentrates on $K$, where $K^c$ denotes the complement of $K$. In addition, by part (a) we have

$$\int_K u(x)\eta^{\varphi_n}(dx,da) = \int_X u(x)v(dx) + \alpha \int_K \int_X u(y)Q(dy|x,a)\eta^{\varphi_n}(dx,da)$$

for each bounded continuous function $u$ on $X$, which together with condition (B4) yields

$$\int_K u(x)\eta(dx,da) = \int_X u(x)v(dx) + \alpha \int_K \int_X u(y)Q(dy|x,a)\eta(dx,da).$$

Hence, by part (a) we see that $\eta \in \mathcal{N}$, and so $\mathcal{N}$ is closed.

To prove the compactness of $\mathcal{N}$, it suffices to show that $\mathcal{N}$ is relatively compact in the $\omega_1$-weak topology. By (8.17) we have

$$\sup_{\eta \in \mathcal{N}} \int_K \omega_1(x)\eta(dx,da) = \sup_{\varphi \in \Phi} \int_K \omega_1(x)\eta^{\varphi}(dx,da)$$

$$\leq \frac{v(\omega_1^2)}{1-\alpha} + \frac{m}{(1-\beta_1)(1-\alpha)} < \infty. \qquad (8.19)$$

On the other hand, from condition (B2), we see that $\varpi_n := \inf\{\omega_1(x) : (x,a) \notin K_n\} \uparrow \infty$. Then, by conditions (B1) and (B3), we have

$$\varpi_n \int_{K_n^c} \omega_1(x)\eta^{\varphi}(dx,da) \leq \int_K \omega_1^2(x)\eta^{\varphi}(dx,da)$$

$$\leq \frac{v(\omega_1^2)}{1-\alpha} + \frac{m}{(1-\beta_1)(1-\alpha)} \qquad (8.20)$$

for all $\varphi \in \Phi$. Thus, by (8.20) we see that for any $\varepsilon > 0$, there exists an integer $N_2 > 0$ such that

$$\sup_{\varphi \in \Phi} \int_{K_{N_2}^c} \omega_1(x) \eta^\varphi(\mathrm{d}x, \mathrm{d}a) \leq \varepsilon. \tag{8.21}$$

Hence, by (8.19), (8.21), and Corollary A.46 in [14, p.424], we conclude that $\mathcal{N}$ is relatively compact in the $\omega_1$-weak topology. Therefore, $\mathcal{N}$ is compact in the $\omega_1$-weak topology. □

Under conditions (B1)–(B5), from Lemma 8.4.5 and (8.18), we define a real-valued function $G$ on $C \times D := C_{\omega_1}(K) \times \mathcal{N}$ as follows:

$$G(c, \eta) := \int_K c(x, a) \eta(\mathrm{d}x, \mathrm{d}a) \quad \text{for } (c, \eta) \in C \times D = C_{\omega_1}(K) \times \mathcal{N}. \tag{8.22}$$

Obviously, the function $G$ defined in (8.22) satisfies (8.1). Moreover, let $D' := \mathcal{N}$.

Now we provide our main result of Sect. 8.4.1 on the existence of constrained-optimal policies for (8.16).

**Proposition 8.4.1.** *Under conditions (B1)–(B6), there exists a constrained-optimal policy $\varphi^* \in \Phi$ for the constrained MDPs in (8.16), that is, $V(r, \varphi^*) = \sup_{\pi \in U_1} V(r, \pi)$.*

*Proof.* We first verify Assumption 8.2.1. From conditions (B1) and (B5), we see that for each $\lambda \geq 0$, the mapping $\eta^\varphi \mapsto \int_K (r(x, a) - \lambda c(x, a)) \eta^\varphi(\mathrm{d}x, \mathrm{d}a)$ is continuous on $\mathcal{N}$. Thus, Assumption 8.2.1(i) follows from the compactness of $\mathcal{N}$. Moreover, by condition (B1) and (8.17), we have

$$\max\{|G(r, \eta)|, |G(c, \eta)|\} \leq \frac{L_1 \nu(\omega_1^2)}{1 - \alpha} + \frac{mL_1}{(1 - \beta_1)(1 - \alpha)} =: M$$

for all $\eta \in \mathcal{N}$, and so Assumption 8.2.1(ii) follows. By conditions (B1) and (B5), we see that Assumption 8.2.1(iii) is obviously true.

Secondly, Assumption 8.2.2 follows from condition (B6) and Lemma 24 in [29, p.141].

Finally, since $G(c, \theta \eta_1 + (1 - \theta) \eta_2) = \theta G(c, \eta_1) + (1 - \theta) G(c, \eta_2)$ for all $\theta \in [0, 1]$ and $\eta_1, \eta_2 \in \mathcal{N}$, Assumption 8.3.1 is obviously true.

Hence, Theorem 8.3.1 gives the existence of $\eta^* \in \mathcal{N}$ such that

$$\int_K r(x, a) \eta^*(\mathrm{d}x, \mathrm{d}a) = \sup_{\eta \in U_o} \int_K r(x, a) \eta(\mathrm{d}x, \mathrm{d}a),$$

which together with Theorem 6.3.7 in [22] implies Proposition 8.4.1. □

### 8.4.2   Constrained MDPs with State-Dependent Discount Factors

In this subsection, we use discrete-time constrained MDPs with state-dependent discount factors to present another application of the constrained optimization problem. Discrete-time unconstrained MDPs with nonconstant discount factors are studied in [26, 34]. Moreover, [36] deals with discrete-time constrained MDPs with state-dependent discount factors in which the costs are assumed to be bounded from below by a convex analytic approach, and here we use Theorem 8.3.1 above to deal with the case in which the costs are allowed to be unbounded from above and from below.

The model of discrete-time constrained MDPs with state-dependent discount factors is as follows:

$$\{X, A, (A(i), i \in X), Q(\cdot|i,a), (\alpha(i), i \in X), r(i,a), c(i,a), \rho\},$$

where the state space $X$ is the set of all positive integers, $\alpha(i) \in (0,1)$ are given discount factors depending on state $i \in X$, and the other components are the same as in (8.15), with $i_k$ here in lieu of $x_k$ in Sect. 8.4.1.

Fix any initial distribution $v$ on $X$. The discounted criteria, $W(r,\pi)$ and $W(c,\pi)$, are defined by

$$W(r,\pi) := E_v^\pi \left[ r(i_0,a_0) + \sum_{n=1}^{\infty} \prod_{k=0}^{n-1} \alpha(i_k) r(i_n,a_n) \right],$$

$$W(c,\pi) := E_v^\pi \left[ c(i_0,a_0) + \sum_{n=1}^{\infty} \prod_{k=0}^{n-1} \alpha(i_k) c(i_n,a_n) \right] \quad \text{for all } \pi \in \Pi.$$

Then, the constrained optimization problem is as follows:

$$\sup_{\pi \in \Pi} W(r,\pi) \text{ subject to } W(c,\pi) \le \rho. \tag{8.23}$$

To ensure the existence of a constrained-optimal policy $\pi^*$ for (8.23) (i.e., $W(r,\pi^*) \ge W(r,\pi)$ for all $\pi$ such that $W(c,\pi) \le \rho$), we consider the following conditions from [34]:

(C1)   There exists a constant $\overline{\alpha} \in (0,1)$ such that $0 < \alpha(i) \le \overline{\alpha}$ for all $i \in X$.

(C2)   There exist constants $L_2 > 0$ and $\beta_2$, with $1 \le \beta_2 < \frac{1}{\overline{\alpha}}$ and a function $\omega_2 \ge 1$ on $X$ such that, for each $(i,a) \in K$,

$$|r(i,a)| \le L_2 \omega_2(i), \quad |c(i,a)| \le L_2 \omega_2(i), \text{ and } \sum_{j \in X} \omega_2(j) Q(j|i,a) \le \beta_2 \omega_2(i).$$

(C3)   $v(\omega_2) := \sum_{i \in X} \omega_2(i) v(i) < \infty.$

(C4)  For each $i \in X$, $A(i)$ is compact.
(C5)  For each $i, j \in X$, the functions $r(i,a)$, $c(i,a)$, $Q(j|i,a)$, and $\sum_{k \in X} \omega_2(k)Q(k|i,a)$

are continuous in $a \in A(i)$.
(C6)  There exists $\widetilde{\pi} \in \Pi$ such that $W(c, \widetilde{\pi}) < \rho$.

*Remark 8.4.7.* Conditions (C1)–(C3) are known as the finiteness conditions. Conditions (C4) and (C5) are the usual continuity-compactness conditions. Condition (C6) is the Slater-like condition.

Under conditions (C1)–(C5), we define a real-valued function $G$ on $C \times D :=$ $C_{\omega_2}(K) \times \Pi$ as follows:

$$G(c, \pi) := W(c, \pi) \quad \text{for } (c, \pi) \in C \times D = C_{\omega_2}(K) \times \Pi. \tag{8.24}$$

Obviously, since the set $\Pi$ is convex, the function $G$ defined in (8.24) satisfies (8.1).

Let $D' := F$. Then, we state our main result of Sect. 8.4.2 on the existence of constrained-optimal policies for (8.23).

**Theorem 8.4.3.** *Suppose that conditions (C1)–(C6) hold. Then there exists a constrained-optimal policy for (8.23), which is either a stationary policy or a randomized stationary policy that randomizes between two stationary policies differing in at most one state; that is, there exist two stationary policies $f^1$, $f^2$, a state $i^* \in X$, and a number $p^* \in [0, 1]$ such that $f^1(i) = f^2(i)$ for all $i \neq i^*$, and, in addition, the randomized stationary policy $\pi^{p^*}(\cdot|i)$ is constrained-optimal, where*

$$\pi^{p^*}(a|i) = \begin{cases} p^* & \text{for } a = f^1(i) \text{ when } i = i^*, \\ 1 - p^* & \text{for } a = f^2(i) \text{ when } i = i^*, \\ 1 & \text{for } a = f^1(i) \text{ when } i \neq i^*. \end{cases}$$

*Remark 8.4.8.* Theorem 8.4.3 extends the corresponding one in [32] for a constant discount factor to the case of state-dependent discount factors. Moreover, we remove the nonnegativity assumption on the costs as in [32].

We will prove Theorem 8.4.3 using Theorem 8.3.2. To do so, we introduce the notation below.

For each $\varphi \in \Phi$, $i \in X$, and $\pi \in \Pi$, define

$$W_r(i, \pi) := E_i^\pi \left[ r(i_0, a_0) + \sum_{n=1}^{\infty} \prod_{k=0}^{n-1} \alpha(i_k) r(i_n, a_n) \right],$$

$$W_c(i, \pi) := E_i^\pi \left[ c(i_0, a_0) + \sum_{n=1}^{\infty} \prod_{k=0}^{n-1} \alpha(i_k) c(i_n, a_n) \right],$$

$$b^\lambda(i, a) := r(i, a) - \lambda c(i, a) \quad \text{for all } (i, a) \in K,$$

$$W^\lambda(i,\pi) := E_i^\pi \left[ b^\lambda(i_0, a_0) + \sum_{n=1}^{\infty} \prod_{k=0}^{n-1} \alpha(i_k) b^\lambda(i_n, a_n) \right],$$

$$W_\lambda^*(i) := \sup_{\pi \in \Pi} W^\lambda(i, \pi),$$

and

$$u(i, \varphi) := \int_{A(i)} u(i, a)\varphi(da|i), \quad \text{for } u(i,a) = b^\lambda(i,a), r(i,a), c(i,a),$$

$$Q(j|i, \varphi) := \int_{A(i)} Q(j|i, a)\varphi(da|i) \quad \text{for } j \in X.$$

Then, we give three lemmas below, which are used to prove Theorem 8.4.3.

**Lemma 8.4.6.** *Under conditions (C1)–(C5), the following assertions hold:*

(a) $|W(r, \pi)| \le \frac{L_2 v(\omega_2)}{1 - \bar{\alpha}\beta_2}$ *and* $|W(c, \pi)| \le \frac{L_2 v(\omega_2)}{1 - \bar{\alpha}\beta_2}$ *for all* $\pi \in \Pi$.

(b) *For each fixed* $\varphi \in \Phi$, $W_u(\cdot, \varphi)$ $(u = r, c)$ *is the unique solution in* $B_{\omega_2}(X)$ *to the following equation:*

$$v(i) = u(i, \varphi) + \alpha(i) \sum_{j \in X} v(j) Q(j|i, \varphi) \quad \text{for all } i \in X.$$

(c) $W(r, f)$ *and* $W(c, f)$ *are continuous in* $f \in F$.

*Proof.* For the proofs of (a) and (b), see Theorem 3.1 in [34].

(c) We only prove the continuity of $W(r, f)$ in $f \in F$ because the other case is similar. Let $f_n \to f$ as $n \to \infty$, and fix any $i \in X$. Choose any subsequence $\{W_r(i, f_{n_m})\}$ of $\{W_r(i, f_n)\}$ converging to some point $v(i)$ as $m \to \infty$. Then, since $X$ is denumerable, the Tychonoff theorem, together with the part (a) and $f_n \to f$, gives the existence of subsequence $\{W_r(j, f_{n_k}), j \in X\}$ of $\{W_r(j, f_{n_m}), j \in X\}$ such that

$$\lim_{k \to \infty} W_r(j, f_{n_k}) =: v'(j), \ v'(i) = v(i), \text{ and } \lim_{k \to \infty} f_{n_k}(j) = f(j) \text{ for all } j \in X. \quad (8.25)$$

Furthermore, by Theorem 3.1 in [34], we have $|v'(j)| \le \frac{L_2 \omega_2(j)}{1 - \bar{\alpha}\beta_2}$ for all $j \in X$, which implies that $v' \in B_{\omega_2}(X)$. On the other hand, for the given $i \in X$ and all $k \ge 1$, by part (b) we have

$$W_r(i, f_{n_k}) = r(i, f_{n_k}) + \alpha(i) \sum_{j \in X} W_r(j, f_{n_k}) Q(j|i, f_{n_k}). \quad (8.26)$$

Then, it follows from (8.25), (8.26), condition (C5), and Lemma 8.3.7 in [23, p.48] that

$$v'(i) = r(i, f) + \alpha(i) \sum_{j \in X} v'(j) Q(j|i, f).$$

Hence, part (b) yields

$$v'(i) = W_r(i, f).  \tag{8.27}$$

Thus, as the above subsequence $\{W_r(i, f_{n_m})\}$ and $i \in X$ are arbitrarily chosen and (by (8.27)) all such subsequences have the same limit $W_r(i, f)$, we have

$$\lim_{n \to \infty} W_r(i, f_n) = W_r(i, f) \quad \text{for all } i \in X.$$

Therefore, from condition (C3) and Theorem A.6 in [22, p.171], we get

$$\lim_{n \to \infty} W(r, f_n) = \sum_{i \in X} \left[ \lim_{n \to \infty} W_r(i, f_n) \right] v(i) = \sum_{i \in X} W_r(i, f) v(i) = W(r, f),$$

which gives the desired conclusion, $W(r, f_n) \to W(r, f)$ as $n \to \infty$.  $\square$

**Lemma 8.4.7.** *Suppose that conditions (C1), (C2), (C4), and (C5) hold. Then we have*

*(a)* $W_\lambda^*$ *is the unique solution of the following equation in* $B_{\omega_2}(X)$:

$$v(i) = \sup_{a \in A(i)} \left\{ b^\lambda(i, a) + \alpha(i) \sum_{j \in X} v(j) Q(j|i, a) \right\} \quad \text{for all } i \in X.  \tag{8.28}$$

*(b)* *There exists a function* $f^* \in F$ *such that* $f^*(i) \in A(i)$ *attains the maximum in (8.28) for each* $i \in X$, *that is,*

$$W_\lambda^*(i) = b^\lambda(i, f^*) + \alpha(i) \sum_{j \in X} W_\lambda^*(j) Q(j|i, f^*) \quad \text{for all } i \in X,  \tag{8.29}$$

*and* $f^* \in F$ *is optimal. Conversely, if* $f^* \in F$ *is optimal, it satisfies (8.29).*

*Proof.* See Theorem 3.2 in [34].  $\square$

*Remark 8.4.9.* Under conditions (C1), (C2), (C4), and (C5), for each $\lambda \geq 0$, let $D_\lambda^*(e) := \{f \in F : W^\lambda(i, f) = W_\lambda^*(i) \text{ for all } i \in X\}$. Then, it follows from Lemma 8.4.7 that $D_\lambda^*(e) \neq \emptyset$, and that $f \in D_\lambda^*(e)$ if and only if $f \in F$ satisfies (8.29).

**Lemma 8.4.8.** *Suppose that conditions (C1)–(C5) hold. Then, for each* $f_1, f_2 \in D_\lambda^*(e)$ *(with any fixed* $\lambda \geq 0$), *and* $0 \leq p \leq 1$, *define a policy* $\pi^p$ *by* $\pi^p(\cdot|i) := p\hat{\delta}_{f_1(i)}(\cdot) + (1-p)\delta_{f_2(i)}(\cdot)$ *for all* $i \in X$. *Then,*

*(a)* $W^\lambda(i, \pi^p) = W_\lambda^*(i)$ *for all* $i \in X$.
*(b)* $W(c, \pi^p)$ *is continuous in* $p \in [0, 1]$.

*Proof.* (a) Since

$$b^\lambda(i, \pi^p) = pb^\lambda(i, f_1) + (1 - p)b^\lambda(i, f_2), \tag{8.30}$$

$$Q(j|i, \pi^p) = pQ(j|i, f_1) + (1 - p)Q(j|i, f_2), \tag{8.31}$$

by Lemma 8.4.7 and the definition of $D_\lambda^*(e)$, we have

$$W_\lambda^*(i) = b^\lambda(i, f_l) + \alpha(i) \sum_{j \in X} W_\lambda^*(j)Q(j|i, f_l) \quad \text{for all } i \in X \text{ and } l = 1, 2,$$

which together with (8.30) and (8.31) gives

$$W_\lambda^*(i) = b^\lambda(i, \pi^p) + \alpha(i) \sum_{j \in X} W_\lambda^*(j)Q(j|i, \pi^p) \quad \text{for all } i \in X. \tag{8.32}$$

Therefore, by Lemma 8.4.6(b) and (8.32), we have $W^\lambda(i, \pi^p) = W_\lambda^*(i)$ for all $i \in X$, and so part (a) follows.

(b) For any $p \in [0, 1]$ and any sequence $\{p_m\}$ in $[0, 1]$ such that $\lim_{m \to \infty} p_m = p$, by Lemma 8.4.6, we have

$$W_c(i, \pi^{p_m}) = c(i, \pi^{p_m}) + \alpha(i) \sum_{j \in X} W_c(j, \pi^{p_m})Q(j|i, \pi^{p_m}) \quad \text{for all } i \in X \text{ and } m \geq 1.$$

$$\tag{8.33}$$

Hence, as in the proof of Lemma 8.4.6, from (8.33), the definition of $\pi^{p_m}$ and Theorem A.6 in [22, p.171], we have

$$\lim_{m \to \infty} W(c, \pi^{p_m}) = W(c, \pi^p),$$

and so $W(c, \pi^p)$ is continuous in $p \in [0, 1]$.  □

*Proof of Theorem* 8.4.3. By Lemmas 8.4.6–8.4.8 and (8.24), we see that Assumptions 8.2.1 and 8.3.1 hold. Moreover, Assumptions 8.2.2 and 8.3.2 follow from condition (C6) and Lemma 8.4.7, respectively. Hence, by Theorem 8.3.2, we complete the proof.  □

### 8.4.3 Continuous-Time Constrained MDPs with Average Criteria

In this subsection, removing the nonnegativity assumption on the cost function as in [20, 35], we will prove that the corresponding results in [20, 35] still hold using Theorem 8.3.2 above.

The model of continuous-time constrained MDPs is of the form [18, 20, 30, 35]:

$$\{X, A, (A(i), i \in X), q(\cdot|i, a), r(i, a), c(i, a), \rho\}, \tag{8.34}$$

where $X$ is assumed to be a denumerable set. Without loss of generality, we assume that $X$ is the set of all positive integers. Furthermore, the transition rates $q(j|i,a)$, which satisfy $q(j|i,a) \geq 0$ for all $(i,a) \in K$ and $j \neq i$. We also assume that the transition rates $q(j|i,a)$ are conservative, i.e., $\sum_{j \in X} q(j|i,a) = 0$ for all $(i,a) \in K$, and stable, which means that $q^*(i) := \sup_{a \in A(i)} -q(i|i,a) < \infty$ for all $i \in X$. In addition, $q(j|i,a)$ is measurable in $a \in A(i)$ for each fixed $i, j \in X$. The other components are the same as in (8.15), with a state $i$ here in lieu of a state $x$ in Sect. 8.4.1.

We denote by $\Pi_m$, $\Phi$ and $F$ the classes of all randomized Markov policies, randomized stationary policies, and stationary policies, respectively; see [18,20,30,35] for details.

To guarantee the regularity of the $Q$-process, we impose the following drift condition from [30]:

(D1) There exists a nondecreasing function $\omega_3 \geq 1$ on $X$ such that $\lim_{i \to \infty} \omega_3(i) = \infty$.

(D2) There exist constants $\gamma_1 \geq \kappa_1 > 0$ and a state $i_0 \in X$ such that

$$\sum_{j \in X} q(j|i,a)\omega_3^2(j) \leq -\kappa_1 \omega_3^2(i) + \gamma_1 I_{i_0}(i) \quad \text{for all } (i,a) \in K,$$

where $I_B(\cdot)$ denotes the indicator function of any set $B$.

Let $\overline{T} := [0,\infty)$, and let $(\Omega, \mathscr{B}(\Omega))$ be the canonical product measurable space with $(X \times A)^{\overline{T}}$ being the set of all maps from $\overline{T}$ to $X \times A$. Fix an initial distribution $\nu$ on $X$. Then, under conditions (D1) and (D2), by Theorem 2.3 in [20, p.14], for each policy $\pi \in \Pi_m$, there exist a unique probability measure $P_\nu^\pi$ on $(\Omega, \mathscr{B}(\Omega))$ and a stochastic process $\{(x(t),a(t)), t \geq 0\}$. The expectation operator with respect to $P_\nu^\pi$ is denoted by $E_\nu^\pi$.

For each $\pi \in \Pi_m$, we define the expected average criteria, $\overline{V}(r,\pi)$ and $\overline{V}(c,\pi)$, as follows:

$$\overline{V}(r,\pi) := \liminf_{T \to \infty} \frac{E_\nu^\pi \left[ \int_0^T r(x(t),a(t))dt \right]}{T},$$

$$\overline{V}(c,\pi) := \limsup_{T \to \infty} \frac{E_\nu^\pi \left[ \int_0^T c(x(t),a(t))dt \right]}{T}.$$

Then, the constrained optimization problem for the average criteria is as follows:

$$\sup_{\pi \in \Pi_m} \overline{V}(r,\pi) \quad \text{subject to} \quad \overline{V}(c,\pi) \leq \rho. \tag{8.35}$$

To guarantee the existence of a constrained-optimal policy $\pi^*$ for (8.35) (i.e., $\overline{V}(r,\pi^*) \geq \overline{V}(r,\pi)$ for all $\pi$ such that $\overline{V}(c,\pi) \leq \rho$), we need the following conditions from [20,30,35]:

(D3) There exists a constant $L_3 > 0$ such that

$$|r(i,a)| \leq L_3 \omega_3(i) \text{ and } |c(i,a)| \leq L_3 \omega_3(i) \text{ for all } (i,a) \in K.$$

(D4) For each $i \in X$, $A(i)$ is compact.
(D5) For each $i \in X$, $q^*(i) \leq \omega_3(i)$.
(D6) $v(\omega_3^2) := \sum_{i \in X} \omega_3^2(i) v(i) < \infty.$
(D7) For each $i, j \in X$, the functions $r(i,a)$, $c(i,a)$, $q(j|i,a)$, and $\sum_{k \in X} \omega_3(k) q(k|i,a)$
are continuous in $a \in A(i)$. (D8) For each $\varphi \in \Phi$, the corresponding Markov process
with transition rates $q(j|i,\varphi)$ is irreducible, where $q(j|i,\varphi) := \int_{A(i)} q(j|i,a)\varphi(da|i)$
for all $i, j \in X$.
(D9) There exists $\varphi' \in \Phi$ such that $\overline{V}(c, \varphi') < \rho.$

From conditions (D1), (D2), and (D8), by Theorem 4.2 in [28], for each $\varphi \in \Phi$,
the corresponding Markov process with transition rates $q(j|i,\varphi)$ has a unique
invariant probability measure $\mu_\varphi$ on $X$. Moreover, under conditions (D1)–(D4),
(D7), and (D8), by Theorem 7.2 in [30] we have

$$\overline{V}(r, \varphi) = \lim_{T \to \infty} \frac{1}{T} E_v^\varphi \left[ \int_0^T r(x(t), a(t)) dt \right] = \sum_{i \in X} r(i, \varphi) \mu_\varphi(i) \qquad (8.36)$$

and

$$\overline{V}(c, \varphi) = \lim_{T \to \infty} \frac{1}{T} E_v^\varphi \left[ \int_0^T c(x(t), a(t)) dt \right] = \sum_{i \in X} c(i, \varphi) \mu_\varphi(i), \qquad (8.37)$$

where

$$r(i, \varphi) := \int_{A(i)} r(i,a)\varphi(da|i) \text{ and } c(i, \varphi) := \int_{A(i)} c(i,a)\varphi(da|i) \text{ for all } i \in X.$$

For each $\varphi \in \Phi$, we define the average occupation measure $\widehat{\mu}_\varphi$ by

$$\widehat{\mu}_\varphi(\{i\} \times B) := \mu_\varphi(i)\varphi(B|i) \text{ for all } i \in X \text{ and } B \in \mathscr{B}(A(i)).$$

The set of all the average occupation measures is denoted by $\mathscr{N}_1$, i.e., $\mathscr{N}_1 := \{\widehat{\mu}_\varphi : \varphi \in \Phi\}$.
Then, we have the following result.

**Lemma 8.4.9.** *Suppose that conditions (D1)–(D8) hold. Then, for any $\pi \in \Pi_m$ with*
$\overline{V}(c, \pi) \leq \rho$, *there exists $\widehat{\mu}_{\varphi_0} \in \mathscr{N}_1$ such that $\overline{V}(r, \varphi_0) \geq \overline{V}(r, \pi)$ and $\overline{V}(c, \varphi_0) \leq \rho$.*

*Proof.* For each fixed $\pi \in \Pi_m$ with $\overline{V}(c, \pi) \leq \rho$ and $n \geq 1$, we define a measure
$\mu_n$ by

$$\mu_n(\{i\} \times B) := \frac{1}{n} \int_0^n E_v^\pi \left[ I_{\{i\} \times B}(x(t), a(t)) \right] dt \text{ for all } i \in X \text{ and } B \in \mathscr{B}(A(i)).$$

$$(8.38)$$

Then, by conditions (D2) and (D6), Lemma 6.3 in [20, p.90], and (8.38), we have

$$
\sum_{i \in X} \int_{A(i)} \omega_3^2(i) \mu_n(i, \mathrm{d}a) = \frac{1}{n} \int_0^n E_v^\pi \left[ \omega_3^2(x(t)) \right] \mathrm{d}t
$$

$$
\leq \frac{1}{n} \int_0^n \sum_{i \in X} \left[ e^{-\kappa_1 t} \omega_3^2(i) + \frac{\gamma_1}{\kappa_1}(1 - e^{-\kappa_1 t}) \right] v(i) \mathrm{d}t
$$

$$
\leq v(\omega_3^2) + \frac{\gamma_1}{\kappa_1} < \infty. \tag{8.39}
$$

On the other hand, from conditions (D1) and (D4), we see that the sets $\{(i,a) \in K : \omega_3^2(i) \leq z\omega_3(i)\}$ are compact in $K$ for each $z \geq 1$. Hence, by (8.39) and Corollary A.46 in [14, p.424], we conclude that the sequence $\{\mu_n\}$ is relatively compact in the $\omega_3$-weak topology. Thus, there exist a subsequence $\{\mu_{n_l}\}$ of $\{\mu_n\}$ and a probability measure $\mu \in \mathcal{M}_{\omega_3}(X \times A)$ such that $\mu_{n_l}$ converges to $\mu$ in the $\omega_3$-weak topology. By condition (D4), we see that $K$ is closed. Then, since $\mu_{n_l}$ weakly converges to $\mu$, by Theorem A.38 in [14, p.420], we have

$$
0 = \liminf_{l \to \infty} \mu_{n_l}(K^c) \geq \mu(K^c) \geq 0,
$$

which implies $\mu(K) = 1$. Moreover, for each bounded function $v$ on $X$, a direct calculation together with condition (D5), the Fubini theorem, the Kolmogorov forward equation, and Theorem 2.3 in [20, p.14] gives

$$
\sum_{i \in X} \int_A \left[ \sum_{j \in X} q(j|i,a)v(j) \right] \mu_{n_l}(i, \mathrm{d}a)
$$

$$
= \frac{1}{n_l} \int_0^{n_l} \sum_{i \in X} \int_A \left[ \sum_{j \in X} q(j|i,a)v(j) \right] P_v^\pi(x(t) = i, a(t) \in \mathrm{d}a) \mathrm{d}t
$$

$$
= \frac{1}{n_l} \int_0^{n_l} \sum_{k \in X} \sum_{i \in X} \int_A \left[ \sum_{j \in X} q(j|i,a)v(j) \right] P_k^\pi(x(t) = i) \pi_t(\mathrm{d}a|i) v(k) \mathrm{d}t
$$

$$
= \frac{1}{n_l} \int_0^{n_l} \sum_{k \in X} \sum_{j \in X} v(j) \left[ \sum_{i \in X} q(j|i,\pi_t) p_\pi(0,k,t,i) \right] v(k) \mathrm{d}t
$$

$$
= \frac{1}{n_l} \int_0^{n_l} \sum_{k \in X} \sum_{j \in X} v(j) \frac{\partial p_\pi(0,k,t,j)}{\partial t} v(k) \mathrm{d}t
$$

$$
= \frac{1}{n_l} \sum_{k \in X} \sum_{j \in X} v(j) \left[ p_\pi(0,k,n_l,j) - p_\pi(0,k,0,j) \right] v(k)
$$

$$
= \frac{1}{n_l} E_v^\pi \left[ v(x(n_l)) \right] - \frac{1}{n_l} \sum_{k \in X} v(k) v(k), \tag{8.40}
$$

where $p_\pi(0,i,t,j)$ denotes the minimal transition function with transition rates $q(j|i,\pi_t) := \int_{A(i)} q(j|i,a)\pi_t(\mathrm{d}a|i)$ for all $i,j \in X$ and $t \geq 0$. Letting $l \to \infty$ in (8.40), by conditions (D5), (D7), and (8.39), we have

$$\sum_{i \in X} \int_A \left[ \sum_{j \in X} q(j|i,a)v(j) \right] \mu(i,\mathrm{d}a) = 0$$

for each bounded function $v$ on $X$, which together with Lemma 4.6 in [30] yields $\mu \in \mathcal{N}_1$. Hence, there exists $\varphi_0 \in \Phi$ such that $\mu = \widehat{\mu}_{\varphi_0}$. Furthermore, by condition (D3), (8.38), and (8.39), we have

$$\overline{V}(c,\pi) \geq \limsup_{n \to \infty} \sum_{i \in X} \int_A c(i,a)\mu_n(i,\mathrm{d}a) \text{ and } \overline{V}(r,\pi) \leq \liminf_{n \to \infty} \sum_{i \in X} \int_A r(i,a)\mu_n(i,\mathrm{d}a),$$

which together with conditions (D3) and (D7) yield

$$\rho \geq \overline{V}(c,\pi) \geq \sum_{i \in X} \int_A c(i,a)\widehat{\mu}_{\varphi_0}(i,\mathrm{d}a) = \overline{V}(c,\varphi_0),$$

and

$$\overline{V}(r,\pi) \leq \sum_{i \in X} \int_A r(i,a)\widehat{\mu}_{\varphi_0}(i,\mathrm{d}a) = \overline{V}(r,\varphi_0).$$

This completes the proof of the lemma.                                                                    □

By Lemma 8.4.9 we see that the constrained optimization problem (8.35) is equivalent to the following form:

$$\sup_{\varphi \in \Phi} \overline{V}(r,\varphi) \text{ subject to } \overline{V}(c,\varphi) \leq \rho. \tag{8.41}$$

Under conditions (D1)–(D8), from (8.41) we define a real-valued function $G$ on $C \times D := C_{\omega_3}(K) \times \Phi$ as follows:

$$G(c,\varphi) := \overline{V}(c,\varphi) \text{ for } (c,\varphi) \in C \times D = C_{\omega_3}(K) \times \Phi. \tag{8.42}$$

Then, by (8.36) and (8.37), we see that the function $G$ defined in (8.42) satisfies (8.1). Moreover, let $D' := F$.

Now we state our main result of Sect. 8.4.3 on the existence of constrained-optimal policies for (8.35).

**Proposition 8.4.2.** *Suppose that conditions (D1)–(D9) hold. Then there exists a constrained-optimal policy for (8.35), which may be a stationary policy or a randomized stationary policy that randomizes between two stationary policies differing in at most one state.*

*Proof.* It follows from Lemma 7.2, Theorem 7.8, and Lemma 12.5 in [20] that Assumptions 8.2.1 and 8.3.2 hold. Obviously, condition (D9) implies Assumption 8.2.2. Finally, from Lemma 12.6 and the proof of Theorem 12.4 in [20], we see that Assumption 8.3.1 holds. Hence, by Theorem 8.3.2, we complete the proof. □

*Remark 8.4.10.* Proposition 8.4.2 shows that the nonnegativity assumption on the costs as in [20, 35] is not required.

### 8.4.4   Continuous-Time Constrained MDPs with Discounted Criteria

In this subsection, we consider the following discounted criteria $J(r,\pi)$ and $J(c,\pi)$ in (8.43) below for the model (8.34), in lieu of the average criteria above. Removing the nonnegativity assumption on the cost function as in [18, 20], we will prove that the corresponding results in [18, 20] still hold using Theorem 8.3.2 above.

With the same components as in the model (8.34), we consider the following drift condition from [18, 20]:

(E1) There exists a function $\omega_4 \geq 1$ on $X$ and constants $\gamma_2 \geq 0$, $\kappa_2 \neq 0$, and $\overline{L} > 0$ such that

$$q^*(i) \leq \overline{L}\omega_4(i) \text{ and } \sum_{j \in X} \omega_4(j)q(j|i,a) \leq \kappa_2\omega_4(i) + \gamma_2 \text{ for all } (i,a) \in K.$$

Fix a discount factor $\alpha > 0$ and an initial distribution $\nu$ on $X$. For each $\pi \in \Pi_m$, we define the discounted criteria, $J(r,\pi)$ and $J(c,\pi)$, as follows:

$$J(r,\pi) := E_\nu^\pi\left[\int_0^\infty e^{-\alpha t} r(x(t),a(t))dt\right], \ J(c,\pi) := E_\nu^\pi\left[\int_0^\infty e^{-\alpha t} c(x(t),a(t))dt\right].$$

$$(8.43)$$

Then, the constrained optimization problem for the discounted criteria is as follows:

$$\sup_{\pi \in \Pi_m} J(r,\pi) \text{ subject to } J(c,\pi) \leq \rho. \tag{8.44}$$

To ensure the existence of a constrained-optimal policy $\pi^*$ for (8.44) (i.e., $J(r,\pi^*) \geq J(r,\pi)$ for all $\pi$ such that $J(c,\pi) \leq \rho$), we consider the following conditions from [18, 20]:

(E2) There exists a constant $L_4 > 0$ such that

$$|r(i,a)| \leq L_4\omega_4(i) \text{ and } |c(i,a)| \leq L_4\omega_4(i) \text{ for all } (i,a) \in K.$$

(E3) The positive discount factor $\alpha$ verifies that $\alpha > \kappa_2$, with $\kappa_2$ as in (E1).

(E4) $v(\omega_4) := \sum\limits_{i \in X} \omega_4(i) v(i) < \infty$.

(E5) For each $i \in X$, $A(i)$ is compact.

(E6) For each $i, j \in X$, the functions $r(i,a)$, $c(i,a)$, $q(j|i,a)$ and $\sum\limits_{k \in X} \omega_4(k) q(k|i,a)$ are continuous in $a \in A(i)$.

(E7) There exist a nonnegative function $\omega'$ on $X$ and constants $\gamma_3 \geq 0$, $\kappa_3 > 0$, and $L' > 0$ such that

$$q^*(i)\omega_4(i) \leq L'\omega'(i) \quad \text{and} \quad \sum_{j \in X} \omega'(j) q(j|i,a) \leq \kappa_3 \omega'(i) + \gamma_3 \quad \text{for all } (i,a) \in K.$$

(E8) There exists $\widehat{\pi} \in \Pi_m$ such that $J(c, \widehat{\pi}) < \rho$.

Note that the set $\Pi_m$ is convex. That is, if $\pi^1$ and $\pi^2$ are in $\Pi_m$, and for any $p \in [0,1]$, $i \in X$, and $t \in [0,\infty)$, $\pi_t^p(\cdot|i) := p\pi_t^1(\cdot|i) + (1-p)\pi_t^2(\cdot|i)$, then $\pi^p \in \Pi$.

Under conditions (E1)–(E6), we define a real-valued function $G$ on $C \times D := C_{\omega_4}(K) \times \Pi_m$ as follows:

$$G(c, \pi) := J(c, \pi), \quad \text{for } (c, \pi) \in C \times D = C_{\omega_4}(K) \times \Pi_m. \tag{8.45}$$

Obviously, the function $G$ defined in (8.45) satisfies (8.1). Moreover, let $D' := F$.

Now we state our main result of Sect. 8.4.4 on the existence of constrained-optimal policies for (8.44).

**Proposition 8.4.3.** *Suppose that conditions (E1)–(E8) hold. Then there exists a constrained-optimal policy for (8.44), which may be a stationary policy or a randomized stationary policy that randomizes between two stationary policies differing in at most one state.*

*Proof.* It follows from Theorems 6.5 and 6.10 and Lemma 11.6 in [20] that Assumptions 8.2.1 and 8.3.2 hold. Obviously, condition (E8) implies Assumption 8.2.2. Finally, from Lemma 11.7 and the proof of Theorem 11.4 in [20], we see that Assumption 8.3.1 holds. Hence, by Theorem 8.3.2, we complete the proof. □

*Remark 8.4.11.* Proposition 8.4.3 shows that the nonnegativity assumption on the costs as in [18, 20] is not required.

# References

1. Aliprantis, C., Border, K. (2007). *Infinite Dimensional Analysis*. Springer-Verlag, New York.
2. Altman, E. (1999). *Constrained Markov Decision Processes*. Chapman and Hall/CRC Press, London.
3. Beutler, F.J., Ross, K.W. (1985). Optimal policies for controlled Markov chains with a constraint. *J. Math. Anal. Appl.* **112**, 236–252.

4.  Borkar, V., Budhiraja, A. (2004). Ergodic control for constrained diffusions: characterization using HJB equations. *SIAM J. Control Optim.* **43**, 1467–1492.
5.  Borkar, V.S., Ghosh, M.K. (1990). Controlled diffusions with constraints. *J. Math. Anal. Appl.* **152**, 88–108.
6.  Borkar, V.S. (1993). Controlled diffusions with constraints II. *J. Math. Anal. Appl.* **176**, 310–321.
7.  Borkar, V.S. (2005). Controlled diffusion processes. *Probab. Surv.* **2**, 213–244.
8.  Budhiraja, A. (2003). An ergodic control problem for constrained diffusion processes: existence of optimal Markov control. *SIAM J. Control Optim.* **42**, 532–558.
9.  Budhiraja, A., Ross, K. (2006). Existence of optimal controls for singular control problems with state constraints. *Ann. Appl. Probab.* **16**, 2235–2255.
10. Feinberg, E.A., Shwartz, A. (1995). Constrained Markov decision models with weighted discounted rewards. *Math. Oper. Res.* **20**, 302–320.
11. Feinberg, E.A., Shwartz, A. (1996). Constrained discounted dynamic programming. *Math. Oper. Res.* **21**, 922–945.
12. Feinberg, E.A., Shwartz, A. (1999). Constrained dynamic programming with two discount factors: applications and an algorithm. *IEEE Trans. Autom. Control.* **44**, 628–631.
13. Feinberg, E.A. (2000). Constrained discounted Markov decision processes and Hamiltonian cycles. *Math. Oper. Res.* **25**, 130–140.
14. Föllmer, H., Schied, A. (2004). *Stochastic Finance: An Introduction in Discrete Time.* Walter de Gruyter, Berlin.
15. González-Hernández, J., Hernández-Lerma, O. (1999). Envelopes of sets of measures, tightness, and Markov control processes. *Appl. Math. Optim.* **40**, 377–392.
16. González-Hernández, J., Hernández-Lerma, O. (2005). Extreme points of sets of randomized strategies in constrained optimization and control problems. *SIAM J. Optim.* **15**, 1085–1104.
17. Guo, X.P. (2000). Constrained denumerable state non-stationary MDPs with expected total reward criterion. *Acta Math. Appl. Sin. (English Ser.)* **16**, 205–212.
18. Guo, X.P., Hernández-Lerma, O. (2003). Constrained continuous-time Markov controlled processes with discounted criteria. *Stochastic Anal. Appl.* **21**, 379–399.
19. Guo, X.P. (2007). Constrained optimization for average cost continuous-time Markov decision processes. *IEEE Trans. Autom. Control.* **52**, 1139–1143.
20. Guo, X.P., Hernández-Lerma, O. (2009). *Continuous-time Markov Decision Processes: Theory and Applications.* Springer-Verlag, Berlin Heidelberg.
21. Guo, X.P., Song, X.Y. (2011). Discounted continuous-time constrained Markov decision processes in Polish spaces. *Ann. Appl. Probab.* **21**, 2016–2049.
22. Hernández-Lerma, O., Lasserre, J.B. (1996). *Discrete-Time Markov Control Processes: basic optimality criteria.* Springer-Verlag, New York.
23. Hernández-Lerma, O., Lasserre, J.B. (1999). *Further Topics on Discrete-Time Markov Control Processes.* Springer-Verlag, New York.
24. Hernández-Lerma, O., González-Hernández, J. (2000). Constrained Markov control processes in Borel spaces: the discounted case. *Math. Methods Oper. Res.* **52**, 271–285.
25. Huang, Y., Kurano, M. (1997). The LP approach in average reward MDPs with multiple cost constraints: the countable state case. *J. Inform. Optim. Sci.* **18**, 33–47.
26. González-Hernández, J., López-Martínez, R.R., Pérez-Hernández, J.R. (2007). Markov control processes with randomized discounted cost. *Math. Methods. Oper. Res.* **65**, 27–44.
27. López-Martínez, R.R., Hernández-Lerma, O. (2003). The Lagrange approach to constrained Markov processes: a survey and extension of results. *Morfismos.* **7**, 1–26.
28. Mey, S.P., Tweedie, R.L. (1993). Stability of Markov processes III: Foster-Lyapunov criteria for continuous-time processes. *Adv. Appl. Prob.* **25**, 518–548.
29. Piunovskiy, A.B. (1997). *Optimal Control of Random Sequences in Problems with Constraints.* Kluwer, Dordrecht.
30. Prieto-Rumeau, T., Hernández-Lerma, O. (2006). Ergodic control of continuous-time Markov chains with pathwise constraints. *SIAM J. Control Optim.* **45**, 51–73.

31. Puterman, M.L. (1994). *Markov Decision Processes: Discrete Stochastic Dynamic Programming*. Wiley, New York.
32. Sennott, L.I. (1991). Constrained discounted Markov chains. *Probab. Engrg. Inform. Sci.* **5**, 463–475.
33. Sennott, L.I. (1993). Constrained average cost Markov chains. *Probab. Engrg. Inform. Sci.* **7**, 69–83.
34. Wei, Q.D., Guo, X.P. (2011). Markov decision processes with state-dependent discount factors and unbounded rewards/costs. *Oper. Res. Lett.* **39**, 369–374.
35. Zhang, L.L., Guo, X.P. (2008). Constrained continuous-time Markov decision processes with average criteria. *Math. Methods Oper. Res.* **67**, 323–340.
36. Zhang, Y. (2011). Convex analytic approach to constrained discounted Markov decision processes with non-constant discount factors. *Top.* doi: 10.1007/s11750-011-0186-8.

# Chapter 9
# Optimal Execution of Derivatives: A Taylor Expansion Approach

**Gerardo Hernandez-del-Valle and Yuemeng Sun**

## 9.1 Introduction

The problem of optimal execution is a very general problem in which a trader who wishes to buy or sell a *large* position $K$ of a given asset $S$—for instance, wheat, shares, derivatives, etc.—is confronted with the dilemma of executing slowly or as quick as possible. In the first case, he/she would be exposed to volatility, and in the second, to the laws of offer and demand. Thus, the trader must hedge between the *market impact* (due to his trade) and the *volatility* (due to the market).

The main *aim* of this chapter is to study and characterize the so-called Markowitz-optimal open-loop execution trajectory of contingent claims.

The problem of minimizing expected overall liquidity costs has been analyzed using different market models by [1, 6, 8], and [2], just to mention a few. However, some of these approaches miss the volatility risk associated with time delay. Instead, [3, 4] suggested studying and solving a mean-variance optimization for sales revenues in the class of deterministic strategies. Further on, [5] allowed for intertemporal updating and proved that this can *strictly* improve the mean-variance performance. Nevertheless, in [9], the authors study the original problem of expected utility maximization with CARA utility functions. Their main result states that for CARA investors there is surprisingly no added utility from allowing for intertemporal updating of strategies. Finally, we mention that the Hamilton-Jacobi-Bellman approach has also recently been studied in [7].

G. Hernandez-del-Valle (✉)
Statistics Department, Columbia University, New York, NY 10027, USA
e-mail: gerardo@stat.columbia.edu

Y. Sun
Cornell University, Theca, NY 14853, USA
e-mail: ys273@cornell.edu

D. Hernández-Hernández and A. Minjárez-Sosa (eds.), *Optimization, Control, and Applications of Stochastic Systems*, Systems & Control: Foundations & Applications, DOI 10.1007/978-0-8176-8337-5_9, © Springer Science+Business Media, LLC 2012

The chapter is organized as follows: in Sect. 9.2, we state the optimal execution contingent claim problem. Next, in Sect. 9.3, we provide its closed form solution. In Sect. 9.4, a numerical example is studied, and finally we conclude in Sect. 9.5 with some final remarks and comments.

## 9.2 The Problem

**The Model.** A trader wishes to execute $K = k_0 + \cdots + k_n$ units of a *contingent claim* $C$ with underlying $S$ by time $T$. The quantity to optimize is given by the so-called execution shortfall, defined as

$$Y = \sum_{j=0}^{n} k_j C_j - KC_0,$$

and the problem is then to find $k_0, \ldots, k_n$ such that attain the minimum

$$\min_{k_0, \ldots, k_n} \left( \mathbf{E}[Y] + \lambda \mathbf{V}[Y] \right),$$

for some $\lambda > 0$. Assuming the derivative $C$ is smooth in terms of its underlying $S$, it follows from the Taylor series expansion that

$$C_j = f(S_0) + f'(S_0)(\tilde{S}_j - S_0) + \frac{1}{2} f''(S_0)(\tilde{S}_j - S_0)^2 + R_3,$$

where $\tilde{S}$ is the effective price and $R_3$ is the remainder which is $o((\tilde{S}_j - S_0)^3)$. Hence,

$$\sum_{j=0}^{n} k_j C_j = \sum_{j=0}^{n} k_j f(S_0) + \sum_{j=0}^{n} f'(S_0) k_j (\tilde{S}_j - S_0) + \frac{1}{2} f''(S_0) \sum_{j=0}^{n} k_j (\tilde{S}_j - S_0)^2 + \sum_{j=0}^{n} k_j R_3$$

$$= KC_0 + f'(S_0) \left( \sum_{j=0}^{n} k_j \tilde{S}_j - KS_0 \right) s + \frac{1}{2} f''(S_0) \sum_{j=0}^{n} k_j (\tilde{S}_j - S_0)^2 + \sum_{j=0}^{n} k_j R_3.$$

That is,

$$Y = \sum_{j=0}^{n} k_j C_j - KC_0$$

$$= f'(S_0) \left( \sum_{j=0}^{n} k_j \tilde{S}_j - KS_0 \right) + \frac{1}{2} f''(S_0) \sum_{j=0}^{n} k_j (\tilde{S}_j - S_0)^2 + \sum_{j=0}^{n} k_j R_3. \qquad (9.1)$$

Note that if we use only the first-order approximation, then our optimization problem has already been solved and corresponds to [4] trading trajectory.

## 9.3 Second-Order Taylor Approximation

In this section, we extend [4] market impact model for the case of a contingent claim. We provide our main result which is the closed form objective function by adapting a second order-Taylor approximation.

### 9.3.1 Effective Price Process

Let $\xi_1, \xi_2, \ldots$ be a sequence of i.i.d. Gaussian random variables with mean zero and variance 1, and let the execution times be equally spaced, that is, $\tau := T/n$. Then, the price and "effective" processes are respectively defined as

$$S_j = S_{j-1} - \tau g\left(\frac{k_j}{\tau}\right) + \sigma\tau^{1/2}\xi_j,$$

$$\tilde{S}_j = S_j - h\left(\frac{k_j}{\tau}\right),$$

and the permanent and temporary market impact will be modeled, for simplicity, as

$$g\left(\frac{k_j}{\tau}\right) = \alpha\frac{k_j}{\tau}, \qquad h\left(\frac{k_j}{\tau}\right) = \beta\frac{k_j}{\tau},$$

for some constant $\alpha$ and $\beta$. Hence, letting

$$x_j := K - \sum_{m=0}^{j} k_m \quad \text{and}$$

$$W_j := \sum_{m=1}^{j} \xi_m, \quad \text{i.e. } W_j \sim \mathbf{N}(0, j), \text{ Cov}(W_j, W_i) = \min(i, j),$$

it follows that

$$\tilde{S}_j - S_0 = \sigma\tau^{1/2}W_j - \alpha(K - x_j) - \frac{\beta}{\tau}k_j. \tag{9.2}$$

### 9.3.2 Second-Order Approximation

From (9.1) and (9.2), the second-order approximation of the execution shortfall $Y$ is given by

$$Y \approx f'(S_0) \sum_{j=0}^{n} k_j \left( \sigma \tau^{1/2} W_j - \alpha(K - x_j) - \frac{\beta}{\tau} k_j \right)$$

$$+ \frac{1}{2} f''(S_0) \sum_{j=0}^{n} k_j \left( \sigma \tau^{1/2} W_j - \alpha(K - x_j) - \frac{\beta}{\tau} k_j \right)^2. \qquad (9.3)$$

Next, expanding the squared term, we get

$$\left( \sigma \tau^{1/2} W_j - \alpha(K - x_j) - \frac{\beta}{\tau} k_j \right)^2 = \sigma^2 \tau W_j^2 + \alpha^2 (K - x_j)^2 + \frac{\beta^2}{\tau^2} k_j^2 - 2 \frac{\beta \sigma}{\tau^{1/2}} k_j W_j$$

$$- 2 \alpha \sigma \tau^{1/2} (K - x_j) W_j + 2 \frac{\alpha \beta}{\tau} k_j (K - x_j),$$

Thus the expected value of $Y$ is approximately

$$\mathbf{E}[Y] = f'(S_0) \sum_{j=0}^{n} k_j \left( -\alpha(K - x_j) - \frac{\beta}{\tau} k_j \right)$$

$$+ \frac{1}{2} f''(S_0) \sum_{j=0}^{n} k_j \left[ \sigma^2 \tau j + \alpha^2 (K - x_j)^2 + \frac{\beta^2}{\tau^2} k_j^2 + 2 \frac{\alpha \beta}{\tau} k_j (K - x_j) \right],$$

$$(9.4)$$

to compute the variance $\mathbf{V}$ of $Y$ we rearrange (9.3) as

$$Y \approx \sum_{j=0}^{n} v_j k_j W_j + \sum_{j=0}^{n} \eta_j k_j W_j^2 + D,$$

where $D$ are all the deterministic terms and

$$v_j := f'(S_0) \sigma \tau^{1/2} - f''(S_0) \left[ \alpha \sigma \tau^{1/2} (K - x_j) + \frac{\beta \sigma}{\tau^{1/2}} k_j \right]$$

$$\eta_j := \frac{1}{2} f''(S_0) \sigma^2 \tau.$$

It follows that the variance of $Y$ is

$$\mathbf{V}[Y] = \mathbf{V} \left[ \sum_{j=0}^{n} v_j k_j W_j \right] + \mathbf{V} \left[ \sum_{j=0}^{n} \eta_j k_j W_j^2 \right] + 2 \operatorname{Cov} \left( \sum_{j=0}^{n} v_j k_j W_j, \sum_{j=0}^{n} \eta_j k_j W_j^2 \right)$$

$$= \sum_{j=0}^{n} v_j^2 k_j^2 j + 2 \sum_{0 \le i < j \le n} v_i k_i v_j k_j i + \sum_{j=0}^{n} \eta_j^2 k_j^2 \cdot 2 j^2 + 2 \sum_{0 \le i < j \le n} \eta_i k_i \eta_j k_j \cdot 2 i^2$$

$$(9.5)$$

and the last term equals zero.

### 9.3.3  Optimal Trading Schedule for the Second-Order Approximation

To find the optimal trading schedule for the second-order approximation of $Y$, we need find the sequence of $k_0, \ldots, k_n$ such that

$$\mathbf{E}[Y] + \lambda \mathbf{V}[Y]$$

is minimized for a given $\lambda$ and where $\mathbf{E}[Y]$ and $\mathbf{V}[Y]$ are as in (9.4) and (9.5), respectively. After some simplification,

$$
\begin{aligned}
\mathbf{E}[Y] + \lambda \mathbf{V}[Y] = {}& f'(S_0) \sum_{j=0}^{n} k_j \left[ \alpha(x_j - K) - \frac{\beta}{\tau} k_j \right] \\
& + \frac{1}{2} f''(S_0) \sum_{j=0}^{n} k_j \left[ \sigma^2 \tau j + \alpha^2 (K - x_j)^2 + \frac{\beta^2}{\tau^2} k_j^2 + \frac{2\alpha\beta}{\tau} (K - x_j) k_j \right] \\
& + \lambda \sum_{j=0}^{n} j k_j \left[ v_j^2 k_j + 2 j k_j \eta_j^2 + 2 v_j \sum_{m=j+1}^{n} v_m k_m + 4 j \eta_j \sum_{m=j+1}^{n} \eta_m k_m \right].
\end{aligned}
$$

## 9.4  Numerical Solution

For $Y$ as in (9.3), the optimization problem we aim to solve is

$$\min_{k_0, k_1, \ldots, k_n} (\mathbf{E}[Y] + \lambda \mathbf{V}[Y])$$

subject to

$$\sum_{j=0}^{n} k_j = K.$$

We solve the problem using *fmincon* in the **Matlab**.

*Example 9.4.1.* For this example let

$$n = 2; \quad K = 1000; \quad \alpha = 0.1; \quad \beta = 0.5; \quad \lambda = 0.4; \quad \tau = 1;$$
$$\delta = f'(S_0) = 0.5; \quad \gamma = f'(S_0) = 0.2; \quad \sigma = 0.5,$$

the optimal trading strategy is

$$k_0 = 333.3348, \quad k_1 = 333.3336, \quad k_2 = 333.3316,$$

and the optimal objective function is $5.5736 \times 10^8$.

*Remark 9.4.1.* The trading trajectory has a downward trend. Intuitively, and on contrast to executing a large size at a single transaction, our result suggests to split the overall position in almost even trades. The linear assumption that we made on the temporary and the permanent impacts seems to explain the almost equal execution quantities.

## 9.5  Concluding Remarks

In this work, we study the Markowitz-optimal execution trajectory of contingent claims. In order to do so, we use a second-order Taylor approximation with respect to the contingent claim $C$ evaluated at the initial value of the underlying $S$. We obtain the closed form objective function given a risk averse criterion. Our approach allows us to obtain the explicit numerical solution and we provide an example.

**Acknowledgements**  The authors were partially supported by Algorithmic Trading Management LLC.

## References

1. A. Alfonsi, A. Fruth and A. Schied (2010). Optimal execution strategies in limit order books with general shape functions, *Quantitative Finance*, **10**, no. 2, 143–157.
2. A. Alfonsi and A. Schied (2010). Optimal trade execution and absence of price manipulations limit order books models, *SIAM J. on Financial Mathematics*, **1**, pp. 490–522.
3. R. Almgren and N. Chriss (1999). Value under liquidation, *Risk*, **12**, pp. 61–63.
4. R. Almgren and N. Chriss (2000). Optimal execution of portfolio transactions, *J. Risk*, **3** (2), pp. 5–39.
5. R. Almgren and J. Lorenz (2007). Adaptive arrival price, *Trading*, no. 1, pp. 59–66.
6. D. Bertsimas and D. Lo (1998). Optimal control of execution costs, *Journal of Financial Markets*, **1**(1), pp. 1–50.
7. P.A. Forsyth (2011). Hamilton Jacobi Bellman approach to optimal trade schedule, *Journal of Applied Numerical Mathematics*, **61**(2), pp. 241–265.
8. A. Obizhaeva, and J. Wang (2006). Optimal trading strategy and supply/demand dynamics. *Journal of Financial Markets*, forthcoming.
9. A. Schied, T. Schöneborn and M. Tehranchi (2010). Optimal basket liquidation for CARA investors is deterministic, *Applied Mathematical Finance*, **17**, pp. 471–489.

# Chapter 10
# A Survey of Some Model-Based Methods for Global Optimization

**Jiaqiao Hu, Yongqiang Wang, Enlu Zhou, Michael C. Fu, and Steven I. Marcus**

## 10.1 Introduction

Global optimization aims at characterizing and computing global optimal solutions to problems with nonconvex, multimodal, or badly scaled objective functions; it has applications in many areas of engineering and science. In general, due to the absence of structural information and the presence of many local extrema, global optimization problems are extremely difficult to solve exactly. There are many different types of methods in the literature on global optimization, which can be categorized based on different criteria. For instance, they can be classified either based on the properties of problems to be solved (combinatorial or continuous, nonlinear, linear, convex, etc.) or by the properties of algorithms that search for new candidate solutions such as *deterministic* or *random search* algorithms. Random search algorithms can further be classified as *instance-based* or *model-*

J. Hu
Department of Applied Mathematics and Statistics, State University at Stony Brook,
Stony Brook, NY 11794, USA
e-mail: jqhu@ams.sunysb.edu

Y. Wang • S.I. Marcus (✉)
Department of Electrical and Computer Engineering & Institute for Systems Research,
University of Maryland, College Park, MD 20742, USA
e-mail: yqwang@umd.edu; marcus@umd.edu

E. Zhou
Department of Industrial and Enterprise Systems Engineering,
University of Illinois at Urbana-Champaign, IL 61801, USA
e-mail: enluzhou@illinois.edu

M.C. Fu
The Robert H. Smith School of Business & Institute for Systems Research,
University of Maryland, College Park, MD 20742, USA
e-mail: mfu@umd.edu

D. Hernández-Hernández and A. Minjárez-Sosa (eds.), *Optimization, Control, and Applications of Stochastic Systems*, Systems & Control: Foundations & Applications,
DOI 10.1007/978-0-8176-8337-5_10, © Springer Science+Business Media, LLC 2012

*based* algorithms according to the mechanism of generating new candidate solutions [46].

Instance-based algorithms maintain a single solution or population of candidate solutions, and the construction of new generate of candidate solutions depends explicitly on the previously generated solutions. Some well-known instance-based algorithms include simulated annealing [25], genetic algorithms [16,36], tabu search [15], nested partitions [35], generalized hill climbing [22, 23], and evolutionary programming [12]. Model-based search algorithms are a class of new solution techniques and were introduced only in recent years [18, 27, 32–34, 42]. In model-based algorithms, new solutions are generated via an intermediate probabilistic model that is updated or induced from the previously generated solutions. Thus, there is only an implicit/indirect dependency among the solutions generated at successive iterations of the algorithm. Specific model-based algorithms include annealing adaptive search (AAS) [31, 41], the cross-entropy (CE) method [32–34], and estimation of distribution algorithms (EDAs) [27, 42]. Instance-based algorithms have been extensively studied in past decades. After briefly reviewing some model-based algorithms, this chapter focuses on several model-based methods that have been developed recently.

## 10.2 Global Optimization and Previous Work

### 10.2.1 Problem Statement

In many engineering design and optimization applications, we are concerned with finding parameter values that achieve the optimum of an objective function. Such problems can be mathematically stated in the generic form:

$$x^* \in \arg \max_{x \in \mathbf{X}} H(x), \tag{10.1}$$

where $x$ is a vector of $n$ decision variables, the solution space $\mathbf{X}$ is a nonempty (often compact) subset of $\Re^n$, and the objective function $H : \mathbf{X} \to \Re$ is a bounded deterministic function.

Throughout this chapter, we assume that there exists a global optimal solution to (10.1), i.e., $\exists x^* \in \mathbf{X}$ such that $H(x) \leq H(x^*) \ \forall x \neq x^*$, $x \in \mathbf{X}$. In practice, this assumption can be justified under fairly general conditions. For example, for continuous optimization problems with compact solution spaces, the existence of an $x^*$ is guaranteed by the well-known Weierstrass theorem, whereas in discrete optimization, the assumption holds trivially when $\mathbf{X}$ is a (nonempty) finite set. Note that no further structural assumptions, such as convexity or differentiability, are imposed on the objective function, and there may exist many locally optimal solutions. In other words, our focus is on general global optimization problems with little known structure. This setting arises in many complex systems of interest, e.g., when the explicit form of $H$ is not readily available and the objective function values can only be assessed via "black-box" evaluations.

## 10.2.2   Previous Work on Random Search Methods

In this section, we review a class of global optimization algorithms collectively known as random search methods. A random search method usually refers to an algorithm that is iterative in nature, and uses some sort of randomized mechanism to generate a sequence of iterates, e.g., candidate solutions or probabilistic models, in order to successively approximate the optimal solution. What type of iterates an algorithm produces and how these iterates are generated are what differentiates approaches. A major advantage of stochastic search methods is that they are robust and easy to implement, because they typically only rely on the objective function values rather than structural information such as convexity and differentiability. This feature makes these algorithms especially prominent in optimization of complex systems with little structure.

From an algorithmic point of view, a random search algorithm can further be classified as being either *instance-based* or *model-based* [46]. In instanced-based algorithms, an iterate comprises a single or a set/population of candidate solution(s), and the construction of new candidate solutions depends explicitly on previously generated solutions. Such algorithms can be represented abstractly by the following framework:

1. Given a set/population of candidate solutions $Y^{(k)}$ (which might be a singleton set), generate a set of new candidate solutions $X^{(k)}$ according to a specified random mechanism.
2. Update the current population $Y^{(k+1)}$ based on population $Y^{(k)}$ and candidate solutions in $X^{(k)}$; increase the iteration counter $k$ by 1 and reiterate from Step 1.

Thus the two major steps in an instance-based algorithm are the generation step that produces a set of candidate solutions, and the selection/update step that determines whether a newly generated solution in $X^{(k)}$ should be included in the next generation. Over the past few decades, a significant amount of research effort has been centered around instance-based methods, with numerous algorithms proposed in the literature and their behaviors relatively well studied and understood. Some well-known examples include simulated annealing [25], genetic algorithms [16,36], tabu search [15], nested partitions [35], generalized hill climbing [22, 23], and evolutionary programming [12].

We focus on model-based methods, which differ from instance-based approaches in that candidate solutions are generated at each iteration by sampling from an intermediate probability distribution model over the solution space. The idea is to iteratively modify the distribution model based on the sampled solutions to bias the future search toward regions containing high-quality solutions. In its most basic from, a model-based algorithm typically consists of the following two steps: let $g_k$ be a probability distribution on $\mathbf{X}$ at the $k$th iteration of an algorithm:

1. Randomly generate a set/population of candidate solutions $X^{(k)}$ from $g_k$.
2. Update $g_k$ based on the sampled solutions in $X^{(k)}$ to obtain a new distribution $g_{k+1}$; increase $k$ by 1 and reiterate from step 1.

The underlying idea is to construct a sequence of iterates (probability distributions) $\{g_k\}$ with the hope that $g_k \to g^*$ as $k \to \infty$, where $g^*$ is a limiting distribution that assigns most of its probability mass to the set of optimal solutions. So it is the probability distribution (as opposed to candidate solutions as in instance-based algorithms) that is propagated from one iteration to the next.

Clearly, the two key questions one needs to address in a model-based algorithm are how to generate samples from a given distribution $g_k$ and how to construct the distribution sequence $\{g_k\}$. In order to address these questions, we provide brief descriptions of three model-based algorithms: annealing adaptive search (AAS) [31, 41], the cross-entropy (CE) method [32–34], and estimation of distribution algorithms (EDAs) [27, 42].

The annealing adaptive search algorithm was originally introduced in Romeijn and Smith [18] as a means to understand the behavior of simulated annealing. The algorithm generates candidate solutions by sampling from a sequence of Boltzmann distributions parameterized by time-dependent temperatures. As the temperature decreases to zero, the sequence of Boltzmann distributions becomes more concentrated on the set of optimal solutions, so that a solution sampled at later iterations will be close to the global optimum with high probability. For the class of Lipschitz optimization problems, it is shown that the expected number of iterations required by AAS to achieve a given level of precision increases at most linearly in the problem dimension [31, 41]. However, the idealized AAS is not intended to be a practically useful algorithm, because the problem of sampling exactly from a given Boltzmann distribution is known to be extremely difficult. This implementation issue has motivated a number of algorithms that approximate AAS, where a primary focus has been on the design and refinement of Markov chain-based sampling techniques embedded within the AAS framework [40, 41].

The CE method was motivated by an adaptive algorithm for estimating probabilities of rare events in complex stochastic networks [32], which involves variance minimization. It was later realized [33] that the method can be modified to solve combinatorial and continuous optimization problems. The CE method uses a family of parameterized probability distributions on the solution space and tries to find the parameter of the distribution that assigns maximum probability to the set of optimal solutions. Implicit in CE is an optimal importance sampling distribution concentrated only on the set of optimal solutions. The key idea is to use an iterative scheme to successively estimate the optimal parameter that minimizes the Kullback-Leibler (KL) divergence between the optimal distribution and the family of parameterized distributions. Although there have been extensive developments regarding implementation and successful practical applications of CE (see [34]), the literature analyzing the convergence properties of the CE method is relatively sparse, with most of the existing results limited to specific settings (see, e.g., [17] for a convergence proof of a variational version of CE in the context of estimation of rare event probabilities, and [7] for probability one convergence proofs of CE for discrete optimization problems). General convergence and asymptotic rate results for CE were recently obtained in [21] by relating the algorithm to recursions of stochastic approximation type (see Sect. 10.6).

EDAs were first introduced in the field of evolutionary computation. They inherit the spirit of the well-known genetic algorithms (GAs), but eliminate the crossover and mutation operators to avoid the disruption of partial solutions. In EDAs, a new population of candidate solutions are generated according to the probability distribution induced or estimated from the promising solutions selected from the previous generation. Unlike CE, EDAs often take into account the interrelations between the underlying decision variables needed to represent the individual candidate solutions. At each iteration of the algorithm, a high-dimensional probabilistic model that better represents the interdependencies between the decision variables is induced; this step constitutes the most crucial and difficult part of the method. We refer the reader to [27] for a review of the way in which different probabilistic models are used as EDA instantiations. A proof of convergence of a class of EDAs, under the idealized infinite population assumption, can be found in [42].

There are many other model-based algorithms proposed for global optimization. Some interesting examples include ant colony optimization (ACO) [9], probability collectives (PCs) [39], and particle swarm optimization (PSO) [24]. We do not provide a comprehensive description of all of them, but instead present some recently developed frameworks and approaches that allow us to view these algorithms in a unified setting. These approaches, including model reference adaptive search (MRAS) [18], the particle-filtering (PF) approach [43], the evolutionary games approach [38], and the stochastic approximation gradient approach [20, 21], will be discussed in detail in the following sections.

## 10.3   Model Reference Adaptive Search

As we have seen from Sect. 10.2, model-based algorithms differ from each other in the choices of the distribution sequence $\{g_k\}$. Examples of the $\{g_k\}$ sequence include (a) Boltzmann distributions, used in AAS; (b) optimal importance sampling measure, primarily used in the CE method; and (c) proportional selection schemes, used in EDAs, ACOs, and PCs.

However, in all the above cases, the construction of $g_k$ often depends on the objective function $H$, whose explicit form may not be available. In addition, since $g_k$ may not have any special structure, sampling exactly from the distribution is in general intractable. To address these computational challenges arising in model-based methods, we have formalized in [18] a general approach called model reference adaptive search (MRAS), where the basic idea is to use a convenient parametric distribution as a surrogate to approximate $g_k$ and then sample candidate solutions from the surrogate distribution. More specifically, the method starts by specifying a family of parameterized distributions $\{f_\theta, \theta \in \Theta\}$ (with $\Theta$ being the parameter space) and then projects $g_k$ onto the family to obtain a sampling distribution $f_{\theta_k}$, where the projection is implemented at each iteration by finding an optimal parameter $\theta_k$ that minimizes the Kullback-Leibler (KL) divergence between $g_k$ and the parameterized family [34], i.e.,

$$\theta_k = \arg\min_{\theta \in \Theta} \mathscr{D}(g_k, f_\theta) := \arg\min_{\theta \in \Theta} \left( \int_{\mathbf{X}} \ln \frac{g_k(x)}{f_\theta(x)} g_k(\mathrm{d}x) \right). \tag{10.2}$$

The idea is that the parameterized family is specified with some structure (e.g., family of normal distributions parameterized by means and variances) so that once its parameter is specified, sampling from the corresponding distribution can be performed relatively easily and efficiently. Another advantage is that the task of constructing the entire surrogate distribution now simplifies to the task of finding its associated parameters. Roughly speaking, each sampling distribution $f_{\theta_k}$ obtained via (10.2) can be viewed as a compact approximation of $g_k$, and consequently the entire sequence $\{f_{\theta_k}\}$ may (hopefully) retain some nice properties of the distribution sequence $\{g_k\}$. Thus, to ensure the convergence of the MRAS method, it is intuitively clear that the sequence $\{g_k\}$ should be chosen in a way so that it can be shown to converge to a limiting distribution concentrated only on the set of optimal solutions. Since the distribution $g_k$ is primarily used to guide the parameter updating process and to express the desired properties of the MRAS method, it is called the *reference* distribution.

We now provide a summary of the MRAS method:

0. Select a sequence of reference distributions $\{g_k\}$ with desired convergence properties and choose a parameterized family $\{f_\theta\}$.
1. Given $\theta_k$, sample $N$ candidate solutions $X_k^1, \ldots, X_k^N$ from $f_{\theta_k}$.
2. Update the parameter $\theta_{k+1}$ by minimizing the KL divergence

$$\theta_{k+1} = \arg\min_{\theta} \mathscr{D}(g_{k+1}, f_\theta);$$

increase $k$ by 1 and reiterate from step 1.

Note that the algorithm above assumes that the expectation/integral involved in the KL divergence (cf. (10.2)) can be evaluated exactly. In practice, it is often estimated by an empirical average based on samples obtained at step 1.

The MRAS framework accommodates many algorithms aforementioned in Sect. 10.2. For example, when Boltzmann distributions are used as reference models, the resulting algorithm becomes AAS with an additional projection step. The algorithm instantiation considered in [18] uses the following recursive procedure to construct the $g_k$ sequence:

$$g_{k+1}(x) = \frac{H(x)g_k(x)}{\int_{\mathbf{X}} H(x)g_k(\mathrm{d}x)}, \tag{10.3}$$

where $g_0(x)$ is a given initial distribution on $\mathbf{X}$ and we have assumed for simplicity that $H(x) > 0$ for all $x \in \mathbf{X}$ to prevent negative probabilities. This form of reference distributions has also been used in a class of EDAs with proportional selection schemes. It weights the new distribution $g_{k+1}$ by the value of the objective function $H(x)$, so that each iteration of (10.3) improves the expected performance in the sense that

$$E_{g_{k+1}}[H(X)] := \int_{\mathbf{X}} H(x)g_{k+1}(dx) = \frac{\int_{\mathbf{X}} H^2(x)g_k(dx)}{\int_{\mathbf{X}} H(x)g_k(dx)} \geq E_{g_k}[H(X)],$$

so solutions with better performance are given more probability under $g_{k+1}$. This results in a $\{g_k\}$ sequence that converges to a degenerate distribution at the optimal solution. Furthermore, it is shown in [18] that the CE method can also be recovered by replacing $g_k$ in the right-hand side of (10.3) with $f_{\theta_k}$. In other words, there is a sequence of reference distributions implicit in CE that takes the form

$$g_{k+1}(x) = \frac{H(x)f_{\theta_k}(x)}{\int_{\mathbf{X}} H(x)f_{\theta_k}(dx)}. \tag{10.4}$$

Since $g_{k+1}$ in (10.4) is obtained by tilting the sampling distribution $f_{\theta_k}$ with the objective function $H$, it improves the expected performance of $f_{\theta_k}$, i.e.,

$$E_{g_{k+1}}[H(X)] = \frac{\int_{\mathbf{X}} H^2(x)f_{\theta_k}(dx)}{\int_{\mathbf{X}} H(x)f_{\theta_k}(dx)} \geq \int_{\mathbf{X}} H(x)f_{\theta_k}(dx) := E_{\theta_k}[H(X)].$$

Therefore, it is reasonable to expect that the projection of $g_{k+1}$ on the parameterized family, $f_{\theta_{k+1}}$, also improves $f_{\theta_k}$, i.e., $E_{\theta_{k+1}}[H(X)] \geq E_{\theta_k}[H(X)]$. This view of CE leads to an important monotonicity property of the method, generalizing that of [34], which is only proved for the one-dimensional case.

### 10.3.1  Convergence Result

For the family of natural exponential distributions (NEFs), the optimization problem involved at step 2 of the MRAS method can be solved analytically in closed form, which makes the approach very convenient to implement in practice. We recall the definition of NEFs.

**Definition 10.3.1.** A parameterized family $\{f_\theta, \theta \in \Theta \subseteq \Re^d\}$ is said to belong to the natural exponential family if there exist mappings $\Gamma : \Re^n \to \Re^d$ and $K : \Re^d \to \Re$ such that each $f_\theta$ in the family can be represented in the form $f_\theta(x) = \exp(\theta^T \Gamma(x) - K(\theta))$, where $K(\theta)$ is a normalization constant given by $K(\theta) = \ln \int_{\mathbf{X}} \exp(\theta^T \Gamma(x))dx$.

The function $K(\theta)$ plays an important role in the theory of NEFs. It is strictly convex in the interior of $\Theta$ with gradient $\nabla_\theta K(\theta) = E_\theta[\Gamma(X)]$ and Hessian matrix $\text{Cov}_\theta[\Gamma(X)]$. We define the mean vector function

$$m(\theta) := E_\theta[\Gamma(X)].$$

Since the Jacobian of $m(\theta)$ is strictly positive definite, we have from the inverse function theorem that $m(\theta)$ is a one-to-one invertible function of $\theta$. Generally speaking, $m(\theta)$ can be viewed as a transformed version of the sufficient statistic $\Gamma(x)$, whose value contains all necessary information to estimate the parameter $\theta$. For example, for the univariate normal distribution $\mathbf{N}(\mu, \sigma^2)$ with mean $\mu$ and variance $\sigma^2$, it can be seen that $\Gamma(x) = (x, x^2)^T$ and $\theta = (\frac{\mu}{\sigma^2}, -\frac{1}{2\sigma^2})^T$. Thus, $m(\theta) = E_\theta[\Gamma(X)]$ becomes $(\mu, \sigma^2 + \mu^2)^T$, which can be uniquely solved for $\mu$ and $\sigma^2$ given the value of $m(\theta)$.

When NEFs are used as the parameterized family, we have the following convergence theorem for the instantiation of MRAS considered in [18].

**Theorem 10.3.1.** *When $\{g_k\}$ in (10.3) are used as reference distributions in MRAS, let $\{\theta_k\}$ be the sequence of parameters generated by the algorithm based on the sampled candidate solutions. Under appropriate assumptions (see [18]),*

$$\lim_{k \to \infty} m(\theta_k) = \Gamma(x^*) \ w.p.1.$$

The interpretation of Theorem 10.3.1 relies on the parameterized family used in MRAS and, in particular, on the specific form of the sufficient statistic $\Gamma(x)$. We consider two special cases of Theorem 10.3.1. (a) In continuous optimization when multivariate normal distributions with mean vector $\mu$ and covariance matrix $\Sigma$ are used as the parameterized family, then it is easy to show that Theorem 10.3.1 implies $\lim_{k \to \infty} \mu_k = x^*$ and $\lim_{k \to \infty} \Sigma_k = 0_{n \times n}$ w.p.1, where $0_{n \times n}$ represents an $n$-by-$n$ zero matrix. In other words, the sequence of sampling distributions $\{f_{\theta_k}\}$ will converge to a delta distribution with all probability mass concentrated on $x^*$. (b) For a discrete optimization problem with feasible domain $\mathbf{X}$ that contains $l$ distinct values denoted by $x_1, \ldots, x_l$, the parameterized family can be specified in terms of an $l$-by-1 probability vector $Q$, whose $i$th entry $q_i$ represents the probability that a (random) solution will take the $i$th value $x_i$. A probability mass function on $\mathbf{X}$, when parameterized by $Q$, can thus be expressed as

$$f_\theta(x) = \prod_{i=1}^{l} q_i^{I\{x=x_i\}} := e^{\theta^T \Gamma(x)},$$

where $I\{\cdot\}$ is the indicator function, $\theta = [\ln q_1, \ldots, \ln q_l]^T$, and the sufficient statistic $\Gamma(x) = [I\{x = x_1\}, \ldots, I\{x = x_l\}]^T$. Therefore, a simple application of Theorem 10.3.1 yields

$$\lim_{k \to \infty} \sum_{x \in \mathbf{X}} \prod_{i=1}^{l} (q_i^k)^{I\{x=x_i\}} I\{x = x_j\} = I\{x^* = x_j\} \ \forall j \ w.p.1,$$

where $q_i^k$ is the $i$th entry of the probability vector $Q_k$ obtained at the $k$th iteration of the algorithm. This in turn implies that $\lim_{k \to \infty} q_i^k = I\{x^* = \mathbf{x}_i\}$ w.p.1., i.e., the sequence of $Q_k$ will convergence to a degenerate probability vector assigning unit mass to $x^*$.

We remark that Theorem 10.3.1 does not address the convergence rate of the algorithm. Moreover, the proof techniques used in [18] cannot be directly carried over to analyze other algorithms such as CE, due to the dependency of $g_k$ on the parameterized family (cf. (10.4)). In Sect. 10.6, we show that with some appropriate modifications of the MRAS method, we can arrive at a general framework linking model-based methods to recursive algorithms of stochastic approximation type, which makes the convergence and convergence rate analysis of these algorithms more tractable.

## 10.4   Particle-Filtering Approach

Filtering refers to the estimation of an unobserved state in a dynamical system based on noisy observations that arrive sequentially in time (c.f. [8] for an introduction). The idea behind the particle-filtering approach is to transform the optimization problem into a filtering problem. Using a novel interpretation, the distribution sequence $\{g_k\}$ in model-based optimization corresponds to the sequence of conditional distributions of the unobserved state given the observation history in filtering, and hence, $\{g_k\}$ is updated from a Bayesian perspective. A class of simulation-based filtering techniques called particle filtering can then be employed to sample from $\{g_k\}$, leading to a framework for model-based optimization algorithms.

More specifically, the optimization problem (10.1) can be transformed into a filtering problem by choosing an appropriate state-space model, such as the following:

$$X_k = X_{k-1}, \ k = 1, 2, \ldots,$$
$$Y_k = H(X_k) - V_k, \ k = 1, 2, \ldots, \tag{10.5}$$

where $X_k \in \mathfrak{R}^n$ is the unobserved state, $Y_k \in \mathfrak{R}$ is the observation, and $\{V_k, k = 1, 2, \ldots\}$ is an i.i.d. sequence of nonnegative random variables that have a p.d.f. $\varphi$. A prior distribution on $X_0$ is denoted by $g_0$. The goal of filtering is to compute the conditional density $g_k$ of the unobserved state $X_k$ given the past observations $\{Y_1 = y_1, \ldots, Y_k = y_k\}$ for $k = 1, 2, \ldots$. Let $\mathcal{F}$ denote the $\sigma$-field of Borel sets of $\mathfrak{R}^n$. Then the conditional density $g_k$ satisfies

$$P(X_k \in A | Y_{1:k} = y_{1:k}) = \int_A g_k(x) dx, \ \forall A \in \mathcal{F},$$

where $Y_{1:k} = \{Y_1, \ldots, Y_k\}$, and $y_{1:k} = \{y_1, \ldots, y_k\}$. Using Bayes rule, the evolution of $g_k(x)$ can be derived as follows:

$$g_k(x) = p(x|y_{0:k-1}, y_k)$$
$$= \frac{p(y_k|x)p(x|y_{0:k-1})}{p(y_k|y_{0:k-1})}$$
$$= \frac{\varphi(H(x) - y_k)g_{k-1}(x)}{\int \varphi(H(x) - y_k)g_{k-1}(x)dx}, \tag{10.6}$$

where the last line uses the density functions induced by (10.5).

The intuition of (10.5) and (10.6) and their connection with optimization can be explained as follows: the unobserved state $\{X_k\}$ is constant with the underlying value being the optimum $x^*$, which needs to be estimated; the observations $\{y_k\}$ are noisy observations of the optimal function value $H(x^*)$ and come from the sample function values in an optimization algorithm; the conditional density $g_k$ is a density estimate of the optimum $x^*$ at iteration $k$ based on the sample function values $\{y_1, \ldots, y_k\}$. Equation (10.6) implies that $g_k$ is tuned the more promising area where $H(x)$ is greater than $y_k$ since $\varphi(H(x) - y_k)$ is positive if $H(x) \geq y_k$ and is zero otherwise. Hence, randomization in the optimization algorithm is brought in by the randomness of $V_k$, and the choice of the p.d.f. of $V_k$, $\varphi$, results in different sample selection or weighting schemes in the algorithm. In order to ensure the resultant optimization algorithm monotonically approaches the optimum, the following general condition (C) on $\varphi$ is imposed:

(C)   The p.d.f. $\varphi(\cdot)$ is positive, strictly increasing, and continuous on its support $[0, \infty)$.

It is shown in [45] that if $\varphi$ satisfies the condition, then for an arbitrary, fixed observation sequence $\{y_1, y_2, \ldots\}$, the estimate of the function value is monotonically increasing, i.e.,

$$E_{g_{k+1}}[H(X)] \geq E_{g_k}[H(X)].$$

Hence, it has the same monotonicity property as MRAS and CE. Furthermore, the estimate of the optimal function value asymptotically converges to the true optimal function value as stated in the following theorem that is also shown in [45].

**Theorem 10.4.2.** *Suppose the following conditions hold:*

(i) *For all $H(x) < H(x^*)$, the set $\{z \in \mathbf{X} : H(z) \geq H(x)\}$ has strictly positive measure with respect to the initial sampling distribution, i.e., $\int_{\{z \in \mathbf{X}: H(z) \geq H(x)\}} g_0(x)dx > 0$.*

(ii) *There is a unique optimum $x^*$, and $H(x)$ is continuous at $x^*$.*

(iii) *$\varphi$ satisfies the condition (C).*

*Then for an arbitrary, fixed observation sequence $\{y_1, y_2, \ldots\}$,*

$$\lim_{k \to \infty} E_{g_k}[H(X)] = H(x^*).$$

The conditions (i) and (ii) ensure that any neighborhood of the optimum always has a positive probability to be sampled. The result implies that the samples drawn from $g_k$ in the limit will be concentrated on the optimum.

## 10.4.1   Algorithms

The distribution sequence $\{g_k\}$ in general does not have a closed-form solution. Various numerical filtering methods (cf. [5] for a recent survey) are available to numerically approximate $\{g_k\}$. However, the most akin to model-based optimization algorithms is the particle-filtering technique, which is a more recent class of approximate filtering methods based on Sequential Monte Carlo (SMC) simulation (cf. the tutorial [1] and the more recent tutorial [11] for a quick reference and the book [10] for a more comprehensive account). Despite its abundant successful applications in many areas, particle filtering has rarely been explored in optimization.

The basic particle filter is a sequential importance sampling resampling algorithm, each iteration of which is composed of an importance sampling step to propagate the particles (i.e., samples) from the previous iteration to the current, a Bayes updating step to update the weights of the particles, and a resampling step to generate new particles in order to prevent sample degeneracy. Applying it to the distribution sequence $\{g_k\}$ specified in (10.6) leads to the particle filtering for optimization (PFO) framework as follows:

0. *Initialization.* Specify $g_0$, and draw i.i.d. samples $\{X_1^i\}_{i=1}^{N_1}$ from $g_0$. Set $k = 1$.
1. *Bayes updating.* Take $y_k$ to be a sample function value $H(X_k^i)$ according to a certain rule. Compute the weight $w_k^i$ for sample $X_k^i$ according to

$$w_k^i \propto \varphi(H(X_k^i) - y_k), i = 1, 2, \ldots, N_k,$$

   and normalize the weights such that they sum up to 1.
2. *Resampling.* Generate i.i.d. samples $\{X_{k+1}^i\}_{i=1}^{N_{k+1}}$ from the weighted samples $\{w_k^i, X_k^i\}_{i=1}^{N_k}$ using regularized method, density projection method, or resample-move method.
3. *Stopping.* If a stopping criterion is satisfied, then stop; else, increase $k$ by 1 and reiterate from step 1.

Note that the simple method of sampling with replacement cannot be used in the resampling step since it does not generate new values for the samples and hence does not explore new candidate solutions for the purpose of optimization. Several other known resampling methods can be used to generate new candidate solutions and can also be easily implemented, including the regularized method [28], the density projection method [44], and the resample-move method [13]. The regularized method draws new i.i.d. samples from a continuous mixture distribution, where each continuous kernel of the mixture distribution is centered at each sample

$X_k^i$ and the weight of that kernel is equal to the probability mass $w_k^i$ of $X_k^i$. The density projection method resembles MRAS and CE in finding a parameterized density $f_{\theta_k}$ by minimizing the KL divergence between the discrete distribution $\{w_k^i, X_k^i\}$ and the parameterized family. The resample-move method applies a Markov chain Monte Carlo (MCMC) step to move the particles after they are generated by sampling with replacement. Depending on the resampling methods, the convergence properties of the different instantiations of PFO are also slightly different, but all readily follow from the existing convergence results of the corresponding particle filters in the literature [6, 14, 44] under suitable assumptions.

We end this section with a final remark that the PFO framework provides a new perspective on CE and MRAS. We will use the truncated selection scheme for sample selection as an illustration. Suppose that the objective function $H(x)$ is bounded by $H_1 \leq H(x) \leq H_2$. In the state-space model (10.5), let the observation noise $V_k$ follow a uniform distribution $U(0, H_2 - H_1)$, and then $\varphi$, the p.d.f. of $V_k$, satisfies

$$\varphi(u) = \begin{cases} \frac{1}{H_2 - H_1}, & \text{if } 0 \leq u \leq H_2 - H_1; \\ 0, & \text{otherwise.} \end{cases} \tag{10.7}$$

Since $y_k$ is a sample function value, the inequality $H(x) - y_k \leq H_2 - H_1$ holds with probability 1, so substituting (10.7) into (10.6) yields

$$g_k(x) = \frac{I\{H(x) \geq y_k\} g_{k-1}(x)}{\int I\{H(x) \geq y_k\} g_{k-1}(x) \mathrm{d}x}.$$

The standard CE method can be viewed as PFO with the above choice of distribution sequence $\{g_k\}$ and the density projection method for resampling, so the samples $\{X_k^i\}$ are generated from $f_{\theta_{k-1}}$ and the weights of the samples are computed according to $w_k^i \propto I\{H(X_k^i) \geq y_k\}$. However, the approximation of $g_{k-1}$ by $f_{\theta_{k-1}}$ introduces an approximation error, which is accumulated to the next iteration. This approximation error can be corrected by taking $f_{\theta_{k-1}}$ as an importance density and hence can be taken care of by the weights of the samples. That is, in the case of MRAS or CE in which the sequence $\{y_k\}$ is monotonically increasing, the weights are computed according to

$$w_k^i = \frac{g_k(X_k^i)}{f_{\theta_{k-1}}(X_k^i)} \propto \frac{I\{H(X_k^i) \geq y_k\}}{f_{\theta_{k-1}}(X_k^i)}.$$

This instantiation of PFO coincides with an instantiation of MRAS. More details on a unifying perspective on EDAs, CE, and MRAS are given in [45].

## 10.5  Evolutionary Games Approach

The main idea of the evolutionary games approach is to formulate the global optimization problem as an evolutionary game and to use dynamics from evolutionary game theory to study the evolution of the candidate solutions. Searching for the optimal solution is carried out through the dynamics of reaching equilibrium points in evolutionary games. Specifically, we establish a connection between evolutionary game theory and optimization by formulating the global optimization problem as an evolutionary game with continuous strategy spaces. We show that there is a strong connection between a particular equilibrium set of the replicator dynamics and the global optimal solutions. By using Lyapunov theory, we also show that the particular equilibrium set is asymptotically stable under mild conditions. Based on the connection between the equilibrium points and global optimal solutions, we develop a model-based evolutionary optimization (MEO) algorithm.

First, we set up an evolutionary game with a continuous strategy space. Let $\mathscr{B}$ be the Borel $\sigma$-field on $\mathbf{X}$, the strategy space of the game; for each $t$, let $\mathbb{P}_t$ be a probability measure defined on $(\mathbf{X}, \mathscr{B})$. Let $\Delta$ denote set of all the strategies (probability measures) on $\mathbf{X}$. Each point $x \in \mathbf{X}$ can be viewed as a pure strategy. Roughly speaking, the fraction of agents playing the pure strategy $x$ at time $t$ is $\mathbb{P}_t(\mathrm{d}x)$. An agent playing the pure strategy $x$ obtains a fitness $\phi(H(x))$, where $\phi(\cdot) : \mathfrak{R} \to \mathfrak{R}^+$ is a strictly increasing function. An appropriate chosen $\phi(\cdot)$ can facilitate the expression of the model updating rule presented later. Let $X$ be a random variable with probability distribution $\mathbb{P}_t$. The fractions of agents adopting different strategies in the continuous game is described by the probability measure $\mathbb{P}_t$ defined on the strategy space $\mathbf{X}$, so the average payoff of the whole population is given by

$$E_{\mathbb{P}_t}[\phi(H(X))] = \int_{\mathbf{X}} \phi(H(x))\mathbb{P}_t(\mathrm{d}x).$$

In evolutionary game theory [29], the evolution of this probability measure is governed by some dynamics such as the so-called replicator dynamics. Let $\mathscr{A}$ be a measurable set in $\mathbf{X}$. If the replicator dynamics with a continuous strategy space is adopted, we have

$$\dot{\mathbb{P}}_t(\mathscr{A}) = \int_{\mathscr{A}} (\phi(H(x)) - E_{\mathbb{P}_t}[\phi(H(X))])\mathbb{P}_t(\mathrm{d}x). \tag{10.8}$$

From (10.8), we can see that if $\phi(H(x))$ outperforms $E_{\mathbb{P}_t}[\phi(H(X))]$ at $x$, the probability measure around $x$ will increase. If there exists a probability density function $p_t$, such that $\mathbb{P}_t(\mathrm{d}x) = p_t\mu(\mathrm{d}x)$, where $\mu(\cdot)$ is the Lebesgue measure defined on $(\mathbf{X}, \mathscr{B})$, then (10.8) becomes

$$\dot{p}_t(x) = (\phi(H(x)) - E_{\mathbb{P}_t}[\phi(H(X))])p_t(x), \tag{10.9}$$

which governs the evolution of the probability density function on the continuous strategy space. When $p_t(x)$ is used as our model to generate candidate solutions for the global optimization problem (10.1), the differential equation (10.9) can be used to update the model $p_t(x)$, with the final goal of making the probability density function $p_t(x)$ converge to a small set containing the global optimal solution. Then, the global optimization problem can be easily solved by sampling from the obtained probability density function.

### 10.5.1 Convergence Analysis

In this section, we study the properties of the equilibrium points of (10.8) and their connection with the global optimal solutions for the optimization problem, by employing the tools of equilibrium analysis in game theory and stability analysis in dynamic systems.

Assume that the optimization problem (10.1) has $m$ global optimal solutions $\{x_i^\star, i = 1, \ldots, m\}$. It is easy to see that $\mathbb{P}^{(x)} = \delta(x - x_i^\star)$ for $i = 1, \ldots, m$ are equilibrium points of (10.8), and we might guess there is a strong connection between the equilibrium points of (10.8) and the optimal solutions of the global optimization problem (10.1). We enforce the following assumption on function $\phi$.

**Assumption 10.5.1** $\phi(\cdot)$ is a continuous and strictly increasing function; there exist constants $\mathscr{L}$ and $\mathscr{M}$ such that $\mathscr{L} \leq \phi(H(x)) \leq \mathscr{M}$ for all $x \in \mathbf{X}$.

The following theorem shows that the overall fitness of the strategy (probability measure) $\mathbb{P}_t$ is monotonically increasing over time.

**Theorem 10.5.3.** Let $\mathbb{P}_t$ be a solution of the replicator dynamics (10.8). Under Assumption 10.5.1, the average payoff of the entire population $E_{\mathbb{P}_t}[\phi(H(X))]$ is monotonically increasing with time $t$. If $\mathbb{P}_t$ is not an equilibrium point of (10.8), then $E_{\mathbb{P}_t}[\phi(H(X))]$ is strictly increasing with time $t$.

To further study the properties of the equilibrium points of the replicator dynamics (10.8), the Prokhorov metric is used to measure the distance between different strategies (probability measures):

$$\rho(\mathbb{P}, \mathbb{Q}) := \inf\{\varepsilon > 0 : \mathbb{Q}(\mathscr{A}) \leq \mathbb{P}(\mathscr{A}^\varepsilon) + \varepsilon \text{ and } \mathbb{P}(\mathscr{A}) \leq \mathbb{Q}(\mathscr{A}^\varepsilon) + \varepsilon, \quad \forall \mathscr{A} \in \mathscr{B}\},$$

where $\mathscr{A}^\varepsilon := \{x : \exists \tilde{y} \in \mathscr{A}, d(\tilde{y}, x) < \varepsilon\}$, in which $d$ is a metric defined on $\mathbf{X}$. Then, the convergence of $\rho(\mathbb{Q}_n, \mathbb{Q}) \to 0$ is equivalent to the weak convergence of $\mathbb{Q}_n$ to $\mathbb{Q}$ [3].

**Definition 10.5.2.** Let $\mathscr{E}$ be a set in $\Delta$. For a point $\mathbb{P} \in \Delta$, define the distance between $\mathbb{P}$ and $\mathscr{E}$ as $\rho(\mathbb{P}, \mathscr{E}) = \inf\{\rho(\mathbb{P}, \mathbb{Q}), \forall \mathbb{Q} \in \mathscr{E}\}$. $\mathscr{E}$ is called Lyapunov stable if for all $\varepsilon > 0$, there exists $\eta > 0$ such that $\rho(\mathbb{P}_0, \mathscr{E}) < \eta \implies \rho(\mathbb{P}_t, \mathscr{E}) < \varepsilon$ for all $t > 0$.

**Definition 10.5.3.** Let $\mathscr{E}$ be a set in $\Delta$. $\mathscr{E}$ is called asymptotically stable if $\mathscr{E}$ is Lyapunov stable and there exists $\eta > 0$ such that $\rho(\mathbb{P}_0, \mathscr{E}) < \eta \implies \rho(\mathbb{P}_t, \mathscr{E}) \to 0$ as $t \to \infty$.

**Definition 10.5.4.** $\Delta_0 \subset \Delta$ is the set containing all $\mathbb{P}_0$ for which there exists a $x_k^\star$ such that $\mathbb{P}_0(\mathscr{A}) > 0$ for any set $\mathscr{A} \in \mathscr{B}$ that contains $x_k^\star$ and has a positive Lebesgue measure $\mu(\mathscr{A}) > 0$. Let $\mathscr{C} = \{\mathbb{P}^\star : \mathbb{P}^\star = \lim_{t \to \infty} \mathbb{P}_t \text{ starting from some } \mathbb{P}_0 \in \Delta_0\}$.

To present the main convergence result, we also need the following assumption.

**Assumption 10.5.2** *There is a finite number of global optimal solutions $\{x_1^\star, \ldots, x_m^\star\}$ for the optimization problem (10.1), where m is a positive integer.*

**Theorem 10.5.4.** *If Assumptions 10.5.1 and 10.5.2 hold, then for any $\mathbb{P}^\star \in \mathscr{C}$, there exist $\alpha_i \geq 0$, for $i = 1, \ldots, m$ with $\sum_{i=1}^m \alpha_i = 1$ such that $\mathbb{P}^{(x)} = \sum_{i=1}^m \alpha_i \delta(x - x_i^\star)$; the set $\mathscr{C}$ can be represented as $\mathscr{C} = \{\mathbb{P}^\star : \mathbb{P}^{(x)} = \sum_{i=1}^m \alpha_i \delta(x - x_i^\star), \text{ for some } \sum_{i=1}^m \alpha_i = 1, \alpha_i \geq 0, \forall i = 1, \ldots, m\}$, and in addition, the set $\mathscr{C}$ is asymptotically stable.*

## 10.5.2   Model-Based Evolutionary Optimization

From the above analysis, we know that the global optimal solutions can be obtained by generating samples from equilibrium distributions of the replicator dynamics (10.8); these equilibrium distributions can be approached by following trajectories of (10.8) starting from $\mathbb{P}_0 \in \Delta_0$. Note that by Theorem 10.5.4, the equilibrium points obtained by starting from $\mathbb{P}_0 \in \Delta_0$ are of the form $\mathbb{P}^{(x)} = \sum_{i=1}^m \alpha_i \delta(x - x_i^\star)$, where $\sum_{i=1}^m \alpha_i = 1$ and $\alpha_i \geq 0$ for $i = 1, \ldots, m$, which suggests using a sum of Dirac functions to approximate $p_t$. Assume a group of candidate solutions $\{y_t^i\}_{i=1}^N$ is generated from $p_t$; then the probability density function $p_t$ can be approximated by $\hat{p}_t(x) = \sum_{i=1}^N w_t^i \delta(x - x_t^i)$, where $\delta$ denotes the Dirac function and $\{w_t^i\}_{i=1}^N$ are weights satisfying $\sum_{i=1}^N w_t^i = 1$. If we use this approximation $\hat{p}_t$ as our probabilistic model and substitute it into (10.9), we have

$$\frac{\partial w_t^i}{\partial t} = \left( \phi(H(x_t^i)) - \sum_{j=1}^N w_t^j \phi(H(x_t^j)) \right) w_t^i, \quad \forall i = 1, \ldots, N. \tag{10.10}$$

The corresponding discrete-time version of (10.10) is

$$w_{k+1}^i = \frac{\phi(H(x_k^i))}{\sum_{j=1}^N w_k^j \phi(H(x_k^j))} w_k^i, \quad \forall i = 1, \ldots, N. \tag{10.11}$$

We can let $\phi(\cdot)$ be an exponential function so that the denominator of the right-hand side of (10.11) is not equal to zero. Although an updated density approximation $\hat{p}_{k+1}(x) = \sum_{i=1}^N w_{k+1}^i \delta(x - x_k^i)$ is obtained, it cannot be used directly to generate new candidate solutions. We construct a new continuous density to approximate $\hat{p}_{k+1}$, which is done by projecting $\hat{p}_{k+1}$ onto some parameterized family of distributions $g_\theta$. The idea of projection onto a parameterized family has also been used in CE and MRAS, as discussed above. Specifically, we minimize the KL divergence between the parameterized distribution $g_\theta$ and $\hat{p}_{k+1}$:

$$\theta_{k+1} = \arg\min_{\theta\in\Theta} \mathscr{D}(\hat{p}_{k+1}, g_\theta), \qquad (10.12)$$

where $\Theta$ is the domain of $\theta$. After some algebraic operations, we can show that solving (10.12) is equivalent to: $\max_{\theta\in\Theta} \sum_{i=1}^{N} w_{k+1}^i \ln g_\theta(y_k^i)$.

All the above analysis is carried out when replicator dynamics, e.g., (10.8) and (10.9), are used. There are some other dynamics in evolutionary game theory such as imitation dynamics, logit dynamics, and Brown-von Neumann-Nash dynamics that can be used to update the weights $\{w_k^i\}$. To present the algorithm in a more general setting, the updating of weights is denoted as

$$w_k^i = D_d\left(\phi(H(x_{k-1}^i))I_{\{H(x_{k-1}^i)\geq\gamma_{k-1}\}}, \sum_{j=1}^{N} w_{k-1}^j \phi(H(x_{k-1}^j))I_{\{H(x_{k-1}^j)\geq\gamma_{k-1}\}}, w_{k-1}^i\right),$$
$$(10.13)$$

where $\gamma_{k-1}$ is a constant that is used to select good candidate solutions; $D_d$ is a function of three variables, which is used to represent the updating rule. For example, when $D_d$ is derived from replicator dynamics, we have

$$w_k^i = \frac{\frac{1}{N}\phi(H(x_{k-1}^i))I_{\{H(x_{k-1}^i)\geq\gamma_{k-1}\}}}{\sum_{j=1}^{N}\frac{1}{N}\phi(H(x_{k-1}^j))I_{\{H(x_{k-1}^j)\geq\gamma_{k-1}\}}} w_{k-1}^i, \quad \forall i=1,\ldots,N.$$

Based on the above analysis, a Monte Carlo simulation version of the MEO algorithm is given as follows.

### Model-Based Evolutionary Optimization Algorithm (MEO)

0. Initialization. Specify $N$ as the total number of candidate solutions generated at each iteration. Choose $\rho\in(0,1]$ and an initial $g_{\theta_0}$ defined on $\mathbf{X}$. Set $k=0$, $w_0^i = 1/N$ for $i=1,\ldots,N$, and $\gamma_0 = -\infty$.
1. Quantile calculation. Generate $N$ candidate solutions $\{x_k^i\}_{i=1}^N$ from $g_{\theta_k}$. Calculate the $1-\rho$ quantile $\gamma_k$ of $\{x_k^i\}_{i=1}^N$. If $\gamma_k < \gamma_{k-1}$ and $k>1$, set $\gamma_k = \gamma_{k-1}$ and $w_{k-1}^i = 1/N$ for $i=1,\ldots,N$. Set $k=k+1$ and go to Step 2.
2. Updating the probabilistic model. The discrete approximation of the model is $\hat{p}_k(x) = \sum_{i=1}^N w_k^i \delta(x-x_{k-1}^i)$, where $\{w_k^i\}$ are updated according to (10.13).
3. Density projection. Construct $g_\theta$ by projecting the density $\hat{p}_k = \sum_{i=1}^N w_k^i \delta(x-x_{k-1}^i)$ onto $g_\theta$: $\theta_k = \arg\max_{\theta\in\Theta}\sum_{i=1}^N w_k^i \ln g_\theta(x_{k-1}^i)$.
4. Stop if some stopping criterion is satisfied; otherwise go to Step 1.

Generally, it is not easy to solve the optimization problem (10.12), which depends on the choice of $g_\theta$. However, for $g_\theta$ in an exponential family, analytical solutions can be obtained. A comprehensive exposition of the evolutionary games approach is given in [37, 38].

## 10.6   Stochastic Approximation Approach

In this section, we present a stochastic approximation framework to study model-based algorithms [21]. The framework is based on the MRAS method presented in Sect. 10.3 and is intended to combine the robust features of model-based algorithms encountered in practice with rigorous convergence guarantees. Specifically, by exploiting a natural connection between model-based algorithms and the well-known stochastic approximation (SA) method [2,4,26,30], we show that, regardless of the type of decision variables involved in (10.1), algorithms conforming to the framework can be equivalently formulated in the form of a generalized stochastic approximation procedure on a transformed continuous parameter space for solving a sequence of stochastic optimization problems with differentiable structures. This viewpoint, which is new to this type of random search algorithms, allows us to study the asymptotic convergence and rate properties of these algorithms by using existing theory and tools from SA.

The key idea that leads to the proposed framework is based on replacing the reference sequence $\{g_k\}$ in the original MRAS method by a more general distribution sequence in the recursive form:

$$\hat{g}_{k+1}(x) = \alpha_k g_{k+1}(x) + (1 - \alpha_k) f_{\theta_k}(x), \ \alpha_k \in (0,1) \ \forall k, \tag{10.14}$$

which is a mixture of the reference distribution $g_{k+1}$ and the sampling distribution $f_{\theta_k}$ obtained at the $k$th iteration. Such a mixture $\hat{g}_{k+1}$ retains the properties of $g_{k+1}$ while, on the other hand, ensures that its difference from $f_{\theta_k}$ is only incremental. Thus, the intuition is that if one were to replace $g_{k+1}$ with $\hat{g}_{k+1}$ in minimizing the KL divergence $\mathscr{D}(\hat{g}_{k+1}, f_\theta)$, then the new sampling distribution $f_{\theta_{k+1}}$ obtained would also stay close to the current sampling distribution $f_{\theta_k}$.

When $\{\hat{g}_k\}$ instead of $\{g_k\}$ is used at step 2 of MRAS to minimize the KL divergence, the following lemma reveals a key link between the two successive mean vector functions of the projected probability distributions [21].

**Lemma 10.6.1.** *If $f_\theta$ belongs to NEFs and the new parameter $\theta_{k+1}$ obtained via minimizing $\mathscr{D}(\hat{g}_{k+1}, f_\theta)$ is an interior point of the parameter space $\Theta$ for all $k$, then*

$$m(\theta_{k+1}) - m(\theta_k) = -\alpha_k \nabla_\theta \mathscr{D}(g_{k+1}, f_\theta)|_{\theta=\theta_k}. \tag{10.15}$$

Basically, Lemma 10.6.1 states that regardless of the specific form of $g_k$, the mean vector function $m(\theta_k)$ (i.e., a one-to-one transformation of $\theta_k$) is updated at each step along the gradient descent direction of the *time-varying* objective function for the minimization problem $\min_\theta \mathscr{D}(g_{k+1}, f_\theta)$. In particular, in the case of the CE method, i.e., when $g_{k+1}$ in (10.15) takes the form $g_{k+1}(x) = \frac{H(x)f_{\theta_k}(x)}{\int_{\mathbf{X}} H(x)f_{\theta_k}(dx)}$ (cf. (10.4)), it can be seen that recursion (10.15) becomes

$$m(\theta_{k+1}) - m(\theta_k) = \alpha_k \nabla_\theta \ln E_\theta[H(X)]|_{\theta=\theta_k}. \qquad (10.16)$$

Hence, $m(\theta_k)$ is updated along the gradient direction of the objective function for the maximization problem $\max_\theta \ln E_\theta[H(X)]$, the optimal solution to which is a sampling distribution $f_{\theta^*}$ that assigns maximum probability to the set of optimal solutions of (10.1). Note that the parameter sequence $\{\alpha_k\}$ turns out to be the gain sequence for the gradient iteration, so that the special case $\alpha_k \equiv 1$ corresponds to the original MRAS method. This suggests that all model-based algorithms that fall under the MRAS framework can be equivalently viewed as gradient-based recursions on the parameter space $\Theta$ for solving a sequence of optimization problems with differentiable structures. This new interpretation of model-based algorithms provides a key insight to understand how these algorithms address hard optimization problems with little structure.

In actual implementation, when integrals/expectations are replaced by sample averages based on Monte Carlo sampling, (10.15) and (10.16) become recursive algorithms of stochastic approximation type with direct gradient estimation. Thus, it is clear that the rich body of tools and results from stochastic approximation can be incorporated into the framework to analyze model-based algorithms.

### 10.6.1 Convergence of the CE Method

The convergence of the CE algorithm has recently been studied in [19,21] by casting a Monte Carlo version of recursion (10.16) in the form of a generalized Robbins-Monro algorithm in terms of the true gradient, bias, and an error term due to random sampling and then following the arguments of the ordinary differential equation (ODE) approach [2,4]. The main convergence results are summarized below, where for notational convenience, we define $\eta := m(\theta)$ and $\eta_k := m(\theta_k)$.

**Theorem 10.6.5.** *(Convergence of CE) Under some regularity conditions (see [21]), the sequence of iterates $\{\eta_k\}$ generated by the CE algorithm converges w.p.1 to a compact connected internally chain recurrent set of the ODE*

$$\frac{d\eta(t)}{dt} = L(\eta),\ t \geq 0, \qquad (10.17)$$

*where $L(\eta) := \nabla_\theta \ln E_\theta[H(X)]|_{\theta=m^{-1}(\eta)}$.*

Theorem 10.6.5 indicates that the long-run behavior (e.g., local/global convergence) of CE is primarily governed by the asymptotic solution of an underlying ODE. This result formalizes our prior observation in [18], which provides counterexamples indicating that CE and its variants are in general local improvement methods. Under the more stringent assumption that the convergence of $\{\eta_k\}$ occurs to a unique limiting point $\eta^*$, the following asymptotic normality result was obtained in [21].

**Theorem 10.6.6.** *(Asymptotic normality of CE) Under some appropriate conditions (see Theorem 4.1 of [21]),*

$$k^{\frac{\tau}{2}}(\eta_k - \eta^*) \xrightarrow{dist} \mathbf{N}(0, \Sigma) \ as \ k \to \infty,$$

*where $\tau \in (0,1)$ is some appropriate constant and $\Sigma$ is a positive definite covariance matrix.*

## 10.6.2  Model-Based Annealing Random Search

To further illustrate the stochastic approximation approach, we present an algorithm instantiation of the framework called model-based annealing random search (MARS) [20]. MARS can essentially be viewed as an implementable version of the annealing adaptive search (AAS) algorithm, in that it provides an alternative approach to address the implementation difficulty of AAS (cf. Sect. 10.2). The basic idea is to use a sequence of NEF distributions to approximate the target Boltzmann distributions and then use the sequence as surrogate distributions to generate candidate points. Thus, by treating Boltzmann distributions as reference distributions, candidate solutions are drawn at each iteration of MARS *indirectly* from a Boltzmann distribution by sampling exactly from its approximation. This is in contrast to Markov chain-based techniques [41] that aim to *directly* sample from the Boltzmann distributions.

The MARS algorithm is conceptually very simple and is summarized below:

0. Choose a parameterized family $\{f_\theta\}$, an annealing schedule used in the Boltzmann distribution, and a gain sequence $\{\alpha_k\}$.
1. Given $\theta_k$, sample $N$ candidate solutions $X_k^1, \ldots, X_k^N$ from $f_{\theta_k}$.
2. Update the parameter $\theta_{k+1} = \arg_\theta \min \mathscr{D}(\tilde{g}_{k+1}, f_\theta)$; increase $k$ by 1 and reiterate from step 1.

At Step 2 of MARS, the reference distribution is given by $\tilde{g}_{k+1}(x) = \alpha_k \bar{g}_{k+1}(x) + (1 - \alpha_k) f_{\theta_k}(x)$, where $\bar{g}_{k+1}$ is an empirical estimate of the true Boltzmann distribution $g_{k+1}(x) := \frac{e^{H(x)/T_k}}{\int_{\mathbf{x}} e^{H(x)/T_k} dx}$ based on the sampled solutions $X_k^1, \ldots, X_k^N$, and $\{T_k\}$ is a sequence of decreasing temperatures that controls how fast the sequence of Boltzmann distributions will degenerate.

Under its equivalent gradient interpretation, Lemma 10.6.1 shows that the mean vector function $m(\theta_{k+1})$ of the new distribution $f_{\theta_{k+1}}$ obtained at step 2 of MARS can be viewed as an iterate generated by a gradient descent algorithm for solving the iteration-varying minimization problem $\min_\theta \mathscr{D}(\bar{g}_{k+1}, f_\theta)$ on the parameter space $\Theta$, i.e.,

$$m(\theta_{k+1}) - m(\theta_k) = -\alpha_k \nabla_\theta \mathscr{D}(\bar{g}_{k+1}, f_\theta)|_{\theta=\theta_k}. \tag{10.18}$$

Note that since the reference distribution $\bar{g}_{k+1}$ may change shape with $k$, a primary difference between MARS and CE is that the gradient in (10.18) is time-varying vs. stationary in (10.16). Stationarity in general only guarantees local convergence, whereas the time-varying feature of MARS provides a viable way to ensure that the algorithm escapes from local optima, leading to global convergence. By the properties of NEFs, recursion (10.18) can be further written as

$$m(\theta_{k+1}) - m(\theta_k) = -\alpha_k \left( m(\theta_k) - E_{g_{k+1}}[\Gamma(X)] + E_{g_{k+1}}[\Gamma(X)] - E_{\bar{g}_{k+1}}[\Gamma(X)] \right)$$
$$= -\alpha_k \nabla_\theta \mathscr{D}(g_{k+1}, f_\theta)|_{\theta=\theta_k} - \alpha_k \left( E_{g_{k+1}}[\Gamma(X)] - E_{\bar{g}_{k+1}}[\Gamma(X)] \right).$$

This becomes a Robbins-Monro-type stochastic approximation algorithm in terms of the true gradient and a noise term due to the approximation error between $g_{k+1}$ and $\bar{g}_{k+1}$. Thus, in light of the existing theories from stochastic approximation, the convergence analysis of MARS essentially boils down to the issue of inspecting whether the Boltzmann distribution $g_{k+1}$ can be closely approximated by its empirical estimate $\bar{g}_{k+1}$. The following results are obtained in [20].

**Theorem 10.6.7.** *(Global convergence of MARS) Under some appropriate conditions (see Theorem 3.1 of [20]),*

$$\lim_{k\to\infty} m(\theta_k) = \Gamma(x^*) \ w.p.1.$$

**Theorem 10.6.8.** *(Asymptotic normality of MARS) Let $\alpha_k = a/k^\alpha$ and the sample size be polynomially increasing $N_k = ck^\beta$ for constants $a > 0$, $c > 0$, $\alpha \in (\frac{1}{2}, 1)$, and $\beta > \alpha$. Under some additional conditions on $\{T_k\}$,*

$$k^{\frac{\alpha+\beta}{2}} \left( m(\theta_k) - \Gamma(x^*) \right) \xrightarrow{\text{dist}} \mathbf{N}(0, \Sigma) \ as \ k \to \infty,$$

*where $\Sigma$ is some positive definite covariance matrix.*

Numerical results on high-dimensional multi-extremal benchmark problems reported in [20] show that MARS may yield high-quality solutions within a modest number of function evaluations and provide superior performance over some of the existing algorithms.

## 10.7   Conclusions

We reviewed several recent contributions to model-based methods for global optimization, including algorithms and convergence results for model reference adaptive search, the particle-filtering approach, the evolutionary games approach, and the stochastic approximation gradient approach. These approaches analyze model-based methods from different perspectives, providing useful tools to explore properties of the updating mechanism of probabilistic models and to facilitate proofs of convergence of model-based algorithms.

**Acknowledgements** This work was supported in part by the National Science Foundation (NSF) under Grants CNS-0926194, CMMI-0856256, CMMI-0900332, CMMI-1130273, CMMI-1130761, EECS-0901543, and by the Air Force Office of Scientific Research (AFOSR) under Grant FA9550-10-1-0340.

## References

1. Arulampalam, S., Maskell, S., Gordon, N.J., Clapp, T.: A tutorial on particle filters for on-line non-linear/non-Gaussian Bayesian tracking. IEEE Transactions on Signal Processing **50**(2), 174–188 (2002)
2. Benaim, M.: A dynamical system approach to stochastic approximations. SIAM Journal on Control and Optimization **34**, 437–472 (1996)
3. Billingsley, P.: Convergence of Probability Measures. John Wiley & Sons, Inc., New York (1999)
4. Borkar, V.: Stochastic approximation: a dynamical systems viewpoint. Cambridge University Press; New Delhi: Hindustan Book Agency (2008)
5. Budhiraja, A., Chen, L., Lee, C.: A survey of numerical methods for nonlinear filtering problems. Physica D: Nonlinear Phenomena **230**, 27–36 (2007)
6. Chopin, N.: Central limit theorem for sequential Monte Carlo and its applications to Bayesian inference. The Annals of Statistics **32**(6), 2385–2411 (2004)
7. Costa, A., Jones, O., Kroese, D.: Convergence properties of the cross-entropy method for discrete optimization. Operations Research Letters (35), 573–580 (2007)
8. Davis, M.H.A., Marcus, S.I.: An introduction to nonlinear filtering. The Mathematics of Filtering and Identification and Applications. Amsterdam, The Netherlands, Reidel (1981)
9. Dorigo, M., Gambardella, L.M.: Ant colony system: A cooperative learning approach to the traveling salesman problem. IEEE Trans. on Evolutionary Computation **1**, 53–66 (1997)
10. Doucet, A., deFreitas, J.F.G., Gordon, N.J. (eds.): Sequential Monte Carlo Methods In Practice. Springer, New York (2001)
11. Doucet, A., Johansen, A.M.: A tutorial on particle filtering and smoothing: Fifteen years later. Handbook of Nonlinear Filtering. Cambridge University Press, Cambridge (2009)
12. Eiben, A., Smith, J.: Introduction to Evolutionary Computing. Natural Computing Series, Springer (2003)
13. Gilks, W., Berzuini, C.: Following a moving target - Monte Carlo inference for dynamic Bayesian models. Journal of the Royal Statistical Society **63**(1), 127–146 (2001)
14. Gland, F.L., Oudjane, N.: Stability and uniform approximation of nonlinear filters using the Hilbert metric and application to particle filter. The Annals of Applied Probability **14**(1), 144–187 (2004)
15. Glover, F.W.: Tabu search: A tutorial. Interfaces **20**, 74–94 (1990)

16. Goldberg, D.E.: Genetic Algorithms in Search, Optimization, and Machine Learning. Addison-Wesley, Boston, MA (1989)
17. Homem-De-Mello, T.: A study on the cross-entropy method for rare-event probability estimation. INFORMS Journal on Computing **19**, 381–394 (2007)
18. Hu, J., Fu, M.C., Marcus, S.I.: A model reference adaptive search method for global optimization. Operations Research **55**(3), 549–568 (2007)
19. Hu, J., Hu, P.: On the performance of the cross-entropy method. In: Proceedings of the 2009 Winter Simulation Conference, pp. 459–468 (2009)
20. Hu, J., Hu, P.: Annealing adaptive search, cross-entropy, and stochastic approximation in global optimization. Naval Research Logistics **58**, 457–477 (2011)
21. Hu, J., Hu, P., Chang, H.: A stochastic approximation framework for a class of randomized optimization algorithms. IEEE Transactions on Automatic Control (forthcoming) (2012)
22. Jacobson, S., Sullivan, K., Johnson, A.: Discrete manufacturing process design optimization using computer simulation and generalized hill climbing algorithms. Engineering Optimization **31**, 247–260 (1998)
23. Johnson, A., Jacobson, S.: A class of convergent generalized hill climbing algorithms. Applied Mathematics and Computation **125**, 359–373 (2001)
24. Kennedy, J., Eberhart, R.: Particle swarm optimization. In Proceedings of IEEE International Conference on Neural Networks. IEEE Press, Piscataway, NJ, pp. 1942–1948 (1995)
25. Kirkpatrick, S., Gelatt, C.D., Vecchi, M.P.: Optimization by simulated annealing. Science **220**, 671–680 (1983)
26. Kushner, H.J., Clark, D.S.: Stochastic Approximation Methods for Constrained and Unconstrained Systems and Applications. Springer-Verlag, New York (1978)
27. Larrañaga, P., Lozano, J. (eds.): Estimation of Distribution Algorithms: A New Tool for Evolutionary Computation. Kluwer Academic Publisher, Boston, MA (2002)
28. Musso, C., Oudjane, N., Gland, F.L.: Sequential Monte Carlo Methods in Practice. Springer-Verlag, New York (2001)
29. Oechssler, J., Riedel, F.: On the dynamics foundation of evolutionary stability in continuous models. Journal of Economic Theory **107**, 223–252 (2002)
30. Robbins, H., Monro, S.: A stochastic approximation method. Annals of Mathematical Statistics **22**, 400–407 (1951)
31. Romeijn, H., Smith, R.: Simulated annealing and adaptive search in global optimization. Probability in the Engineering and Informational Sciences **8**, 571–590 (1994)
32. Rubinstein, R.Y.: Optimization of computer simulation models with rare events. European Journal of Operations Research **99**, 89–112 (1997)
33. Rubinstein, R.Y.: The cross-entropy method for combinatorial and continuous optimization. Methodology and Computing in Applied Probability **2**, 127–190 (1999)
34. Rubinstein, R.Y., Kroese, D.P.: The Cross-Entropy Method: A Unified Approach to Combinatorial Optimization, Monte-Carlo Simulation, and Machine Learning. Springer, New York, NY (2004)
35. Shi, L., Olafsson, S.: Nested partitions method for global optimization. Operations Research **48**(3), 390–407 (2000)
36. Srinivas, M., Patnaik, L.M.: Genetic algorithms: A survey. IEEE Computer **27**, 17–26 (1994)
37. Wang Y., Fu, M.C., Marcus, S.I.: Model-based evolutionary optimization. In Proceedings of the 2010 Winter Simulation Conference. IEEE Press, Piscataway, NJ, pp. 1199–1210 (2010)
38. Wang, Y., Fu, M.C., Marcus, S.I.: An evolutionary game approach for model-based optimization. Working paper (2011)
39. Wolpert, D.H. Finding bounded rational equilibria part i: Iterative focusing. In Proceedings of the Eleventh International Symposium on Dynamic Games and Applications, T. Vincent (Editor), Tucson AZ, USA (2004)
40. Zabinsky, Z., Smith, R., McDonald, J., Romeijn, H., Kaufman, D.: Improving hit-and-run for global optimization. Journal of Global Optimization **3**, 171–192 (1993)
41. Zabinsky, Z.B.: Stochastic Adaptive Search for Global Optimization. Kluwer, The Netherlands (2003)

42. Zhang, Q., Mühlenbein, H.: On the convergence of a class of estimation of distribution algorithm. IEEE Trans. on Evolutionary Computation **8**, 127–136 (2004)
43. Zhou, E., Fu, M.C., Marcus, S.I.: A particle filtering framework for randomized optimization algorithms. In Proceedings of the 2008 Winter Simulation Conference. IEEE Press, Piscataway, NJ, pp. 647–654 (2008)
44. Zhou, E., Fu, M.C., Marcus, S.I.: Solving continuous-state POMDPs via density projection. IEEE Transactions on Automatic Control **55**(5), 1101–1116 (2010)
45. Zhou, E., Fu, M.C., Marcus, S.I.: Particle filtering framework for a class of randomized optimization algorithms. Under review (2012)
46. Zlochin, M., Birattari, M., Meuleau, N., Dorigo, M.: Model-based search for combinatorial optimization: A critical survey. Annals of Operations Research **131**, 373–395 (2004)

# Chapter 11
# Constrained Optimality for First Passage Criteria in Semi-Markov Decision Processes

**Yonghui Huang and Xianping Guo**

## 11.1 Introduction

In the field of Markov decision problems (MDPs), the control horizon is usually a fixed finite interval $[0, T]$ or the infinite interval $[0, +\infty)$. In many real applications, however, the control horizon may be a *random duration* $[0, \tau]$, where the terminal time $\tau$ is a random variable at which the state of the controlled system changes critically and the control beyond $\tau$ may no longer be meaningful or necessary. For example, in the insurance systems [27], the control after the time when the company is bankrupt becomes unnecessary. Therefore, it makes better sense to consider the problem in $[0, \tau]$, where $\tau$ represents the bankruptcy time of the company. Such situations motivate *first passage* problems in MDPs [13,15,19,21,22,28], for which one generally aims at maximizing/minimizing the expected reward/cost over a first passage time to some target set.

This chapter is devoted to studying constrained optimality for first passage criteria, for which the dynamic of a system is described by semi-Markov decision processes (SMDPs). The state space is assumed to be denumerable, while the action set is general. Both reward and cost rates are possibly *unbounded*. A key feature of our model is that the discount rate is state-action dependent, and furthermore, the undiscounted case is allowed. This feature makes our model more general since the state-action-dependent discount rate exactly characterizes the practical cases such as the interest rate in economic and financial systems [2,9,17,23,26], which can be adjusted according to the real circumstances. We aim to maximize the expected reward obtained during a first passage time to some target set, subject to that the associated expected cost over this first passage time does not exceed a given constraint. An interesting special case is that in which the reward rates are uniformly

Y. Huang • X. Guo (✉)
Sun Yat-Sen University, Guangzhou 510275, China
e-mail: hyongh5@mail.sysu.edu.cn; mcsgxp@mail.sysu.edu.cn

D. Hernández-Hernández and A. Minjárez-Sosa (eds.), *Optimization, Control, and Applications of Stochastic Systems*, Systems & Control: Foundations & Applications, DOI 10.1007/978-0-8176-8337-5_11, © Springer Science+Business Media, LLC 2012

equal to one, which corresponds to a stochastic time optimal control problem with a target set; see Remark 11.2.4(d) for details.

Previously, Beutler and Ross [3] consider constrained SMDPs with the *long-run average* criteria. They suppose that the state space of the SMDP is finite, and the action space compact metric. A Lagrange multiplier formulation involving a dynamic programming equation is utilized to relate the constrained optimization to an unconstrained optimization parametrized by the multiplier. This approach leads to a proof for the existence of a semi-simple optimal constrained policy. That is, there is at most one state for which the action is randomized between two possibilities; at all other states, an action is uniquely chosen for each state. Feinberg [4] further investigates *constrained average reward* SMDPs with finite state and action sets. They develop a technique of state-action renewal intensities and provide linear programming algorithms for the computation of optimal policies. On the other hand, Feinberg [5] deals with constrained *infinite horizon discounted* SMDPs. Compared with the existing works above, however, our main interest in this chapter is to analyze the constrained optimality for *first passage* criteria in SMDPs, which, to best of our knowledge, is an issue not yet explored.

To obtain the existence of a constrained *first passage* optimal policy, we first give suitable conditions and then employ the so-called Lagrange multiplier technique to analyze the constrained control problem. Based on the Lagrange multiplier technique, we transform the constrained control problem to an unconstrained one, prove that a constrained optimal policy exists, and show that the constrained optimal policy randomizes between two stationary policies differing in at most one state.

The rest of this chapter is organized as follows. In Sect. 11.2, we formulate the control model, followed by the optimality conditions and the main results on the existence of constrained optimal policies. In Sect. 11.3, some technique preliminaries are given, and the proof of the main result is presented in Sect. 11.4.

## 11.2   The Control Model

The model of constrained SMDPs considered in this chapter is specified by the eight objects

$$\{E, B, (A(i) \subset A, i \in E), Q(\cdot, \cdot \mid i, a), r(i, a), c(i, a), \alpha(i, a), \gamma\}, \tag{11.1}$$

where $E$ is the *state space*, a denumerable set; $B \subset E$ is the given *target set*, such as the set of all bad states or of good states of a system; $A$ is the *action space*, a Borel space endowed with the Borel $\sigma$-field $\mathscr{A}$; and $A(i) \in \mathscr{A}$ is the set of *admissible actions* at state $i \in E$. The transition mechanism of the SMDPs is defined by the *semi-Markov kernel* $Q(\cdot, \cdot \mid i, a)$ on $R_+ \times E$ given $K$, where $R_+ = [0, +\infty)$, and $K = \{(i, a) \mid i \in E, a \in A(i)\}$ is the set of feasible state-action pairs. It is assumed that (1) $Q(\cdot, j \mid i, a)$ (for any fixed $j \in E$ and $(i, a) \in K$) is a nondecreasing, right continuous real function on $R_+$ such that $Q(0, j \mid i, a) = 0$; (2) $Q(t, \cdot \mid \cdot, \cdot)$ (for each fixed $t \in R_+$) is a sub-stochastic kernel on $E$ given $K$; and (3) $P(\cdot \mid \cdot, \cdot) := Q(\infty, \cdot \mid \cdot, \cdot)$ is a stochastic kernel on $E$ given $K$. If action $a \in A(i)$ is selected in state $i$, then $Q(t, j \mid i, a)$ is the joint probability that the sojourn time in state $i$ is not greater than $t \in R_+$, and the next state is $j$. Moreover, $r(i,a)$ and $c(i,a)$ in (11.1) denote the *reward* and *cost rate functions* on $K$ valued in $R = (-\infty, +\infty)$, respectively, which are both assumed to be measurable on $A(i)$ for each fixed $i \in E$. In addition, $\alpha(i,a)$ represents the *discount rate*, which is a measurable function from $K$ to $R_+$. Finally, $\gamma$ is a given *constraint constant*.

*Remark 11.2.1.* Compared with the models of the standard constrained discounted and average criteria [3–5], in this model (11.1), we introduce a target set $B \subset E$ of the controlled system, and furthermore, the discount rate $\alpha(i,a)$ here is state-action dependent and may be equal to zero (i.e., the undiscounted case is allowed).

To state the constrained SMDPs we are concerned with, we need to introduce the classes of policies. For each $n \geq 0$, let $H_n$ be the family of admissible histories up to the $n$th jump (decision epoch), that is, $H_n = (R_+ \times K)^n \times (R_+ \times E)$, for $n = 0, 1, \ldots$.

**Definition 11.2.1.** A randomized history-dependent policy, or simply a policy, is a sequence $\pi = \{\pi_n, n \geq 0\}$ of stochastic kernels $\pi_n$ on $A$ given $H_n$ satisfying

$$\pi_n(A(i_n) \mid h_n) = 1 \quad \forall\, h_n = (t_0, i_0, a_0, \ldots, t_{n-1}, i_{n-1}, a_{n-1}, t_n, i_n) \in H_n, \quad n = 0, 1, \ldots.$$

The class of all policies is denoted by $\Pi$. To distinguish the subclasses of $\Pi$, we let $\Phi$ be the family of all stochastic kernels $\varphi$ on $A$ given $E$ such that $\varphi(A(i) \mid i) = 1$ for all $i \in E$, and $\mathbb{F}$ the set of all functions $f : E \to A$ such that $f(i)$ is in $A(i)$ for every $i \in E$. A policy $\pi = \{\pi_n\} \in \Pi$ is said to be *randomized Markov* if there is a sequence $\{\varphi_n\}$ of $\varphi_n \in \Phi$ such that $\pi_n(\cdot \mid h_n) = \varphi_n(\cdot \mid i_n)$ for every $h_n \in H_n$ and $n \geq 0$. We denote such a policy by $\pi = \{\varphi_n\}$. A randomized Markov policy $\pi = \{\varphi_n\}$ is said to be *randomized stationary* if every $\varphi_n$ is independent of $n$. In this case, we write $\pi = \{\varphi, \varphi, \ldots\}$ as $\varphi$ for simplicity. Further, a randomized Markov policy $\pi = \{\varphi_n\}$ is said to be *deterministic Markov* if there is a sequence $\{f_n\}$ of $f_n \in \mathbb{F}$ such that $\varphi_n(\cdot \mid i)$ is the Dirac measure at $f_n(i)$ for all $i \in E$ and $n \geq 0$. We write such a policy as $\pi = \{f_n\}$. In particular, a deterministic Markov policy $\pi = \{f_n\}$ is said to be (deterministic) *stationary* if $f_n$ are all independent of $n$. Similarly, we write $\pi = \{f, f, \ldots\}$ as $f$ for simplicity. We denote by $\Pi_{\mathrm{RM}}, \Pi_{\mathrm{RS}}, \Pi_{\mathrm{DM}}$, and $\Pi_{\mathrm{DS}}$ the families of all randomized Markov, randomized stationary, deterministic Markov, and stationary policies, respectively. Obviously, $\Phi = \Pi_{\mathrm{RS}} \subset \Pi_{\mathrm{RM}} \subset \Pi$ and $\mathbb{F} = \Pi_{\mathrm{DS}} \subset \Pi_{\mathrm{DM}} \subset \Pi$.

Let $\mathbb{P}(E)$ denote the set of all the probability measures on $E$. For each $(s, \mu) \in R_+ \times \mathbb{P}(E)$ and $\pi \in \Pi$, by the well-known Tulcea's theorem ([10, Proposition C.10]), there exist a unique probability space $(\Omega, \mathscr{F}, P^\pi_{(s,\mu)})$ and a stochastic process $\{T_n, J_n, A_n, n \geq 0\}$ such that, for each $i, j \in E, t \in R_+, C \in \mathscr{A}$ and $n \geq 0$,

$$P^\pi_{(s,\mu)}(T_0 = s, J_0 = i) = \mu(i), \tag{11.2}$$

$$P^\pi_{(s,\mu)}(A_n \in C \mid h_n) = \pi_n(C \mid h_n), \tag{11.3}$$

$$P^\pi_{(s,\mu)}(T_{n+1} - T_n \leq t, J_{n+1} = j \mid h_n, a_n) = Q(t, j \mid i_n, a_n), \tag{11.4}$$

where $T_n, J_n$, and $A_n$ denote the $n$th decision epoch, the state, and the action chosen at the $n$th decision epoch, respectively. The expectation operator with respect to $P^\pi_{(s,\mu)}$ is denoted by $E^\pi_{(s,\mu)}$. In particular, if $\mu$ is the Dirac measure $\delta_i(\cdot)$ concentrated at some state $i \in E$, we write $P^\pi_{(s,\mu)}$ and $E^\pi_{(s,\mu)}$ as $P^\pi_{(s,i)}$ and $E^\pi_{(s,i)}$, respectively. For simplicity, $P^\pi_{(0,\mu)}$ and $E^\pi_{(0,\mu)}$ is denoted by $P^\pi_\mu$ and $E^\pi_\mu$, respectively. Without loss of generality, in the following, we always set the initial decision epoch $T_0 = 0$ and omit it.

*Remark 11.2.2.* (a) The construction of the probability measure space $(\Omega, \mathscr{F}, P^\pi_{(s,\mu)})$ and the above properties (11.2)–(11.4) follow from those in Limnios and Oprisan [18, p.33] and Puterman [24, p.534–535].

(b) Let $X_0 := 0$, $X_n := T_n - T_{n-1}$ $(n \geq 0)$ denote the sojourn times between decision epochs (jumps). Then, the stochastic process $\{T_n, J_n, A_n, n \geq 0\}$ may be rewritten as the one $\{X_n, J_n, A_n, n \geq 0\}$.

To avoid the possibility of an infinite number of decision epochs within finite time, we make the following assumption that the system does not have accumulation points.

**Assumption 11.2.1** *For all $\mu \in \mathbb{P}(E)$ and $\pi \in \Pi$, $P^\pi_\mu(\{\lim_{n\to\infty} T_n = \infty\}) = 1$.*

To verify Assumption 11.2.1, we can use a sufficient condition below.

**Condition 11.2.2** *There exist constants $\delta > 0$ and $\varepsilon > 0$ such that*

$$Q(\delta, E \mid i, a) \leq 1 - \varepsilon \quad \forall (i,a) \in K.$$

*Remark 11.2.3.* In fact, Condition 11.2.2 is the standard regular condition widely used in SMDPs [5, 16, 20, 24, 25], which exactly implies Assumption 11.2.1 above.

Under Assumption 11.2.1, we define an underlying continuous-time state-action process $\{Z(t), W(t), t \in R_+\}$ corresponding to the stochastic process $\{T_n, J_n, A_n\}$ by

$$Z(t) = J_n, \quad W(t) = A_n, \text{ for } T_n \leq t < T_{n+1}, \quad t \in R_+ \text{ and } n \geq 0.$$

**Definition 11.2.2.** The stochastic process $\{Z(t), W(t)\}$ is called a (continuous-time) SMDP.

For the target set $B \subset E$, we consider the random variable

$$\tau_B := \inf\{t \geq 0 \mid Z(t) \in B\} \quad (\text{with } \inf\emptyset := \infty),$$

which is the first passage time into the set $B$ of the process $\{Z(t), t \in R_+\}$. Now, fix an initial distribution $\mu \in \mathbb{P}(E)$. For each $\pi \in \Pi$, the expected first passage reward and cost criteria are defined as follows:

$$V_r(\mu, \pi) := E^\pi_\mu \left[ \int_0^{\tau_B} e^{-\int_0^t \alpha(Z(u), W(u)) du} r(Z(t), W(t)) dt \right], \tag{11.5}$$

$$V_c(\mu, \pi) := E^\pi_\mu \left[ \int_0^{\tau_B} e^{-\int_0^t \alpha(Z(u), W(u)) du} c(Z(t), W(t)) dt \right]. \tag{11.6}$$

To introduce the constrained problem, for the constraint constant $\gamma$ in (11.1), let

$$U := \{\pi \in \Pi \mid V_c(\mu, \pi) \leq \gamma\}$$

be the set of "constrained" policies. We assume that $U \neq \emptyset$ throughout the following. Then, the optimization problem we are interested in is to maximize the expected first passage reward $V_r(\mu, \pi)$ over the set $U$, and our goal is to find a *constrained optimal* policy as defined below.

**Definition 11.2.3.** A policy $\pi^* \in U$ is called constrained optimal if

$$V_r(\mu, \pi^*) = \sup_{\pi \in U} V_r(\mu, \pi).$$

*Remark 11.2.4.* (a) It is worthwhile to point out that the expected first passage reward criterion $V_r(\mu, \pi)$ defined in (11.5) is different from the usual discounted reward criteria [11, 12, 24] and the total reward criteria without discount [6, 11, 24]. In fact, the former concerns with the performance of the system during a first passage time to some target set, while the latter concern with the performance of the system over an infinite horizon. However, if the target set $B = \emptyset$ (and thus $\tau_B \equiv \infty$) and, furthermore, the discount factor $\alpha(i,a)$ is state-action independent (say, $\alpha(i,a) \equiv \alpha$), then the expected first passage reward criterion (11.5) above will be directly reduced to the standard infinite horizon expected discounted reward criteria or expected total reward criteria [6, 11, 12, 14, 24].

(b) Note that the case without discount, that is, $\alpha(i,a) \equiv 0$, is allowed in the context of this chapter; see Remark 11.2.5 for further details.

(c) When the constraint constant $\gamma$ in (11.1) is sufficiently large so that $U = \Pi$, then the constrained first passage optimization problem (recall Definition 11.2.3) is reduced to the usual unconstrained first passage optimization problems [13, 15, 19, 21, 22, 28].

(d) In real situations, the target set $B$ usually represents the set of failure states of a system, and thus $\tau_B$ denotes the working life (functioning life) of the system. Therefore, our aim is to maximize the expected rewards $V_r(\mu, \pi)$ obtained before the system fails, subject to the associated costs $V_c(\mu, \pi)$ incurred before the failure of the system is not more than some constraint constant $\gamma$. In particular, if the reward function rate $r(i,a) \equiv 1$ and the discount factor $\alpha(i,a) \equiv 0$, our aim is then reduced to maximizing the expected working life of the system, subject to the associated costs $V_c(\mu, \pi)$ incurred before the failure of the system are not more than some constraint constant $\gamma$.

To obtain the existence of a constrained optimal policy, we need several sets of conditions.

**Assumption 11.2.3** *There exist constants $M > 0$, $0 < \beta < 1$, and a weight function $w \geq 1$ on $E$ such that for every $i \in B^c := E - B$,*

*(a)* $\sup_{a \in A(i)} |\tilde{r}(i,a)| \leq M w(i)$, *and* $\sup_{a \in A(i)} |\tilde{c}(i,a)| \leq M w(i)$ *for all* $a \in A(i)$, *where*

$$\widetilde{r}(i,a) : = r(i,a) \int_0^\infty \mathrm{e}^{-\alpha(i,a)t}(1 - D(t \mid i,a))\mathrm{d}t,$$

$$\widetilde{c}(i,a) : = c(i,a) \int_0^\infty \mathrm{e}^{-\alpha(i,a)t}(1 - D(t \mid i,a))\mathrm{d}t, \ \ and$$

$$D(t \mid i,a) : = Q(t, E \mid i,a).$$

(b) $\sup_{a\in A(i)} \sum_{j\in B^c} w(j)m(j \mid i,a) \le \beta w(i)$, where $m(j \mid i,a) := \int_0^\infty \mathrm{e}^{-\alpha(i,a)t} Q(\mathrm{d}t, j \mid i,a)$.

*Remark 11.2.5.* (a) In fact, Assumption 11.2.3 is a condition that ensures the first passage criteria (11.5) and (11.6) to be finite and the dynamic programming operators to be contracting; see Lemmas 11.3.1–11.3.2 below.

(b) Assumption 11.2.3(a) shows that the cost function is allowed to have neither upper nor lower bounds, while the ones in the existing works [3–5, 7, 8, 12] for the standard constrained expected discount criteria are assumed to be bounded or nonnegative (bounded below).

(c) Note that the case without discount, that is, "$\alpha(i,a) \equiv 0$", is allowed in Assumption 11.2.3. In this case, Assumption 11.2.3(b) is reduced to that there exists a constant $0 < \beta < 1$ such that

$$\sup_{a\in A(i)} \sum_{j\in B^c} w(j)P(j \mid i,a) \le \beta w(i) \ \forall \, i \in B^c \ (\text{with } P(j \mid i,a) := Q(\infty, j \mid i,a)),$$

(11.7)

which can be still verified. This fact is due to that the restrictions in Assumption 11.2.3(b) are imposed on the data of the set $B^c$ rather than the entire space $E$. However, if the restrictions in Assumption 11.2.3(b) are imposed on the data of the entire space $E$, that is, there exists a constant $0 < \beta < 1$ such that

$$\sup_{a\in A(i)} \sum_{j\in E} w(j)P(j \mid i,a) \le \beta w(i) \ \ \forall \, i \in E,$$

(11.8)

then (11.8) fails to hold itself. Indeed, by taking $\inf_i w(i)$ in the two sides of (11.8), we can conclude from (11.8) that "$\beta \ge 1$", which leads to a contradiction with "$0 < \beta < 1$".

**Assumption 11.2.4** (a) *For each $i \in B^c$, $A(i)$ is compact.*

(b) *The functions $\widetilde{r}(i,a)$, $\widetilde{c}(i,a)$, and $m(j \mid i,a)$ defined in Assumption 11.2.3 are continuous in $a \in A(i)$ for each fixed $i, j \in B^c$, respectively.*

(c) *The function $\sum_{j\in B^c} w(j)m(j \mid i,a)$ is continuous in $a \in A(i)$, with $w$ as in Assumption 11.2.3.*

*Remark 11.2.6.* Assumption 11.2.4 is the compactness-continuity conditions for the first passage criteria, which is similar to the standard compactness-continuity conditions for discount and average criteria; see, for instance, Beutler and Ross [3], Guo and Hernández-Lerma [7, 8]. The difference between them lies in that the former only imposes restrictions on the data of the set $B^c$, while the latter focus on the data of the entire space $E$.

**Assumption 11.2.5** *(a)* $\sum_{j \in B^c} w(j)\mu(j) < \infty$.
*(b)* $U_0 := \{\pi \in \Pi \mid V_c(\mu, \pi) < \gamma\} \neq \emptyset$.

*Remark 11.2.7.* (a) Assumption 11.2.5(a) is a condition on the "tails" of the initial distribution $\mu$, whereas Assumption 11.2.5(b) is a Slater-like hypothesis, typical for constrained problems; see, for instance, Beutler and Ross [3], Guo and Hernández-Lerma [7, 8], and Zhang and Guo [29].

(b) It should be noted that the conditions in Assumptions 11.2.3–11.2.5 are all imposed on the data of the set $B^c$ rather than the entire space $E$ and thus can be fulfilled in greater generality.

Our main result is stated as following.

**Theorem 11.2.1.** *Suppose that Assumptions 11.2.1–11.2.5 hold. Then there exists a constrained optimal policy that may be a stationary policy or a randomized stationary policy which differ in at most one state; that is, there exist two stationary policies $f^1, f^2$, a state $i^* \in B^c$, and a number $p \in [0, 1]$ such that $f^1(i) = f^2(i)$ for all $i \neq i^*$ and, in addition, the randomized stationary policy $\varphi^p(\cdot \mid i)$ is constrained optimal, where*

$$\varphi^p(a \mid i) = \begin{cases} p, & \text{if } a = f^1(i^*), \\ 1 - p, & \text{if } a = f^2(i^*), \\ 1, & \text{if } a = f^1(i) = f^2(i), i \neq i^*. \end{cases} \tag{11.9}$$

*Proof.* See Sect. 11.4. □

## 11.3 Technical Preliminaries

This section provides some technical preliminaries necessary for the proof of Theorem 11.2.1 in Sect. 11.4.

To begin with, we define the $w$-norm for every real-valued function $u$ on $E$ by

$$\|u\|_w := \sup_{i \in E} |u(i)|/w(i),$$

where $w$ is the so-called weight function on $E$ as in Assumption 11.2.3. Let

$$\mathbb{B}_w(E) := \{u : \|u\|_w < \infty\}$$

be the space of $w$-bounded functions on $E$.

**Lemma 11.3.1.** *Suppose that Assumptions 11.2.1 and 11.2.3 hold. Then:*

*(a) For each $i \in E$ and $\pi \in \Pi$,*

$$|V_r(i, \pi)| \leq Mw(i)/(1 - \beta), \quad |V_c(i, \pi)| \leq Mw(i)/(1 - \beta).$$

*Hence, $V_r(\cdot, \pi) \in \mathbb{B}_w(E)$, and $V_c(\cdot, \pi) \in \mathbb{B}_w(E)$.*

(b) *For all $i \in E$, $\pi \in \Pi$, and $u \in \mathbb{B}_w(E)$,*

$$\lim_{n \to \infty} E_i^\pi \left[ \prod_{k=0}^{n-1} e^{-\alpha(J_k, A_k)X_{k+1}} \mathbb{1}_{\{J_0 \in B^c, \dots, J_n \in B^c\}} u(J_n) \right] = 0,$$

*where $\mathbb{1}_D$ is the indicator function on a set $D$.*

*Proof.* (a) By the definition of $V_r(i, \pi)$, we see that $V_r(i, \pi)$ can be expressed as below:

$V_r(i, \pi)$

$$= E_i^\pi \left[ \int_0^{\tau_B} e^{-\int_0^t \alpha(Z(u), W(u))du} r(Z(t), W(t))dt \right]$$

$$= E_i^\pi \left[ \int_0^\infty e^{-\int_0^t \alpha(Z(u), W(u))du} \mathbb{1}_{\{\tau_B > t\}} r(Z(t), W(t))dt \right]$$

$$= E_i^\pi \left[ \sum_{n=0}^\infty \int_{T_n}^{T_{n+1}} e^{-\int_0^t \alpha(Z(u), W(u))du} dt\, \mathbb{1}_{\{J_0 \in B^c, \dots, J_n \in B^c\}} r(J_n, A_n) \right]$$

$$= E_i^\pi \left[ \sum_{n=0}^\infty \int_0^{X_{n+1}} e^{-\int_0^{T_n + t} \alpha(Z(u), W(u))du} dt\, \mathbb{1}_{\{J_0 \in B^c, \dots, J_n \in B^c\}} r(J_n, A_n) \right]$$

$$= E_i^\pi \left[ \sum_{n=0}^\infty e^{-\int_0^{T_n} \alpha(Z(u), W(u))du} \mathbb{1}_{\{J_0 \in B^c, \dots, J_n \in B^c\}} r(J_n, A_n) \right.$$
$$\left. \int_0^{X_{n+1}} e^{-\int_{T_n}^{T_n + t} \alpha(Z(u), W(u))du} dt \right]$$

$$= E_i^\pi \left[ \sum_{n=0}^\infty \prod_{k=0}^{n-1} e^{-\alpha(J_k, A_k)X_{k+1}} \mathbb{1}_{\{J_0 \in B^c, \dots, J_n \in B^c\}} r(J_n, A_n) \int_0^{X_{n+1}} e^{-\alpha(J_n, A_n)t} dt \right]$$

$$= E_i^\pi \left[ \sum_{n=0}^\infty E_i^\pi \left[ \prod_{k=0}^{n-1} e^{-\alpha(J_k, A_k)X_{k+1}} \mathbb{1}_{\{J_0 \in B^c, \dots, J_n \in B^c\}} r(J_n, A_n) \right.\right.$$
$$\left.\left. \times \int_0^{X_{n+1}} e^{-\alpha(J_n, A_n)t} dt \middle| X_0, J_0, A_0, \dots, X_n, J_n, A_n \right] \right]$$

$$= E_i^\pi \left[ \sum_{n=0}^\infty \prod_{k=0}^{n-1} e^{-\alpha(J_k, A_k)X_{k+1}} \mathbb{1}_{\{J_0 \in B^c, \dots, J_n \in B^c\}} r(J_n, A_n) \right.$$
$$\left. \times E_i^\pi \left[ \int_0^{X_{n+1}} e^{-\alpha(J_n, A_n)t} dt \middle| X_0, J_0, A_0, \dots, X_n, J_n, A_n \right] \right]$$

$$= E_i^\pi \left[ \sum_{n=0}^\infty \prod_{k=0}^{n-1} e^{-\alpha(J_k, A_k)X_{k+1}} \mathbb{1}_{\{J_0 \in B^c, \dots, J_n \in B^c\}} r(J_n, A_n) \right]$$

$$\times \int_0^\infty e^{-\alpha(J_n,A_n)t}(1 - D(t \mid J_n,A_n))dt\Bigg]$$

$$= E_i^\pi\Bigg[\sum_{n=0}^\infty \prod_{k=0}^{n-1} e^{-\alpha(J_k,A_k)X_{k+1}} \mathbb{1}_{\{J_0\in B^c,\ldots,J_n\in B^c\}}\tilde{r}(J_n,A_n)\Bigg], \qquad (11.10)$$

where the third equality follows from Assumption 11.2.1 and the ninth equality is due to the property (11.4).

We now show that for each $n = 0, 1, \ldots,$

$$E_i^\pi\Bigg[\prod_{k=0}^{n-1} e^{-\alpha(J_k,A_k)X_{k+1}} \mathbb{1}_{\{J_0\in B^c,\ldots,J_n\in B^c\}}w(J_n)\Bigg] \le \beta^n w(i). \qquad (11.11)$$

Indeed, (11.11) is trivial for $n = 0$. Now, for $n \ge 1$, it follows from the property (11.4) and Assumption 11.2.3(b) that

$$E_i^\pi\Bigg[\prod_{k=0}^{n-1} e^{-\alpha(J_k,A_k)X_{k+1}} \mathbb{1}_{\{J_0\in B^c,\ldots,J_n\in B^c\}}w(J_n)\Bigg]$$

$$= E_i^\pi\Bigg[E_i^\pi\Bigg[\prod_{k=0}^{n-1} e^{-\alpha(J_k,A_k)X_{k+1}} \mathbb{1}_{\{J_0\in B^c,\ldots,J_n\in B^c\}}w(J_n)$$

$$\mid T_0, J_0, A_0, \ldots, T_{n-1}, J_{n-1}, A_{n-1}]\Bigg]$$

$$= E_i^\pi\Bigg[\prod_{k=0}^{n-2} e^{-\alpha(J_k,A_k)X_{k+1}} \mathbb{1}_{\{J_0\in B^c,\ldots,J_{n-1}\in B^c\}}E_i^\pi\Big[e^{-\alpha(J_{n-1},A_{n-1})X_n} \mathbb{1}_{\{J_n\in B^c\}}w(J_n)$$

$$\mid T_0, J_0, A_0, \ldots, T_{n-1}, J_{n-1}, A_{n-1}\Big]\Bigg]$$

$$= E_i^\pi\Bigg[\prod_{k=0}^{n-2} e^{-\alpha(J_k,A_k)X_{k+1}} \mathbb{1}_{\{J_0\in B^c,\ldots,J_{n-1}\in B^c\}}$$

$$\times \sum_{j\in B^c}\int_0^\infty e^{-\alpha(J_{n-1},A_{n-1})t}w(j)Q(dt,j \mid J_{n-1},A_{n-1})\Bigg]$$

$$= E_i^\pi\Bigg[\prod_{k=0}^{n-2} e^{-\alpha(J_k,A_k)X_{k+1}} \mathbb{1}_{\{J_0\in B^c,\ldots,J_{n-1}\in B^c\}}\sum_{j\in B^c}w(j)m(j \mid J_{n-1},A_{n-1})\Bigg]$$

$$\le \beta E_i^\pi\Bigg[\prod_{k=0}^{n-2} e^{-\alpha(J_k,A_k)X_{k+1}} \mathbb{1}_{\{J_0\in B^c,\ldots,J_{n-1}\in B^c\}}w(J_{n-1})\Bigg]. \qquad (11.12)$$

Iterating (11.12) yields (11.11).

Moreover, observe that Assumption 11.2.3(a) and (11.11) gives

$$E_i^\pi \left[ \prod_{k=0}^{n-1} e^{-\alpha(J_k,A_k)X_{k+1}} \mathbb{1}_{\{J_0 \in B^c,\ldots,J_n \in B^c\}} |\widetilde{r}(J_n,A_n)| \right]$$

$$\leq M E_i^\pi \left[ \prod_{k=0}^{n-1} e^{-\alpha(J_k,A_k)X_{k+1}} \mathbb{1}_{\{J_0 \in B^c,\ldots,J_n \in B^c\}} w(J_n) \right]$$

$$\leq M\beta^n w(i),$$

which together with (11.10) yields

$$|V_r(i,\pi)| \leq \sum_{n=0}^\infty E_i^\pi \left[ \prod_{k=0}^{n-1} e^{-\alpha(J_k,A_k)X_{k+1}} \mathbb{1}_{\{J_0 \in B^c,\ldots,J_n \in B^c\}} |\widetilde{r}(J_n,A_n)| \right]$$

$$\leq \sum_{n=0}^\infty M\beta^n w(i) = Mw(i)/(1-\beta).$$

Thus, we get

$$\sup_{i \in E} |V_r(i,\pi)|/w(i) \leq Mw(i)/(1-\beta),$$

which shows that $V_r(\cdot,\pi) \in \mathbb{B}_w(E)$.

Similarly, the conclusion for $V_c(\cdot,\pi)$ can be obtained.

(b) Since $|u(i)| \leq \|u\|_w w(i)$ for all $i \in E$, it follows from (11.11) above that

$$E_i^\pi \left[ \prod_{k=0}^{n-1} e^{-\alpha(J_k,A_k)X_{k+1}} \mathbb{1}_{\{J_0 \in B^c,\ldots,J_n \in B^c\}} |u(J_n)| \right]$$

$$\leq \|u\|_w E_i^\pi \left[ \prod_{k=0}^{n-1} e^{-\alpha(J_k,A_k)X_{k+1}} \mathbb{1}_{\{J_0 \in B^c,\ldots,J_n \in B^c\}} w(J_n) \right] \leq \|u\|_w \beta^n w(i),$$

and so part (b) follows.                                                                 □

*Remark 11.3.8.* In fact, Lemma 11.3.1 here for first passage criteria in SMDPs is similar to Lemma 3.1 in Huang and Guo [15]. The main difference between them is due to that the discount factor $\alpha(i,a)$ here is state-action dependent, and the reward rate here is possibly unbounded (while the ones in Huang and Guo [15] are not).

For $\varphi \in \Phi$, we define the dynamic programming operators $H^\varphi$ and $H$ on $\mathbb{B}_w(E)$ as follows: for $u \in \mathbb{B}_w(E)$, if $i \in B^c$,

$$H^{\varphi}u(i) := \int_{a\in A(i)} \left[ \widetilde{r}(i,a) + \sum_{j\in B^c} u(j)m(j\mid i,a) \right] \varphi(da\mid i), \qquad (11.13)$$

$$Hu(i) := \sup_{a\in A(i)} \left[ \widetilde{r}(i,a) + \sum_{j\in B^c} u(j)m(j\mid i,a) \right], \qquad (11.14)$$

and if $i \in B$, $H^{\varphi}u(i) = Hu(i) := 0$.

**Lemma 11.3.2.** *Suppose that Assumptions 11.2.1 and 11.2.3 hold. Then:*

(a) *For each $\varphi \in \Phi$, $V_r(\cdot,\varphi)$ is the unique solution in $\mathbb{B}_w(E)$ to the equation*

$$V_r(i,\varphi) = H^{\varphi}V_r(i,\varphi) \; \forall i \in E.$$

(b) *If, in addition, Assumption 11.2.4 also holds, $V_r^*(i) := \sup_{\pi\in\Pi} V_r(i,\pi)$ is the unique solution in $\mathbb{B}_w(E)$ to equation*

$$V_r^*(i) = HV_r^*(i) \; \forall i \in E.$$

*Moreover, there exists an $f^* \in \mathbb{F}$ such that $V_r^*(i) = H^{f^*}V_r^*(i)$, and such a policy $f^* \in \mathbb{F}$ satisfies $V_r(i,f^*) = V_r^*(i)$ for every $i \in E$.*

*Proof.* (a)  From Lemma 11.3.1, it is clear that $V_r(\cdot,\varphi) \in \mathbb{B}_w(E)$. We now establish the equation $V_r(i,\varphi) = H^{\varphi}V_r(i,\varphi)$. It is obviously true when $i \in B$, and for $i \in B^c$, by (11.10), we have

$V_r(i,\varphi)$

$$= E_i^{\varphi}\left[ \sum_{n=0}^{\infty}\prod_{k=0}^{n-1} e^{-\alpha(J_k,A_k)X_{k+1}} \mathbb{1}_{\{J_0\in B^c,\dots,J_n\in B^c\}} \widetilde{r}(J_n,A_n) \right]$$

$$= E_i^{\varphi}\left[ E_i^{\varphi}\left[ \sum_{n=0}^{\infty}\prod_{k=0}^{n-1} e^{-\alpha(J_k,A_k)X_{k+1}} \mathbb{1}_{\{J_0\in B^c,\dots,J_n\in B^c\}} \widetilde{r}(J_n,A_n) \mid T_0,J_0,A_0,T_1,J_1 \right] \right]$$

$$= E_i^{\varphi}\left[ \mathbb{1}_{\{J_0\in B^c\}}\widetilde{r}(J_0,A_0) + e^{-\alpha(J_0,A_0)X_1}\mathbb{1}_{\{J_0\in B^c,J_1\in B^c\}}E_i^{\varphi}\left[ \sum_{n=1}^{\infty}\prod_{k=1}^{n-1} e^{-\alpha(J_k,A_k)X_{k+1}} \right. \right.$$

$$\left. \left. \mathbb{1}_{\{J_2\in B^c,\dots,J_n\in B^c\}}\widetilde{r}(J_n,A_n) \mid T_0,J_0,A_0,T_1,J_1 \right] \right]$$

$$= \int_{a\in A(i)} \varphi(da\mid i)\left[ \widetilde{r}(i,a) + \sum_{j\in B^c}\int_0^{\infty} e^{-\alpha(i,a)t}Q(dt,j\mid i,a)E_i^{\varphi}\left[ \sum_{n=1}^{\infty}\prod_{k=1}^{n-1} e^{-\alpha(J_k,A_k)X_{k+1}} \right. \right.$$

$$\left. \left. \mathbb{1}_{\{J_0\in B^c,\dots,J_n\in B^c\}}\widetilde{r}(J_n,A_n) \mid T_0=0,J_0=i,A_0=a,T_0=t,J_1=j \right] \right]$$

$$
= \int_{a \in A(i)} \varphi(da \mid i) \left[ \tilde{r}(i,a) + \sum_{j \in B^c} m(j \mid i,a) E_j^{\varphi} \left[ \sum_{n=0}^{\infty} \prod_{k=0}^{n-1} e^{-\alpha(J_k, A_k) X_{k+1}} \right. \right.
$$

$$
\left. \left. \mathbb{1}_{\{J_0 \in B^c, \dots, J_n \in B^c\}} \tilde{r}(J_n, A_n) \right] \right]
$$

$$
= \int_{a \in A(i)} \varphi(da \mid i) \left[ \tilde{r}(i,a) + \sum_{j \in B^c} m(j \mid i,a) V_r(j, \varphi) \right],
$$

where the fifth equality is due to the properties (11.2)–(11.4) and that policy $\varphi$ is Markov. Hence, we obtain that $V_r(i, \varphi) = H^{\varphi} V_r(i, \varphi)$, $i \in E$.

To complete the proof, we need only show that $H^{\varphi}$ is a contraction from $\mathbb{B}_w(E)$ to $\mathbb{B}_w(E)$, and thus $H^{\varphi}$ has a unique fixed point in $\mathbb{B}_w(E)$. Indeed, for an arbitrary function $u \in \mathbb{B}_w(E)$, by Assumption 11.2.3 it is clear that

$$
|H^{\varphi} u(i)|
$$
$$
= \left| \int_{a \in A(i)} \left[ \tilde{r}(i,a) + \sum_{j \in B^c} u(j) m(j \mid i,a) \right] \varphi(da \mid i) \right|
$$
$$
\leq \int_{a \in A(i)} \left[ |\tilde{r}(i,a)| + \sum_{j \in B^c} |u(j)| m(j \mid i,a) \right] \varphi(da \mid i)
$$
$$
\leq \int_{a \in A(i)} \left[ Mw(i) + \|u\|_w \beta w(i) \right] \varphi(da \mid i)
$$
$$
= (M + \beta \|u\|_w) w(i) \ \forall i \in B^c,
$$

which implies that $H^{\varphi} u \in \mathbb{B}_w(E)$, that is, $H^{\varphi}$ maps $\mathbb{B}_w(E)$ to itself. Moreover, for any $u, u' \in \mathbb{B}_w(E)$, we have

$$
|H^{\varphi} u(i) - H^{\varphi} u'(i)|
$$
$$
= \left| \int_{a \in A(i)} \left[ \sum_{j \in B^c} (u(j) - u'(j)) m(j \mid i,a) \right] \varphi(da \mid i) \right|
$$
$$
\leq \int_{a \in A(i)} \left[ \sum_{j \in B^c} |u(j) - u'(j)| m(j \mid i,a) \right] \varphi(da \mid i)
$$
$$
\leq \int_{a \in A(i)} \left[ \|u - u'\|_w \beta w(i) \right] \varphi(da \mid i)
$$
$$
= \beta \|u - u'\|_w w(i) \ \forall i \in B^c.
$$

Hence, $\|H^{\varphi} u - H^{\varphi} u'\|_w \leq \beta \|u - u'\|_w$, and thus $H^{\varphi}$ is a contraction from $\mathbb{B}_w(E)$ to itself. By Banach's Fixed Point Theorem, $H^{\varphi}$ has a unique fixed point in $\mathbb{B}_w(E)$, and so the proof is achieved.

(b)  Under Assumption 11.2.4, using a similar manner to the proof of part (a) yields
    that $H$ is a contraction from $\mathbb{B}_w(E)$ to itself, and thus, by Banach's Fixed Point
    Theorem, $H$ has a unique fixed point $u^*$ in $\mathbb{B}_w(E)$, that is, $Hu^* = u^*$. Hence, to
    prove part (b), we need to show that: $(b_1)$ $V_r^* \in \mathbb{B}_w(E)$, with $w$-norm $\|V_r^*\| \leq$
    $M/(1-\beta)$. $(b_2)$ $V_r^* = u^*$.
    In fact, $(b_1)$ is an immediate result of Lemma 11.3.1(a). Thus, it remains to prove
$(b_2)$. To this end, we show that $u^* \leq V_r^*$ and $u^* \geq V_r^*$ as below, respectively. It is
clear that $u^*(i) = V_r^*(i) = 0$ for every $i \in B$. Hence, in the following, we restrict our
argument to the case of $i \in B^c$.

(i)  This part is to show that $u^* \leq V_r^*$. By the equality $u^* = Hu^*$ and the measurable
    selection theorem [10, Proposition D.5, p.182], there exists an $f \in \mathbb{F}$ such that

$$u^*(i) = \tilde{r}(i,f) + \sum_{j \in B^c} u^*(j)m(j \mid i,f) \ \forall i \in B^c. \tag{11.15}$$

    Iteration of (11.15) yields

$$u^*(i) = E_i^f \left[ \sum_{m=0}^{n-1} \prod_{k=0}^{m-1} e^{-\alpha(J_k,A_k)X_{k+1}} \mathbb{1}_{\{J_0 \in B^c,\ldots,J_m \in B^c\}} \tilde{r}(J_m,f) \right]$$

$$+ E_i^f \left[ \prod_{k=0}^{n-1} e^{-\alpha(J_k,A_k)X_{k+1}} \mathbb{1}_{\{J_0 \in B^c,\ldots,J_n \in B^c\}} u^*(J_n) \right] \ \forall i \in B^c, \ n = 1,2\ldots,$$

    and letting $n \to \infty$ we get, by Lemma 11.3.1(b),

$$u^*(i) = E_i^f \left[ \sum_{m=0}^{\infty} \prod_{k=0}^{m-1} e^{-\alpha(J_k,A_k)X_{k+1}} \mathbb{1}_{\{J_0 \in B^c,\ldots,J_m \in B^c\}} \tilde{r}(J_m,f) \right] = V_r(i,f) \ \forall i \in B^c.$$

    Thus, by the definition of $V_r^*$, we see that $u^* \leq V_r^*$.

(ii)  This part is to show that $u^* \geq V_r^*$. Note that $u^* = Hu^*$ implies that

$$u^*(i) \geq \tilde{r}(i,a) + \sum_{j \in B^c} u^*(j)m(j \mid i,a) \ \forall i \in B^c, \ a \in A(i), \tag{11.16}$$

    which gives

$$\mathbb{1}_{\{J_n \in B^c\}} u^*(J_n) \geq \mathbb{1}_{\{J_n \in B^c\}} \tilde{r}(J_n,A_n) + \mathbb{1}_{\{J_n \in B^c\}} \sum_{j \in B^c} u^*(j)m(j \mid J_n,A_n) \ \forall n \geq 0.$$

$$\tag{11.17}$$

    Hence, for any initial state $i \in B^c$ and policy $\pi \in \Pi$, using properties (11.2)–
    (11.4) yields

$$\mathbb{1}_{\{J_n \in B^c\}} u^*(J_n) \geq E_i^\pi \left[ \mathbb{1}_{\{J_n \in B^c\}} \widetilde{r}(J_n, A_n) + e^{-\alpha(J_n, A_n) X_{n+1}} \mathbb{1}_{\{J_n \in B^c, J_{n+1} \in B^c\}} \right.$$

$$\left. \times u^*(J_{n+1}) \mid T_0, J_0, A_0, \ldots, T_n, J_n, A_n \right] \forall n = 0, 1 \ldots,$$

which gives

$$\prod_{k=0}^{n-1} e^{-\alpha(J_k, A_k) X_{k+1}} \mathbb{1}_{\{J_0 \in B^c, \ldots, J_n \in B^c\}} u^*(J_n)$$

$$\geq E_i^\pi \left[ \prod_{k=0}^{n-1} e^{-\alpha(J_k, A_k) X_{k+1}} \mathbb{1}_{\{J_0 \in B^c, \ldots, J_n \in B^c\}} \widetilde{r}(J_n, A_n) \right.$$

$$\left. + \prod_{k=0}^{n} e^{-\alpha(J_k, A_k) X_{k+1}} \mathbb{1}_{\{J_0 \in B^c, \ldots, J_{n+1} \in B^c\}} u^*(J_{n+1}) \mid T_0, J_0, A_0, \ldots, T_n, J_n, A_n \right]$$

$\forall n = 0, 1 \ldots.$

Therefore, taking expectation $E_i^\pi$ and summing over $m = 0, 1 \ldots, n - 1$, we obtain

$$u^*(i) \geq E_i^\pi \left[ \sum_{m=0}^{n-1} \prod_{k=0}^{m-1} e^{-\alpha(J_k, A_k) X_{k+1}} \mathbb{1}_{\{J_0 \in B^c, \ldots, J_m \in B^c\}} \widetilde{r}(J_m, A_m) \right]$$

$$+ E_i^\pi \left[ \prod_{k=0}^{n-1} e^{-\alpha(J_k, A_k) X_{k+1}} \mathbb{1}_{\{J_0 \in B^c, \ldots, J_n \in B^c\}} u^*(J_n) \right], \quad \forall n = 1, 2 \ldots.$$

Finally, letting $n \to \infty$ in the latter inequality and using Lemma 11.3.1(b), it follows that

$$u^*(i) \geq V_r(i, \pi)$$

so that, as $i$ and $\pi$ were arbitrary, we conclude that $u^* \geq V_r^*$.

Combining (i) with (ii) yields that $u^* = V_r^*$, and thus we have $V_r^* = HV_r^*$.

Finally, it follows from $V_r^* = HV_r^*$ and the measurable selection theorem that there exists an $f^* \in \mathbb{F}$ such that $V_r^* = H^{f^*} V_r^*$. This fact together with part (a) implies that $V_r^{f^*} = V_r^*$. $\qquad \square$

*Remark 11.3.9.* Note that Lemma 11.3.2 also holds for the case of the expected first passage cost $V_c$ accordingly.

Note that $\mathbb{F}$ can be written as the product space $\mathbb{F} = \prod_{i \in E} A(i)$. Hence, by Assumption 11.2.4(a) and Tychonoff's theorem, $\mathbb{F}$ is a compact metric space.

**Lemma 11.3.3.** *Suppose that Assumptions 11.2.1–11.2.4 and 11.2.5(a) hold. Then the functions $V_r(\mu, f)$ and $V_c(\mu, f)$ are continuous in $f \in \mathbb{F}$.*

*Proof.* We only prove the continuity of $V_r(\mu, f)$ in $f \in \mathbb{F}$ because the other case is similar. Let $f_n \to f$ as $n \to \infty$ and fix any $i \in E$. Let $v(i) := \limsup_{n \to \infty} V_r(i, f_n)$.

Then, by Theorem 4.4 in [1], there exists a subsequence $\{V_r(i, f_{n_m})\}$ (depending on $i$) of $\{V_r(i, f_n)\}$ such that $V_r(i, f_{n_m}) \to v(i)$ as $m \to \infty$. Then, by Lemma 11.3.1(a), we have $V_r(j, f_{n_m}) \in [-Mw(j)/(1-\beta), Mw(j)/(1-\beta)]$ for all $j \in E$ and $m \geq 1$, and so $V_r(\cdot, f_{n_m})$ is in the product space $\prod_{j \in E}[-Mw(j)/(1-\beta), Mw(j)/(1-\beta)]$ for each $m \geq 1$. Since $E$ is denumerable, the Tychonoff theorem shows that the space $\prod_{j \in E}[-Mw(j)/(1-\beta), Mw(j)/(1-\beta)]$ is compact, and thus there exists a subsequence $\{V_r(\cdot, f_{n_k})\}$ of $\{V_r(\cdot, f_{n_m})\}$ converging to some point $u$ in $\prod_{j \in E}[-Mw(j)/(1-\beta), Mw(j)/(1-\beta)]$, that is, $\lim_{k \to \infty} V_r(j, f_{n_k}) = u(j)$ for all $j \in E$, which, together with $f_n \to f$ and $\lim_{m \to \infty} V_r(i, f_{n_m}) = v(i)$, implies that

$$v(i) = u(i), \ \lim_{k \to \infty} V_r(j, f_{n_k}) = u(j), \ \text{and} \lim_{k \to \infty} f_{n_k}(j) = f(j), \ \text{for all } j \in E. \quad (11.18)$$

Moreover, by Lemma 11.3.1(a), we have

$$|u(j)| \leq Mw(j)/(1-\beta), \quad \text{for all } j \in E, \quad (11.19)$$

which implies that $u \in \mathbb{B}_w(E)$.

On the other hand, for $k \geq 1$ and the given $i \in B^c$, by Lemma 11.3.2(a), we have

$$V_r(i, f_{n_k}) = \widetilde{r}(i, f_{n_k}) + \sum_{j \in B^c} V_r(j, f_{n_k}) m(j \mid i, f_{n_k}). \quad (11.20)$$

Then, under Assumptions 11.2.3 and 11.2.4, from (11.18)–(11.20) and Lemma 8.3.7 (i.e., the Generalized Dominated Convergence Theorem) in [11], we get

$$u(i) = \widetilde{r}(i, f) + \sum_{j \in B^c} u(j) m(j \mid i, f). \quad (11.21)$$

Thus, by Lemma 11.3.2(a) and (11.18), we conclude that

$$\limsup_{n \to \infty} V_r(i, f_n) = v(i) = u(i) = V_r(i, f). \quad (11.22)$$

Similarly, we can prove that

$$\liminf_{n \to \infty} V_r(i, f_n) = V_r(i, f),$$

which together with (11.22) implies that

$$\limsup_{n \to \infty} V_r(i, f_n) = \liminf_{n \to \infty} V_r(i, f_n) = V_r(i, f),$$

and so

$$\lim_{n \to \infty} V_r(i, f_n) = V_r(i, f). \quad (11.23)$$

Moreover, noting that $V_r(i, f_n) = V_r(i, f) = 0$ for every $i \in B$ and $n \geq 0$, it follows from Assumption 11.2.5(a) and Lemma 8.3.7 in [11] again that

$$\lim_{n \to \infty} V_r(\mu, f_n) = \lim_{n \to \infty} \sum_{i \in E} [V_r(i, f_n)] \mu(i) = \lim_{n \to \infty} \sum_{i \in B^c} [V_r(i, f_n)] \mu(i)$$

$$= \sum_{i \in B^c} [\lim_{n \to \infty} V_r(i, f_n)] \mu(i) = \sum_{i \in B^c} V_r(i, f) \mu(i) = V_r(\mu, f), \quad (11.24)$$

which gives the stated result: $V_r(\mu, f_n) \to V_r(\mu, f)$, as $n \to \infty$.  $\square$

To analyze the constrained control problem (recall Definition 11.2.3), we introduce a Lagrange multiplier $\lambda \geq 0$ as follows. For each $i \in E$ and $a \in A(i)$, let

$$b^\lambda(i, a) := r(i, a) - \lambda c(i, a). \quad (11.25)$$

Furthermore, for each policy $\pi \in \Pi$ and $i \in E$, let

$$V_{b^\lambda}(i, \pi) := E_i^\pi \left[ \int_0^{\tau_B} e^{-\int_0^t \alpha(Z(u), W(u)) du} b^\lambda(Z(t), W(t)) dt \right], \quad (11.26)$$

$$V_{b^\lambda}(\mu, \pi) := \sum_{j \in E} V_{b^\lambda}(j, \pi) \mu(j), \quad (11.27)$$

$$V_{b^\lambda}^*(i) := \sup_{\pi \in \Pi} V_{b^\lambda}(i, \pi), V_{b^\lambda}^*(\mu) := \sup_{\pi \in \Pi} V_{b^\lambda}(\mu, \pi). \quad (11.28)$$

*Remark 11.3.10.* Notice that, for each $i \in B$, $V_{b^\lambda}(i, \pi) = 0$. Therefore, we have

$$V_{b^\lambda}(\mu, \pi) = \sum_{j \in B^c} V_{b^\lambda}(j, \pi) \mu(j), \qquad V_{b^\lambda}^*(\mu) = \sum_{j \in B^c} V_{b^\lambda}^*(j) \mu(j).$$

Under Assumptions 11.2.1–11.2.4, by Lemma 11.3.2(b), we have

$$V_{b^\lambda}^*(i) = \begin{cases} 0, & \text{for } i \in B, \\ \sup_{a \in A(i)} \left[ \widetilde{b^\lambda}(i, a) + \sum_{j \in B^c} V_{b^\lambda}^*(j) m(j \mid i, a) \right], & \text{for } i \in B^c, \end{cases} \quad (11.29)$$

where $\widetilde{b^\lambda}(i, a) := b^\lambda(i, a) \int_0^\infty e^{-\alpha t} (1 - D(t \mid i, a)) dt$. Moreover, for each $i \in E$, the maximum in (11.29) is realized by some $a \in A(i)$, that is,

$$A_\lambda^*(i) := \begin{cases} A(i), & \text{for } i \in B, \\ \left\{ a \in A(i) \mid V_{b^\lambda}^*(i) = \widetilde{b^\lambda}(i, a) + \sum_{j \in B^c} V_{b^\lambda}^*(j) m(j \mid i, a) \right\}, & \text{for } i \in B^c \end{cases}$$

(11.30)

is *nonempty*. In other words, the following sets

$$\mathbb{F}_\lambda^* := \left\{ f \in \mathbb{F} \mid f(i) \in A_\lambda^*(i) \ \forall i \in E \right\} \tag{11.31}$$

and

$$\Phi^\lambda := \left\{ \varphi \in \Phi \mid \varphi(A_\lambda^*(i) \mid i) = 1 \ \forall i \in E \right\} \tag{11.32}$$

are *nonempty*.

Next lemma reveals that $\Phi^\lambda$ is convex.

**Lemma 11.3.4.** *Under Assumptions 11.2.1–11.2.4, the set $\Phi^\lambda$ is convex.*

*Proof.* For each $\varphi_1, \varphi_2 \in \Phi^\lambda$, and $p \in [0,1]$, let

$$\varphi^p(\cdot \mid i) := p\varphi_1(\cdot \mid i) + (1-p)\varphi_2(\cdot \mid i), \ \forall i \in E. \tag{11.33}$$

Hence, $\varphi^p(A_\lambda^*(i) \mid i) = p\varphi_1(A_\lambda^*(i) \mid i) + (1-p)\varphi_2(A_\lambda^*(i) \mid i) = 1$, and so $\Phi^\lambda$ is convex. $\qquad\square$

**Notation.** For each $\lambda \geq 0$, we take an arbitrary, but fixed policy $f^\lambda \in \mathbb{F}_\lambda^*$, and denote $V_r(\mu, f^\lambda)$, $V_c(\mu, f^\lambda)$, and $V_{b\lambda}(\mu, f^\lambda)$ by $V_r(\lambda)$, $V_c(\lambda)$, and $V_b(\lambda)$, respectively. By Lemma 11.3.2, we have that $V_{b\lambda}(i, f) = V_{b\lambda}^*(i)$ for all $i \in E$ and $f \in \mathbb{F}_\lambda^*$. Hence, $V_b(\lambda) = V_{b\lambda}(\mu, f^\lambda) = V_{b\lambda}^*(\mu)$.

**Lemma 11.3.5.** *If Assumptions 11.2.3–11.2.4 and 11.2.5(a) hold, then $V_c(\lambda)$ is nonincreasing in $\lambda \in [0, \infty)$.*

*Proof.* By (11.5), (11.6), and (11.25)–(11.26) for each $\pi \in \Pi$, we obtain

$$V_{b\lambda}(\mu, \pi) = V_r(\mu, \pi) - \lambda V_c(\mu, \pi) \ \forall \lambda \geq 0.$$

Moreover, noting that $V_b(\lambda) = V_{b\lambda}(\mu, f^\lambda) = V_{b\lambda}^*(\mu)$ for all $\lambda \geq 0$ and $f^\lambda \in \mathbb{F}_\lambda^*$, we have, for any $h > 0$,

$$\begin{aligned}
-hV_c(\lambda) &= V_{b\lambda+h}(\mu, f^\lambda) - V_b(\lambda) \\
&\leq V_b(\lambda + h) - V_b(\lambda) \\
&\leq V_b(\lambda + h) - V_{b\lambda}(\mu, f^{\lambda+h}) = -hV_c(\lambda + h),
\end{aligned}$$

which gives that

$$-hV_c(\lambda) \leq -hV_c(\lambda + h).$$

Hence, $V_c(\lambda)$ is nonincreasing in $\lambda \in [0, \infty)$. $\qquad\square$

**Lemma 11.3.6.** *Suppose that Assumptions 11.2.1–11.2.4 hold. If $\lim\limits_{k \to \infty} \lambda_k = \lambda$, and $f^{\lambda_k} \in \mathbb{F}_{\lambda_k}^*$ is such that $\lim\limits_{k \to \infty} f^{\lambda_k} = f$, then $f \in \mathbb{F}_\lambda^*$.*

*Proof.* Since $f^{\lambda_k} \in \mathbb{F}^*_{\lambda_k}$, for each $i \in B^c$ and $\pi \in \Pi$, we have

$$V^*_{b^{\lambda_k}}(i) = V_r(i, f^{\lambda_k}) - \lambda_k V_c(i, f^{\lambda_k}) \geq V_{b^{\lambda_k}}(i, \pi) = V_r(i, \pi) - \lambda_k V_c(i, \pi). \quad (11.34)$$

Letting $k \to \infty$ in (11.34) and by Lemma 11.3.3, we obtain

$$V_{b^{\lambda}}(i, f) \geq V_{b^{\lambda}}(i, \pi) \ \forall i \in B^c \ \text{and} \ \pi \in \Pi,$$

which together with the fact that $A^*_{\lambda}(i) = A(i)$ for each $i \in B$ implies that $f \in \mathbb{F}^*_{\lambda}$. $\square$

Under Assumptions 11.2.1–11.2.4 and 11.2.5(a), it follows from Lemma 11.3.5 that the following nonnegative constant

$$\overline{\lambda} := \inf\{\lambda \geq 0 \mid V_c(\lambda) \leq \gamma\} \quad (11.35)$$

is well defined.

**Lemma 11.3.7.** *Suppose that Assumptions 11.2.1–11.2.5 hold. Then the constant $\overline{\lambda}$ in (11.35) is finite, that is, $\overline{\lambda} \in [0, \infty)$.*

*Proof.* Suppose that $\overline{\lambda} = \infty$. By Assumption 11.2.5(b), there exists a policy $\pi' \in \Pi$ such that $V_c(\mu, \pi') < \gamma$. Let $d := \gamma - V_c(\mu, \pi') > 0$. Then, for any $\lambda > 0$, we have

$$V_{b^{\lambda}}(\mu, \pi') = V_r(\mu, \pi') - \lambda V_c(\mu, \pi') = V_r(\mu, \pi') - \lambda(\gamma - d). \quad (11.36)$$

As $\overline{\lambda} = \infty$, by (11.35) and Lemma 11.3.5, we obtain $V_c(\lambda) > \gamma$ for all $\lambda > 0$. Therefore, $V_b(\lambda) = V_r(\lambda) - \lambda V_c(\lambda) < V_r(\lambda) - \lambda\gamma$. Since $V_b(\lambda) = V^*_{b^{\lambda}}(\mu) \geq V_{b^{\lambda}}(\mu, \pi')$, from (11.36), we have

$$V_r(\lambda) - \lambda\gamma > V_b(\lambda) \geq V_{b^{\lambda}}(\mu, \pi') = V_r(\mu, \pi') - \lambda(\gamma - d) \ \forall \lambda > 0, \quad (11.37)$$

which gives

$$V_r(\lambda) > V_r(\mu, \pi') + \lambda d \ \ \forall \lambda > 0. \quad (11.38)$$

On the other hand, by Lemma 11.3.1 and Assumption 11.2.5(a), we have

$$\max\left\{|V_r(\mu, \pi')|, |V_r(\lambda)|\right\} \leq M\left[\sum_{j \in B^c} w(j)\mu(j)\right]/(1 - \beta) := \tilde{M} < \infty \quad (11.39)$$

for all $\lambda > 0$. The latter inequality together with (11.38) gives that

$$2\tilde{M} > \lambda d \ \forall \lambda > 0, \quad (11.40)$$

which is clearly a contradiction; for instance, take $\lambda = 3\tilde{M}/d > 0$. Hence, $\overline{\lambda}$ is finite. $\square$

## 11.4   Proof of Theorem 11.2.1

In this section, we prove Theorem 11.2.1 by using the Lagrange approach and the following lemma.

**Lemma 11.4.8.** *If there exist $\lambda_0 \geq 0$ and $\pi^* \in U$ such that*

$$V_c(\mu, \pi^*) = \gamma \ and \ V_{b\lambda_0}(\mu, \pi^*) = V_{b\lambda_0}^*(\mu),$$

*then $\pi^*$ is constrained optimal.*

*Proof.* For any $\pi \in U$, since $V_{b\lambda_0}(\mu, \pi^*) = V_{b\lambda_0}^*(\mu) \geq V_{b\lambda_0}(\mu, \pi)$, we have

$$V_r(\mu, \pi^*) - \lambda_0 V_c(\mu, \pi^*) \geq V_r(\mu, \pi) - \lambda_0 V_c(\mu, \pi). \tag{11.41}$$

Noting that $V_c(\mu, \pi^*) = \gamma$ and $V_c(\mu, \pi) \leq \gamma$ (by $\pi \in U$), from (11.41), we get

$$V_r(\mu, \pi^*) \geq V_r(\mu, \pi) + \lambda_0(\gamma - V_c(\mu, \pi)) \geq V_r(\mu, \pi) \ \forall \pi \in U,$$

which means that $\pi^*$ is constrained optimal.   □

*Proof of Theorem 11.2.1.* By Lemma 11.3.7, the constant $\overline{\lambda} \in [0, \infty)$. Thus, we shall consider the two cases: $\overline{\lambda} = 0$ and $\overline{\lambda} > 0$.

**The case of $\overline{\lambda} = 0$:** By (11.35), there exists a sequence $f^{\lambda_k} \in \mathbb{F}_{\lambda_k}^*$ such that $\lambda_k \downarrow 0$ as $k \to \infty$. Because $\mathbb{F}$ is compact, we may assume that $f^{\lambda_k} \to \tilde{f}$ without loss of generality. Thus, by Lemma 11.3.5, we have $V_c(\mu, f^{\lambda_k}) \leq \gamma$ for all $k \geq 1$, and then it follows from Lemma 11.3.3 that $\tilde{f} \in U$. Moreover, for each $\pi \in U$, we have that $V_b(\lambda_k) = V_{b\lambda_k}(\mu, f^{\lambda_k}) \geq V_{b\lambda_k}(\mu, \pi)$. Hence, by Lemma 11.3.1(a) and (11.39),

$$V_r(\mu, f^{\lambda_k}) - V_r(\mu, \pi) \geq \lambda_k[V_c(\mu, f^{\lambda_k}) - V_c(\mu, \pi)] \geq -2\lambda_k \tilde{M}. \tag{11.42}$$

Letting $k \to \infty$ in (11.42), by Lemma 11.3.3, we have

$$V_r(\mu, \tilde{f}) - V_r(\mu, \pi) \geq 0 \ \forall \pi \in U,$$

which means that $\tilde{f}$ is a constrained optimal stationary policy.

**The case of $\overline{\lambda} > 0$:** First, if there is some $\lambda' \in (0, \infty)$ satisfying $V_c(\lambda') = \gamma$, then there exist an associated $f^{\lambda'} \in \mathbb{F}_{\lambda'}^*$ such that $V_c(\lambda') = V_c(\mu, f^{\lambda'}) = \gamma$, and $V_{b\lambda'}^*(\mu) = V_{b\lambda'}(\mu, f^{\lambda'})$. Thus, by Lemma 11.4.8, $f^{\lambda'}$ is a constrained optimal stationary policy.

Now, suppose that $V_c(\lambda) \neq \gamma$ for all $\lambda \in (0, \infty)$. Then, as $\overline{\lambda}$ is in $(0, \infty)$, there exist two nonnegative sequences $\{\lambda_k\}$ and $\{\delta_k\}$ such that $\lambda_k \uparrow \overline{\lambda}$ and $\delta_k \downarrow \overline{\lambda}$, respectively. Since $\mathbb{F}$ is compact, we may take $f^{\lambda_k} \in \mathbb{F}_{\lambda_k}^*$ and $f^{\delta_k} \in \mathbb{F}_{\delta_k}^*$ such that $f^{\lambda_k} \to f^1 \in \mathbb{F}$ and $f^{\delta_k} \to f^2 \in \mathbb{F}$, without loss of generality. By Lemma 11.3.6, we have that $f^1, f^2 \in$

$\mathbb{F}_{\frac{*}{\lambda}}$. By Lemmas 11.3.3 and 11.3.4, we obtain that $V_c(\mu, f^1) \geq \gamma$ and $V_c(\mu, f^2) \leq \gamma$. If $V_c(\mu, f^1) = \gamma$ ( or $V_c(\mu, f^2) = \gamma$), by Lemma 11.4.8, we have that $f^1$ (or $f^2$) is a constrained optimal stationary policy. Hence, to complete the proof, we shall consider the following case:

$$V_c(\mu, f^1) > \gamma \quad \text{and} \quad V_c(\mu, f^2) < \gamma. \tag{11.43}$$

Now using $f^1$ and $f^2$, we construct a sequence of stationary policies $\{f_n\}$ as follows. For each $n \geq 1$ and $i \in E$, let

$$f_n(i) := \begin{cases} f^1(i), & \text{if } i < n; \\ f^2(i), & \text{if } i \geq n, \end{cases}$$

where, without loss of generality, the denumerable state space is assumed to be the set $\{1, 2, \dots\}$. Obviously, $f_1 = f^2$ and $\lim_{n \to \infty} f_n = f^1$. Hence, by Lemma 11.3.3, $\lim_{n \to \infty} V_c(\mu, f_n) = V_c(\mu, f^1)$. Since $f^1, f^2 \in \mathbb{F}_{\frac{*}{\lambda}}$ (just mentioned), by (11.31), we see that $f_n \in \mathbb{F}_{\frac{*}{\lambda}}$ for all $n \geq 1$. As $f_1 = f^2$, by (11.43), we have $V_c(\mu, f_1) < \gamma$. If there exists $n^*$ such that $V_c(\mu, f_{n^*}) = \gamma$, then by Lemma 11.4.8 and $f_n \in \mathbb{F}_{\frac{*}{\lambda}}$, $f_{n^*}$ a constrained optimal stationary policy. Thus, in the remainder of this section, we may assume that $V_c(\mu, f_n) \neq \gamma$ for all $n \geq 1$. If $V_c(\mu, f_n) < \gamma$ for all $n \geq 1$, $\lim_{n \to \infty} V_c(\mu, f_n) = V_c(\mu, f^1) \leq \gamma$, which is a contradiction to (11.43). Thus, there exists some $n \geq 1$ such that $V_c(\mu, f_n) > \gamma$, which together with $V_c(\mu, f_1) < \gamma$ gives the existence of some $\tilde{n}$ such that

$$V_c(\mu, f_{\tilde{n}}) < \gamma \quad \text{and} \quad V_c(\mu, f_{\tilde{n}+1}) > \gamma. \tag{11.44}$$

Obviously, the stationary policies $f_{\tilde{n}}$ and $f_{\tilde{n}+1}$ differ in at most the state $\tilde{n}$. Here, it should be pointed out that $\tilde{n}$ must be in $B^c$. Indeed, if $\tilde{n} \in B$, we have $V_c(\tilde{n}, f_{\tilde{n}}) = V_c(\tilde{n}, f_{\tilde{n}+1}) = 0$, which implies that $V_c(\mu, f_{\tilde{n}}) = V_c(\mu, f_{\tilde{n}+1})$ and thus leads to a contradiction to (11.44).

For any $p \in [0, 1]$, using the stationary policies $f_{\tilde{n}}$ and $f_{\tilde{n}+1}$, we construct a randomized stationary policy $\varphi^p$ as follows. For each $i \in E$,

$$\varphi^p(a \mid i) = \begin{cases} p, & \text{if } a = f_{\tilde{n}}(\tilde{n}) \text{ when } i = \tilde{n}, \\ 1 - p, & \text{if } a = f_{\tilde{n}+1}(\tilde{n}) \text{ when } i = \tilde{n}, \\ 1, & \text{if } a = f_{\tilde{n}}(i) \text{ when } i \neq \tilde{n}. \end{cases} \tag{11.45}$$

Since $f_{\tilde{n}}, f_{\tilde{n}+1} \in \mathbb{F}_{\frac{*}{\lambda}}$, by Lemma 11.3.4, we have $V_{b\bar{\lambda}}(\mu, \varphi^p) = V^*_{b\bar{\lambda}}(\mu)$ for all $p \in [0, 1]$. We also have that $V_c(\mu, \varphi^p)$ is continuous in $p \in [0, 1]$. Indeed, for any $p \in [0, 1]$ and any sequence $\{p_m\}$ in $[0, 1]$ such that $\lim_{n \to \infty} p_m = p$, as in the proof of Lemma 11.3.2, we have

$$V_c(i, \varphi^{Pm}) = \sum_{a \in A(i)} \varphi^{Pm}(a \mid i) \left[ \widetilde{c}(i,a) + \sum_{j \in B^c} V_c(j, \varphi^{Pm}) m(j \mid i,a) \right] \forall i \in B^c. \quad (11.46)$$

Hence, as in the proof of Lemma 11.3.3, from (11.45) and (11.46), we obtain

$$\lim_{n \to \infty} V_c(\mu, \varphi^{Pm}) = V_c(\mu, \varphi^P),$$

and so $V_c(\mu, \varphi^P)$ is continuous in $p \in [0, 1]$.

Finally, let $p_0 = 0$ and $p_1 = 1$. Then, $V_c(\mu, \varphi^{P_0}) = V_c(\mu, f_{\tilde{n}+1}) > \gamma$ and $V_c(\mu, \varphi^{P_1}) = V_c(\mu, f_{\tilde{n}}) < \gamma$. Therefore, by the continuity of $V_c(\mu, \varphi^P)$ in $p \in [0, 1]$ there exists a $p^* \in (0, 1)$ such that $V_c(\mu, \varphi^{p^*}) = \gamma$. Since $V_{b\overline{\lambda}}(\mu, \varphi^{p^*}) = V_{b\overline{\lambda}}^*(\mu)$, by Lemma 11.4.8, we have that $\varphi^{p^*}$ is a constrained optimal stationary policy, which randomizes between the two stationary policies $f_{\tilde{n}}$ and $f_{\tilde{n}+1}$ that differ in at most the state $\tilde{n} \in B^c$.                                                    □

# References

1. Aliprantis, C.D., Burkinshaw O.: Principles of real analysis. Third edition. Academic Press, Inc., San Diego, CA (1998)
2. Berument, H., Kilinc, Z., Ozlale, U.: The effects of different inflation risk premiums on interest rate spreads. Phys. A **333**, 317–324 (2004)
3. Beutler, F.J., Ross, K.W.: Time-average optimal constrained semi-Markov decision processes. Adv. in Appl. Probab. **18**, 341–359 (1986)
4. Feinberg, E.A.: Constrained semi-Markov decision processes with average rewards. Z. Oper. Res. **39**, 257–288 (1994)
5. Feinberg, E.A.: Continuous time discounted jump Markov decision processes: a discrete-event approach. Math. Oper. Res. **29**, 492–524 (2004)
6. Guo, X.P.: Constrained denumerable state non-stationary MDPs with expected total reward criterion. Acta Math. Appl. Sinica (English Ser.) **16**, 205–212 (2000)
7. Guo, X.P., Hernández-Lerma, O.: Constrained continuous-time Markov control processes with discounted criteria. Stochastic Anal. Appl. **21**, 379–399 (2003)
8. Guo, X.P., Hernández-Lerma, O.: Continuous-Time Markov Decision Processes: Theory and Applications. Springer-Verlag, Berlin Heidelberg (2009)
9. Haberman, S., Sung, J.: Optimal pension funding dynamics over infinite control horizon when stochastic rates of return are stationary. Insur. Math. Econ. **36**, 103–116 (2005)
10. Hernández-Lerma, O., Lasserre, J.B.: Discrete-Time Markov Control Processes: Basic Optimality Criteria. Springer-Verlag, New York (1996)
11. Hernández-Lerma, O., Lasserre, J.B.: Further Topics on Discrete-Time Markov Control Processes. Springer-Verlag, New York (1999)
12. Hernández-Lerma, O., González-Hernández, J.: Constrained Markov control processes in Borel spaces: the discounted case. Math. Methods Oper. Res. **52**, 271–285 (2000)
13. Huang, Y.H, Guo, X.P.: Optimal risk probability for first passage models in semi-Markov decision processes. J. Math. Anal. Appl. **359**, 404–420 (2009)
14. Huang, Y.H, Guo, X.P.: Discounted semi-Markov decision processes with nonnegative costs. Acta. Math. Sinica (Chinese Series) **53**, 503–514 (2010)
15. Huang, Y.H, Guo, X.P.: First passage models for denumerable semi-Markov decision processes with nonnegative discounted costs. Acta. Math. Appl. Sinica **27**, 177–190 (2011)

16. Huang, Y.H, Guo, X.P.: Finite horizon semi-Markov decision processes with application to maintenance systems. European. J. Oper. Res. **212**, 131–140 (2011)
17. Lee, P., Rosenfield, D.B.: When to refinance a mortgage: a dynamic programming approach. European. J. Oper. Res. **166**, 266–277 (2005)
18. Limnios, N., Oprisan, J.: Semi-Markov Processes and Reliability. Birkhäuser, Boston (2001)
19. Lin, Y.L.: Continuous time first arrival target models (1)- discounted moment optimal models. Acta. Math. Appl. Sinica-Chinese Serias **14**, 115–124 (1991)
20. Lippman, S.A.: Semi-Markov decision processes with unbounded rewards. Management Science **19**, 717–731 (1973)
21. Liu, J.Y., Huang S.M.: Markov decision processes with distribution function criterion of first-passage time. Appl. Math. Optim. **43**, 187–201 (2001)
22. Liu, J.Y., Liu, K.: Markov decision programming—the first passage model with denumerable state space. Systems Sci. Math. Sci. **5**, 340–351 (1992)
23. Newell R. G. and Pizer W. A. Discounting the distant future: how much do uncertain rates increase valuation. J. Environ. Econ. Manage **46**, 52–71 (2003)
24. Puterman, M.L.: Markov Decision Processes: Discrete Stochastic Dynamic Programming. John Wiley & Sons. Inc., New York (1994)
25. Ross, S.M.: Average cost semi-Markov decision processes. J. Appl. Probab. **7**, 649–656 (1970)
26. Sack, B., Wieland, V.: Interest-rate smoothing and optimal monetary policy: a review of recent empirical evidence. J. Econ. Bus. **52**, 205–228 (2000)
27. Yong, J.M., Zhou, X.Y.: Stochastic Controls—Hamiltonian Systems and HJB Equations. Springer-Verlag, New York (1999)
28. Yu, S.X., Lin, Y.L., Yan, P.F.: Optimization models for the first arrival target distribution function in discrete time. J. Math. Anal. Appl. **225**, 193–223 (1998)
29. Zhang, L.L., Guo, X.P.: Constrained continuous-time Markov control processes with average criteria. Math. Meth. Oper. Res. **67**, 323–340 (2008)

# Chapter 12
# Infinite-Horizon Optimal Control Problems for Hybrid Switching Diffusions

**Héctor Jasso-Fuentes and George Yin**

## 12.1 Introduction

Owing to the emerging applications arising in manufacturing and production planning, biological and ecological systems, communication networks, and financial engineering, resurgent attention has been drawn to the study of regime-switching diffusion systems, their asymptotic properties, and the associated control and optimization problems. In such processes, continuous dynamics and discrete events coexist and interact. The use of these hybrid models stems from their ability to provide more realistic formulation for real-world applications in which the usual stochastic differential equation models alone are no longer adequate. The regime-switching diffusions blend both continuous and discrete characteristics, where the discrete event process is modeled as a random switching process to represent random environment and other random influences. In this chapter, we focus on controlled dynamic systems whose dynamics are given by the aforementioned switching diffusion processes. We focus on optimal controls of such systems in an infinite horizon. This chapter is dedicated to Professor Onésimo Hernández-Lerma on the occasion of his 65th birthday, who has made many important contributions to infinite-horizon optimal controlled diffusions.

Before proceeding further, we briefly review some relevant literature. Discounted and average reward optimality has been extensively studied for different classes of dynamic systems: for Markov decision processes (MDPs), see [20, 43, 45] and the

H. Jasso-Fuentes (✉)
Departamento de Matemáticas, CINVESTAV-IPN, México D.F. 07000, México
e-mail: hjasso@math.cinvestav.mx

G. Yin
Department of Mathematics, Wayne State University, Detroit, MI 48202, USA
e-mail: gyin@math.wayne.edu

D. Hernández-Hernández and A. Minjárez-Sosa (eds.), *Optimization, Control, and Applications of Stochastic Systems*, Systems & Control: Foundations & Applications, DOI 10.1007/978-0-8176-8337-5_12, © Springer Science+Business Media, LLC 2012

references therein. For continuous-time Markov chains, we can mention the works [16, 17]. In [52, 53], the authors treated singularly perturbed MDPs in discrete and continuous time. Concerning continuous time, in [19, 40], the analysis was given for general controlled Markov processes, whereas the papers in [2, 4–6, 30, 31, 49] considered controlled diffusion process, and [14, 15] for switching diffusions. Dealing with bias optimality, discrete-time MDPs were treated in [21, 22, 57]. From another angle, [38], studied this criterion in the context of continuous-time Markov chains. Other related papers are [26, 27], where the authors considered controlled diffusion processes. The recent papers [12, 29] study this criteria for switching diffusions. Strong overtaking optimality was originally introduced in [44]. This notion was relaxed in [1] and in [51]. Many classes of controlled systems both deterministic and/or stochastic were considered subsequently; see [7, 55] for an overview. Other related papers of overtaking optimality include [9, 21, 33] for discrete-time problems, [38] for continuous-time MDP, [47, 48] for continuous-time deterministic systems, and [26, 34] for controlled diffusion processes. For switching diffusions, the recent papers [12, 29] are the first works dealing with bias and overtaking optimality for this class of dynamic systems. Blackwell optimality was introduced in [3]. This concept has been studied extensively for MDPs by [8, 10, 11, 24, 25, 32, 50]. For continuous-time models, there exists only a handful of papers. For instance, [28, 42] deal with controlled diffusion processes, whereas [39] consider continuous-time controlled Markov chains. To the best of our knowledge, the only paper dealing with hybrid switching diffusions is [29].

In this chapter, we begin with the so-called basic criteria including discounted reward and average reward per unit time criteria. Nevertheless, it is known that the basic criteria have drawbacks. The discounted rewards ignore the future activities and actions, whereas average reward per unit time pays no attention to finite horizons. Thus, the so-called advanced criteria become popular. Not only is the consideration of advance criteria interesting from a mathematical point of view, but it is a practically useful consideration. It would be ideal to document the results obtained so that it will be beneficial to both researchers and practitioners. Since the initiation of the study of stochastic control, there have been numerous papers published in this subject concentrating on infinite horizon control systems. Nevertheless, the results for advanced criteria are still scarce. There have been only a handful of papers focusing on advanced criteria. The rest of this chapter is organized as follows: Sect. 12.2 introduces the control model together with the main assumptions and ergodicity. Section 12.3 concerns the study of basic optimality criteria including the $\rho$-discounted reward criterion and the ergodic criterion. Section 12.4 focuses on the advanced or selective criteria, including bias optimality, overtaking optimality, and sensitive discount optimality. Finally, Sect. 12.5 concludes this chapter.

## 12.2   Controlled Switching Diffusions and Ergodicity

Suppose that $W(\cdot)$ is a $d$-dimensional Wiener process and that $\alpha(\cdot)$ is a continuous-time Markov chain with a finite state space $M = \{1,2,\ldots,l\}$ and a generator $Q = (q_{ij})$. Throughout this chapter, as a standing assumption, we assume that $\alpha(\cdot)$ and $W(\cdot)$ are independent and that $\alpha(\cdot)$ is irreducible. Let $b : \mathbb{R}^n \times M \times U \to \mathbb{R}^n$ and $\sigma : \mathbb{R}^n \times M \to \mathbb{R}^{n \times d}$. Consider the controlled hybrid diffusion process given by

$$dx(t) = b(x(t), \alpha(t), u(t))dt + \sigma(x(t), \alpha(t))dW(t), \qquad (12.1)$$

with $x(0) = x$, $\alpha(0) = i$, and $t \geq 0$. The set $U \subset \mathbb{R}^m$ is called the *control* (or *action*) *set*, and $u(\cdot)$ is a $U$-valued stochastic process representing the controller's action at each time $t \geq 0$.

**Notation.**   For vectors and matrices, we use the usual Euclidean norms $|x|^2 := \sum_i x_i^2$ and $|A|^2 := \mathrm{Tr}(AA') = \sum_{i,j} A_{ij}^2$, where $A'$ is the *transpose* of $A = (A_{ij})$ and $\mathrm{Tr}(B)$ denotes the *trace* of the square matrix $B$, respectively. As usual, the gradient and the Hessian matrix of a function $\varphi$ are represented by $\nabla \varphi$ and $H_\varphi$, respectively. The following assumption ensures the existence of a unique strong solution of (12.1).

**Assumption 12.2.1.**   *(a) The control set $U$ is compact.*
*(b) For each $i \in M$, $b(\cdot, i, \cdot)$ and $\sigma(\cdot, i)$ are continuous on $\mathbb{R}^n \times U$ and on $\mathbb{R}^n$, respectively. Moreover, there exist positive constants $K_{i,1}$ and $K_{i,2}$ such that for each $x$ and $y$ in $\mathbb{R}^n$,*

$$\sup_{u \in U} |b(x,i,u) - b(y,i,u)| \leq K_{i,1}|x-y| \qquad (12.2)$$

*and*

$$|\sigma(x,i) - \sigma(y,i)| \leq K_{i,2}|x-y|. \qquad (12.3)$$

*(c) There exists a positive constant $K_{i,3}$ such that for each $i \in M$, the matrix $a := \sigma\sigma'$ satisfies*

$$x'a(y,i)x \geq K_{i,3}|x|^2 \text{ for all } x, y \in \mathbb{R}^n \quad (uniform\ ellipticity). \qquad (12.4)$$

**Control Policies and Extended Generator.**   Throughout this chapter, we consider only (non-randomized) *Markov control policies*, also named simply *Markov policies*, which consist of all of $U$-valued measurable functions on $[0,\infty) \times \mathbb{R}^n \times M$. A special case of this class is the so-called *stationary Markov policy* or simply *stationary policy* consisting of all $U$-valued measurable functions on $\mathbb{R}^n \times M$. Observe that a control policy $u(t)$, $t \geq 0$ in (12.1) becomes $u(t) := f(t,x(t),\alpha(t))$ (or $u(t) := f(x(t),\alpha(t))$, when we consider Markov (or stationary) policies. By a slight abuse of terminology, we call $f$ itself a Markov policy (or stationary policy). We will denote by $\mathbb{M}$ and by $\mathbb{F}$ the families of Markov and stationary control policies, respectively.

Let $C^2(\mathbb{R}^n \times M)$ be the space of real-valued function $h$ on $\mathbb{R}^n \times M$, such that $h(x,i) \in C^2(\mathbb{R}^n)$ for each $i \in M$. For each $x \in \mathbb{R}^n$, $i \in M$, $u \in U$, and $h \in C^2(\mathbb{R}^n \times M)$, let $\mathscr{Q}h(x,i) := \sum_{k=1}^{l} q_{ik}h(x,k)$, and

$$L^u h(x,i) := \langle b(x,i,u), \nabla h(x,i) \rangle + \frac{1}{2}\mathrm{Tr}\,[H_h(x,i)a(x,i)] + \mathscr{Q}h(x,i). \qquad (12.5)$$

Moreover, for each $f \in \mathbb{F}$, $x \in \mathbb{R}^n$, and $i \in M$, we define

$$L^f h(x,i) := L^{f(x,i)} h(x,i). \qquad (12.6)$$

**Existence and Uniqueness of Solutions.** Under Assumption 12.2.1, for each Markov policy $f \in \mathbb{M}$, there exists a unique strong solution $(x^f(\cdot), \alpha(\cdot))$, which is a Markov-Feller process. For $f \in \mathbb{F}$, the operator in (12.6) corresponds to the infinitesimal generator for $(x^f(\cdot), \alpha(\cdot))$, and its corresponding transition probability and conditional expectation are given by $P^f_{x,i}(t,B) := P((x^f(t), \alpha(t)) \in B | x^f(0) = x, \alpha(0) = i)$, for every Borel set $B \subset \mathbb{R}^n \times M$, and $x \in \mathbb{R}^n$, $i \in M$, and $E^f_{x,i}[\varphi(x(t), \alpha(t))] := \sum_{k=1}^{l} \int_{\mathbb{R}^n} \varphi(y,k)P^f_{x,i}(t,dy \times \{k\})$, respectively, where $\varphi$ is a real-valued function on $\mathbb{R}^n \times M$ such that $E^f_{x,i}[\varphi] < \infty$ (for more details of all these statements, we refer [36, 54]). As was in the previous paragraphs, we sometimes write $(x^f(\cdot), \alpha(\cdot))$ instead of $(x(\cdot), \alpha(\cdot))$ to emphasize the dependence of $f$ on the system (12.1).

**Positive Recurrence and Ergodicity.** We begin by providing conditions that guarantee positive recurrence and exponential ergodicity of the controlled process.

**Assumption 12.2.2.** *There exists a function $w \geq 1$ in $C^2(\mathbb{R}^n \times M)$ and constants $d \geq c > 0$ such that for each $i \in M$,*

(a) $\lim_{|x| \to \infty} w(x,i) = +\infty$.
(b) *For all $x \in \mathbb{R}^n$, and $u \in U$,*

$$L^u w(x,i) \leq -cw(x,i) + d. \qquad (12.7)$$

Based on [13, 37, 56], Assumption 12.2.2 ensures that, for each $f \in \mathbb{F}$, the hybrid diffusion $(x^f(\cdot), \alpha(\cdot))$ is Harris positive recurrent with a *unique* invariant probability measure $\mu_f$ for which

$$\mu_f(w) := \sum_{k=1}^{l} \int_{\mathbb{R}^n} w(y,k)\mu_f(dy,k) < \infty. \qquad (12.8)$$

We now introduce the concept of a $w$-weighted norm, where $w$ is the function in Assumption 12.2.2.

**Definition 12.2.3.** Denote by $B_w(\mathbb{R}^n \times M)$ the normed linear space of real-valued measurable functions $v$ on $\mathbb{R}^n \times M$ with finite $w$-norm defined as $\| v \|_w :=$

$\sup_{i\in M}\sup_{x\in\mathbb{R}^n}\frac{|v(x,i)|}{w(x,i)}$. Use $\mathbb{M}_w(\mathbb{R}^n\times M)$ to denote the normed linear space of finite signed measures $\mu$ on $\mathbb{R}^n\times M$ such that

$$\|\mu\|_w:=\sum_{k=1}^{l}\int_{\mathbb{R}^n}w(y,k)\,|\mu(dy,k)|<\infty,\tag{12.9}$$

where $|\mu|:=\mu^++\mu^-$ denotes the total variation of $\mu$.

The next result provides a bound of $E_x^f w(x(t),\alpha(t))$ in the sense of (12.10); see Propositions 2.5 and 2.7 in [29] and also [12, 26, 27].

**Proposition 12.2.4.** *Suppose that Assumption 12.2.2(b) holds. Then:*

*(a) For every $f\in\mathbb{F}$, $x\in\mathbb{R}^n$, $i\in M$, and $t\geq 0$,*

$$E_{x,i}^f w(x(t),\alpha(t))\leq e^{-ct}w(x,i)+\frac{d}{c}(1-e^{-ct}).\tag{12.10}$$

*(b) For every $v\in B_w(\mathbb{R}^n\times M)$, $x\in\mathbb{R}^n$, $i\in M$, and $f\in\mathbb{F}$,*

$$\lim_{T\to\infty}\frac{1}{T}E_{x,i}^f[v(x(T),\alpha(T))]=0.\tag{12.11}$$

Assumption 12.2.5 concerns *w-exponential ergodicity* for the process $(x(\cdot),\alpha(\cdot))$. Sufficient conditions for Assumption 12.2.5 are given, for instance, in [12, 29].

**Assumption 12.2.5.** *For each $f\in\mathbb{F}$, the process $(x^f(\cdot),\alpha(\cdot))$ is $w$-exponentially ergodic, that is, there exist positive constants $\eta$ and $\delta$ such that*

$$\sup_{f\in\mathbb{F}}\left|E_{x,i}^f v(x(t),\alpha(t))-\mu_f(v)\right|\leq\eta e^{-\delta t}\,\|v\|_w\,w(x,i)\tag{12.12}$$

*for all $x\in\mathbb{R}^n$, $i\in M$, $v\in B_w(\mathbb{R}^n\times M)$, and $t\geq 0$, where $\mu_f(v):=\sum_{k=1}^{l}\int_{\mathbb{R}^n}v(y,k)\,\mu_f(dy,k)$.*

## 12.3  Basic Optimality Criteria

Basic optimality criteria consists of the $\rho$-discounted criterion and the ergodic (or average) reward criterion. For related references, see for instance, [4, 14, 16, 19, 20, 30, 43, 45, 49, 52, 53] for the discounted case and [5, 6, 15, 17, 19, 20, 31, 43, 45, 52, 53] for the ergodic case.

In this section, we give a brief account of this two basic criteria. All of our results are based mainly from the following papers [12, 14, 15, 29]. For the average reward criterion, our main tool is the dynamic programming method, based on the Hamilton–Jacobi–Bellman equations (12.24) and (12.25).

**Discounted Reward.** Define the *reward rate* $r$ as a real-valued function on $\mathbb{R}^n \times M \times U$, whose properties are detailed in Assumption 12.3.1 below. Define the set $B_R := \{x \in \mathbb{R}^n \mid |x| < R, \; R > 0\}$ and denote by $\bar{B}_R$ its closure.

**Assumption 12.3.1.** *(a) For each $i \in M$, the reward rate $r(x,i,u)$ is continuous on $\mathbb{R}^n \times U$, locally Lipschitz in $x$, and uniformly in $u \in U$; that is, for each $R > 0$, there exists a constant $K_i(R)$ such that*

$$\sup_{u \in U} |r(x,i,u) - r(y,i,u)| \leq K_i(R)|x-y| \quad \text{for all } |x|, |y| \leq R. \tag{12.13}$$

*(b) $r(\cdot, \cdot, u)$ is in $B_w(\mathbb{R}^n \times M)$ uniformly in $u$; that is, there exists $C > 0$ such that, for all $x \in \mathbb{R}^n$,*

$$\sup_{u \in U} |r(x,i,u)| \leq Cw(x,i). \tag{12.14}$$

For each Markov policy $f \in \mathbb{M}$, $t \geq 0$, $x \in \mathbb{R}^n$, and $i \in M$, we write $r(t,x,i,f) := r(x,i,f(t,x,i))$, which reduces to $r(x,i,f) := r(x,i,f(x,i))$ if $f \in \mathbb{F}$. Using this notation, we define the expected $\rho$-discounted reward as follows.

**Definition 12.3.2.** Given the discount factor $\rho > 0$, let

$$V_\rho(x,i,f) := E_{x,i}^f \left[ \int_0^\infty e^{-\rho t} r(t,x(t),\alpha(t),f) \, dt \right] \tag{12.15}$$

be the *$\rho$-discounted reward* using the policy $f \in \mathbb{M}$, given the initial data $x \in \mathbb{R}^n$ and $i \in M$. The corresponding optimal value function is defined by $V_\rho^*(x,i) := \sup_{f \in \mathbb{M}} V_\rho(x,i,f)$, and a policy $f^* \in \mathbb{M}$ is said to be *$\rho$-discounted optimal* if $V_\rho(x,i,f^*) = V_\rho^*(x,i)$ for all $x \in \mathbb{R}^n$ and $i \in M$.

The next proposition establishes that $V_\rho$ is bounded in an appropriate sense; see [29, Proposition 3.3] for a proof.

**Proposition 12.3.3.** *Suppose that Assumptions 12.2.1, 12.2.2, 12.2.5, and 12.3.1 are satisfied. Then $V_\rho(\cdot, \cdot, f)$ is in $B_w(\mathbb{R}^n \times M)$ for all $f \in \mathbb{F}$, in fact, for every $x \in \mathbb{R}^n$ and $i \in M$,*

$$\sup_{f \in \mathbb{F}} V_\rho(x,i,f) \leq \bar{C}w(x,i), \quad \text{with } \bar{C} := \frac{C(\rho+d)}{\rho c}. \tag{12.16}$$

The following theorem concerns the existence of $\rho$-discounted optimal policies; see [14] for a proof.

**Theorem 12.3.4.** *Suppose that conditions 12.2.1, 12.2.2, 12.2.5, 12.3.1 are satisfied. Then, for fixed $\rho > 0$, the set of $\rho$-discounted optimal policies is nonempty.*

**Ergodic Reward.** From Sect. 12.2, for each stationary policy, $f \in \mathbb{F}$, $(x^f(\cdot), \alpha(\cdot))$ is positive recurrent and has a *unique* invariant measure $\mu_f$. The next definition is concerned with the ergodic criterion together with average optimal policies.

**Definition 12.3.5.** For each $f \in \mathbb{M}$, $x \in \mathbb{R}^n$, $i \in M$, and $T > 0$, let

$$J_T(x,i,f) := E_{x,i}^f \left[ \int_0^T r(t,x(t),\alpha(t),f) dt \right]. \tag{12.17}$$

The *long-run average reward*, also known as *ergodic reward* or *gain* of $f$, given the initial conditions $x(0) = x$ and $\alpha(0) = i$, is

$$J(x,i,f) := \liminf_{T \to \infty} \frac{1}{T} J_T(x,i,f). \tag{12.18}$$

The function

$$J^*(x,i) := \sup_{f \in \mathbb{M}} J(x,i,f) \quad \text{for } x \in \mathbb{R}^n, i \in M \tag{12.19}$$

is referred to as the *optimal gain* or *optimal average reward*. If there is a policy $f^* \in \mathbb{M}$ for which $J(x,i,f^*) = J^*(x,i)$ for all $x \in \mathbb{R}^n$ and $i \in M$, then $f^*$ is called *average optimal*.

Denote by $\mathbb{F}_{ao}$ the set of all average optimal policies; later on, we will see that this set is nonempty. Given $f \in \mathbb{F}$, we define

$$\bar{r}(f) := \mu_f(r(\cdot,f)) = \sum_{k=1}^l \int_{\mathbb{R}^n} r(y,k,f) \mu_f(dy,k). \tag{12.20}$$

The following result shows that, under our ergodic results, the gain in (12.18) becomes a constant and that it is *uniformly* bounded; see Proposition 3.6 in [29] for a proof (see also [12]).

**Proposition 12.3.6.** *Under Assumptions 12.2.1, 12.2.2, 12.2.5, and 12.3.1, we have:*

*(a) $J(x,i,f) = \bar{r}(f)$, for all $x \in \mathbb{R}^n$, $i \in M$.*
*(b) $\bar{r}(f)$ is uniformly bounded by*

$$\bar{r}(f) \leq C \frac{d}{c}, \tag{12.21}$$

*where c, d, and C are the constants in (12.7) and in (12.14), respectively. Therefore,*

$$r^* := \sup_{f \in \mathbb{F}} \bar{r}(f) < \infty. \tag{12.22}$$

To find average optimal policies, we restrict ourselves to the set $\mathbb{F}$ instead of considering the whole space $\mathbb{M}$. The reason is that the optimal gain (12.19) is attained in the set $\mathbb{F}$ of stationary policies, which also coincides with the constant $r^*$ in (12.22), that is,

$$\sup_{f \in \mathbb{M}} J(x,i,f) = \sup_{f \in \mathbb{F}} J(x,i,f) = r^* \quad \text{for all } x \in \mathbb{R}^n, i \in M. \tag{12.23}$$

For more details, see [6, 34, 35] for the diffusion case and [15] for the hybrid switching diffusion case. The following theorem, established in [29], provides the existence and uniqueness of a solution of the average reward HJB equation defined in (12.24).

**Theorem 12.3.7.** *If Assumptions 12.2.1, 12.2.2, 12.2.5, and 12.3.1 are satisfied, then the following assertions hold:*

(a) *There exists a unique pair $(r^*, h)$ consisting of the constant $r^*$ in (12.22) and of a function $h$ of class $C^2(\mathbb{R}^n \times M) \cap B_w(\mathbb{R}^n \times M)$, with $h(0, i_0) = 0$, such that it satisfies the average reward HJB equation*

$$r^* = \max_{u \in U} \left[ r(x, i, u) + L^u h(x, i) \right] \quad \text{for all } x \in \mathbb{R}^n \text{ and } i \in M. \tag{12.24}$$

(b) *There exists a policy $f \in \mathbb{F}$ which attains the maximum in (12.24), that is,*

$$r^* = r(x, i, f) + L^f h(x, i) \quad \text{for all } x \in \mathbb{R}^n \text{ and } i \in M, \tag{12.25}$$

*this class of policies are so-called canonical policies.*
(c) *A policy is average optimal if and only if it is canonical.*
(d) *There exists an average optimal policy.*

## 12.4  Advanced Criteria

We have mentioned in the introduction that the average reward in (12.18) is undesirable in that it practically ignores the finite-horizon total reward. To overcome this difficulty, we need to consider more selective criteria. In this section, we study bias optimality including some of its characterizations and sensitive discount optimality which allows us to avoid the problem that average reward criteria face. We will give conditions for the existence and characterizations of bias and $m$-discount optimal policies for $-1 \leq m \leq +\infty$. We also give further characterizations of these policies. Our characterizations lead to the relation with the concept of overtaking optimality. This concept was defined separately and independently.

*Remark 12.4.1.* Throughout the rest of this chapter, we assume that Assumptions 12.2.1, 12.2.2, 12.2.5 and 12.3.1 hold.

**Bias Optimality.** In this section, we study bias optimality and some of its characterizations which lead to the concept of overtaking optimality. Our approach to work is the well-known dynamic programming method. All of the results established here will be only stated. Their correspondent proofs can be seen in the following two papers [12, 29].

We next establish the concept of bias of each stationary policy. For each $f \in \mathbb{F}$, we define the *bias* of $f$ as the function

$$h_f(x,i) := E_{x,i}^f \left[ \int_0^\infty [r(x(t), \alpha(t), f) - \bar{r}(f)]\, dt \right] \quad \text{for all } x \in \mathbb{R}^n \text{ and } i \in M. \quad (12.26)$$

Note that we can interpret the bias as the expected total difference between the immediate reward $r(x(t), \alpha(t), f)$ and the ergodic reward $\bar{r}(f)$.

*Remark 12.4.2.* It is easy to see that $h_f$ is in $B_w(\mathbb{R}^n \times M)$ for each $f \in \mathbb{F}$. Namely, by the exponential ergodicity property in (12.12), we have

$$\left| h_f(x,i) \right| \leq \|r(\cdot,\cdot,f)\|_w w(x,i) \eta \int_0^\infty e^{-\delta t} dt$$

$$\leq \delta^{-1} C w(x,i) \eta \quad \text{(by (12.14))}. \quad (12.27)$$

The following lemma establishes that the pair $(\bar{r}(f), h_f)$ consisting of the constant $\bar{r}(f)$ in (12.20) and the bias (12.26) is the solution of the so-called *Poisson equation*; see [18, 29].

**Lemma 12.4.3.** *For each $f \in \mathbb{F}$, the pair $(\bar{r}(f), h_f)$ is the unique solution of the Poisson equation*

$$\bar{r}(f) = r(x,i,f) + L^f h_f(x,i) \quad \text{for all } x \in \mathbb{R}^n \text{ and } i \in M, \quad (12.28)$$

*with the condition*

$$\sum_{k=1}^l \int_{\mathbb{R}^n} h_f(y,k) \mu_f(dy,k) = 0. \quad (12.29)$$

*Moreover, $h_f$ belongs to the space $C^2(\mathbb{R}^n \times M) \cap B_w(\mathbb{R}^n \times M)$.*

The following proposition relates the bias $h_f$ to the solution $h$ of the average reward HJB equation (12.24) when $f$ is average optimal; see Proposition 4.5 in [29].

**Proposition 12.4.4.** *If a policy $\hat{f} \in \mathbb{F}$ is average optimal, then its bias $h_{\hat{f}}$ and any function $h$ in the HJB equation (12.24) coincide up to an additive constant. In fact,*

$$h_{\hat{f}}(x,i) = h(x,i) - \mu_{\hat{f}}(h) \quad \text{for all } x \in \mathbb{R}^n \text{ and } i \in M. \quad (12.30)$$

Fixed $(x,i) \in \mathbb{R}^n \times M$, we define the set

$$U^0(x,i) := \{ u \in U \mid r^* = r(x,i,u) + L^u h(x,i) \}. \quad (12.31)$$

*Remark 12.4.5.* (i) From Theorem 12.3.7, $U^0$ is well defined.
(ii) It is easy to see that $f \in \mathbb{F}_{ao}$ if and only if $f(x,i) \in U^0(x,i)$ for all $x \in \mathbb{R}^n$ and $i \in M$.

The following result establishes some properties of $U^0$. A complete guide of multifunctions and selectors is given, for instance, in [20, Appendix D].

**Proposition 12.4.6.** *Under Assumptions 12.2.1 and 12.3.1, the multifunction $(x,i) \mapsto U^0(x,i)$ satisfies the following properties:*

*(a) For each $(x,i) \in \mathbb{R}^n \times M$, $U^0(x,i)$ is a nonempty compact set.*
*(b) For each $i \in M$, $U^0(\cdot,i)$ is continuous.*
*(c) $U^0(\cdot,i)$ is Borel measurable, for each $i \in M$.*

The proofs of (a) and (b) follow from the continuity of $u \mapsto r(x,i,u) + L^u h(x,i)$ and from the compactness of $U$, whereas a proof of part (c) is given, for instance, in [23, 46].

**Existence of Bias Optimal Policies.** The main objective of this part is to show the existence of bias optimal policies, defined in (12.33). We will use the dynamic programming method based on the *bias optimality HJB equations* described in (12.36)–(12.37), below. To begin with, we state the concept of *bias optimality* as follows.

**Definition 12.4.7.** Consider the function $\hat{h}(x,i) \in C^2(\mathbb{R}^n \times M)$ defined by

$$\hat{h}(x,i) := \sup_{f \in \mathbb{F}_{ao}} h_f(x,i) \quad \text{for all } x \in \mathbb{R}^n \text{ and } i \in M. \tag{12.32}$$

Then $\hat{h}$ is called the *optimal bias function*. Moreover, if there exists a policy $f^* \in \mathbb{F}_{ao}$ for which

$$h_{f^*}(x,i) = \hat{h}(x,i) \quad \text{for all } x \in \mathbb{R}^n \text{ and } i \in M, \tag{12.33}$$

then $f^*$ is referred to as a bias optimal policy.

*Remark 12.4.8.* (a) Definition 12.4.7 above suggests to maximize the bias $h_f$ defined in (12.26) over the class of stationary policies $f \in \mathbb{F}$. This definition seems not to be useful at first sight since we are maximizing a kind of error!— see expression (12.26). It will help us, however, to obtain this class of policies in order to get another family of policies so-called overtaking optimal policies that we are going to analyze later on.
(b) Denote by $\mathbb{F}_{bias}$ the set of all bias optimal policies. Suppose that $\mathbb{F}_{bias}$ is nonempty. Then Lemma 12.4.3 ensures that the optimal bias function $\hat{h}$ belongs to $C^2(\mathbb{R}^n \times M) \cap B_w(\mathbb{R}^n \times M)$.

Using (12.30), we can rewrite (12.32) by

$$\hat{h}(x,i) = h(x,i) + \sup_{f \in \mathbb{F}_{ao}} \mu_f(-h) \quad \text{for all } x \in \mathbb{R}^n \text{ and } i \in M. \tag{12.34}$$

This last relation indicates that to search for bias optimal policies, we need only look for policies $f \in \mathbb{F}_{ao}$ that maximizes the second term on the right-hand side of (12.34). If we define a new control problem, which is referred to as the "bias problem" with:

- The control system (12.1)
- The action set $U^0(\cdot,\cdot)$ defined in (12.31)          $\left.\begin{array}{l}\\\\\end{array}\right\}$        (12.35)
- The reward rate $\tilde{r}(x,i,u) := -h(x,i)$ for all $x \in \mathbb{R}^n$ and $i \in M$

then, to find bias optimal policies, we need only solve the bias problem (12.35) with average criterion given by $\mu_f(-h)$.

Our next theorem, whose proof is provided, for instance, in [29, Theorem 4.11] or in [12, Theorem 7.2], ensures both the existence of solutions to (12.36)–(12.37) and the existence of a policy $f \in \mathbb{F}_{ao}$ that maximizes the right-hand side of these equations. This result is based on the dynamic programming method that uses the bias optimal HJB equations (12.36)–(12.37).

**Theorem 12.4.9.** *(a) The triplet $(r^*, \hat{h}, \phi)$ consisting of the constant (12.22), the optimal bias function (12.32), and some function $\phi \in C^2(\mathbb{R}^n \times M)$ solves the bias optimal HJB equation*

$$r^* = \max_{u \in U} \{r(x,i,u) + L^u h(x,i)\}, \quad and \quad (12.36)$$

$$h(x,i) = \max_{u \in U^0(x,i)} \{L^u \phi(x,i)\}. \quad (12.37)$$

*(b) There exists a policy $f \in \mathbb{F}_{ao}$ that maximizes the right-hand side of (12.36)–(12.37). That is,*

$$r^* = r(x,i,f) + L^f h(x,i), \quad and \quad (12.38)$$

$$h(x,i) = L^f \phi(x,i). \quad (12.39)$$

*(c) A policy f satisfies (12.38)–(12.39) if and only if it is bias optimal.*
*(d) The set of optimal bias policies is nonempty.*

**Bias Optimality vs Overtaking Optimality.** One of the most important features of bias optimality is its closed relation with overtaking optimality, also known as catching up optimality. This concept was introduced in [44] and later through the papers [1, 51]. The idea of overtaking optimality arose from problems of economic growth or capital accumulations. Nowadays, there has been a considerable work dealing with this concept by using different type of dynamic models; see, for instance, [1, 12, 26, 29, 38, 41, 47, 48]. An overview of results can be found in [7, 55]. Let us define now the concept of overtaking optimality.

**Definition 12.4.10.** A policy $f^*$ is said to be *overtaking optimal* in $\mathbb{M}$ if for every $\varepsilon > 0$, $x \in \mathbb{R}^n$, $i \in M$, and $f \in \mathbb{M}$, there exists $T_\varepsilon = T_\varepsilon(x,i,f^*,f)$ such that

$$J_T(x,i,f^*) \geq J_T(x,i,f) - \varepsilon, \quad \text{for all } T \geq T_\varepsilon. \quad (12.40)$$

If $\varepsilon = 0$, then $f^*$ is said to be *strongly overtaking optimal* in $\mathbb{M}$.

By using Ito's formula for hybrid switching diffusions [36] combined with (12.28), we can easily obtain the relation

$$J_T(x,i,f) = T\bar{r}(f) + h_f(x,i) - E_{x,i}^f h_f(x(T),\alpha(T)) \tag{12.41}$$

for each $f \in \mathbb{F}$, $x \in \mathbb{R}^n$, $i \in M$, and $T > 0$. Also, the $w$-exponential ergodicity (12.12) implies

$$E_{x,i}^f h_f(x(T),\alpha(T)) \to \mu_f(h_f) = 0. \tag{12.42}$$

These two facts in (12.41)–(12.42) are key points for the equivalence of bias optimality and overtaking optimality. The following theorem establishes this fact. For more details, we refer [29, Theorem 4.10] or [12, Theorem 6.4].

**Theorem 12.4.11.** *A policy $f^* \in \mathbb{F}$ is bias optimal if and only if it is overtaking optimal.*

**Sensitive Discount Optimality.** We have studied the concept of bias optimality and its equivalence with overtaking optimality. Each of these concepts constitutes a refinement on the set of average optimal policies because any (stationary) policy corresponding to some of these criteria (e.g., bias and overtaking) turns out to be average optimal, and, in addition, it has some other special feature, depending on what context we are dealing with. In this section, we introduce another refinement of the average reward criterion called *sensitive discount optimality*, which includes *m-discount optimality* for $m \geq -1$, and Blackwell optimality for the case $m = +\infty$. Sensitive discount optimality is a very useful tool since it gives interesting connections to some of the optimality criterion we have analyzed so far. Its characterizations are given by means of a nested system of "bias-like" optimal HJB equations.

***m*-Discounted Optimality.** We begin this part by recalling the $\rho$-discounted reward $V_\rho$ given in Definition 12.3.2. Indeed, for $\rho > 0$, $x \in \mathbb{R}^n$, $i \in M$, and $f \in \mathbb{M}$, we have defined $V_\rho(x,i,f) := E_{x,i}^f [\int_0^\infty e^{-\rho t} r(t,x(t),\alpha(t),f) dt]$ and its associated value function by $V_\rho^*(x,i) = \sup_{f \in \mathbb{M}} V_\rho(x,i,f)$. The following definition introduces the concept of *m*-discount optimality for each $m \geq -1$.

**Definition 12.4.12.** Let $m \geq -1$ be an integer. A policy $f^* \in \mathbb{M}$ is said to be *m*-discount optimal if for all $x \in \mathbb{R}^n$, $i \in M$, and $f \in \mathbb{M}$,

$$\liminf_{\rho \downarrow 0} \rho^{-m} \left[ V_\rho(x,i,f^*) - V_\rho(x,i,f) \right] \geq 0. \tag{12.43}$$

We now proceed to generalize the basic criteria (12.15) and (12.18) by using reward rates different from the function $r$ in Assumption 12.3.1. We do need, however, functions satisfying a growth rate similar to Assumption 12.3.1(b). A formal statement of this functions is given next.

**Definition 12.4.13.** Let $w$ be the function defined in Assumption 12.2.2. We define $B_w(\mathbb{R}^n \times M \times U)$ as the space of real-valued functions $v$ on $\mathbb{R}^n \times M \times U$ such that

$$\sup_{u \in U} \{v(x,i,u)\} \leq C_v w(x,i), \qquad (12.44)$$

where $C_v$ is a positive constant dependent on $v$.

*Remark 12.4.14.* (a) Observe that $B_w(\mathbb{R}^n \times M)$ is a subset of $B_w(\mathbb{R}^n \times M \times U)$ because we can write $v \in B_w(\mathbb{R}^n \times M)$ as $v(x,i) = v(x,i,u)$ for all $u \in U$.
(b) By Assumption 12.3.1(b), the reward rate $r$ is in $B_w(\mathbb{R}^n \times M \times U)$.

For each $f \in \mathbb{M}$, we shall write $v(t,x,i,f) := v(x,i,f(t,x,i))$. Similarly, $v(x,i,f) := v(x,i,f(x,i))$ for $f \in \mathbb{F}$. The following definition is a generalization of the basic criteria (12.15) and (12.18) when we use $v$ rather than $r$.

**Definition 12.4.15.** Given $f \in \mathbb{M}$, $x \in \mathbb{R}^n$, $i \in M$, and $v \in B_w(\mathbb{R}^n \times M \times U)$, we define:

(a) For each $\rho > 0$, the $\rho$-discounted $v$-reward by

$$V_\rho(x,i,f,v) := E_{x,i}^f \left[ \int_0^\infty e^{-\rho t} v(t,x(t),\alpha(t),f) dt \right]. \qquad (12.45)$$

(b) The $v$-gain of $f$ is given by

$$J(x,i,f,v) := \liminf_{T \to \infty} \frac{1}{T} E_{x,i}^f \left[ \int_0^T v(t,x(t),\alpha(t),f) dt \right]. \qquad (12.46)$$

Also, given $f \in \mathbb{F} \subset \mathbb{M}$, we define the constant

$$\bar{v}(f) := \mu_f(v(\cdot,\cdot,f)) = \sum_{k=1}^l \int_{\mathbb{R}^n} v(y,k,f) \mu_f(dy,k), \qquad (12.47)$$

where $\mu_f$ is the invariant probability measure of $(x^f(\cdot), \alpha(\cdot))$. Under Assumptions 12.2.1, 12.2.2, 12.2.5, and 12.3.1, for $v \in B_w(\mathbb{R}^n \times M \times U)$, we can follow the same analysis as given in [29, Proposition 3.6] to conclude that the $v$-gain (12.46) turns out to be the constant (12.47); i.e., $\liminf_{T \to \infty} \frac{1}{T} E_{x,i}^f \left[ \int_0^T v(x(t),\alpha(t),f) dt \right] = \bar{v}(f)$, for $f \in \mathbb{F}$. Furthermore, as in the proof of Proposition 3.6 (b) in [29], we can verify that $\bar{v}$ is uniformly bounded; in fact, $\sup_{f \in \mathbb{F}} \bar{v}(f) \leq C_v \frac{d}{c}$.

For each $f \in \mathbb{F}$, consider the operator $A_f$ on $B_w(\mathbb{R}^n \times M \times U)$ defined as follows:

$$A_f v(x,i) := \int_0^\infty \left[ E_{x,i}^f v(x(t),\alpha(t),f) - \bar{v}(f) \right] dt, \qquad (12.48)$$

where $\bar{v}$ is the constant in (12.47) (compare with the bias $h_f$ in (12.26)).

By (12.12), the operator $A_f$ is bounded by

$$|A_f v(x,i)| \leq \delta^{-1} \eta C_v w(x,i); \tag{12.49}$$

in other words, $A_f$ is uniformly bounded in the norm $\|\cdot\|_w$ in that

$$\|A_f v\|_w \leq \delta^{-1} \eta C_v. \tag{12.50}$$

Also observe that for each $f \in \mathbb{F}$, $A_f$ maps the space $B_w(\mathbb{R}^n \times M \times U)$ into $B_w(\mathbb{R}^n \times M)$.

On the other hand, by using the properties of the invariant measure $\mu_f$, we can easily deduce that

$$\mu_f(A_f v) = 0, \quad \text{for all } f \in \mathbb{F} \text{ and } v \in B_w(\mathbb{R}^n \times M \times U). \tag{12.51}$$

Moreover, for any $v \in B_w(\mathbb{R}^n \times M \times U)$, $f \in \mathbb{F}$, and $k = 0, 1, \ldots$, we define the function $h_f^k v \in B_w(\mathbb{R}^n \times M)$ by

$$h_f^k v(x,i) := (-1)^k A_f^{k+1} v(x,i) \quad \text{for all } x \in \mathbb{R}^n, i \in M, \tag{12.52}$$

where $A_f^{k+1}$ is the $(k+1)$ composition of $A_f$ with itself. In addition, we consider the special case when $v \equiv r$. In this case, we simply write $h_f^k r := h_f^k$. Hence, the definition of the bias function in (12.26) coincides with the term $h_f^0$ in (12.52). Furthermore, $h_f^1 = A_f(-h_f^0)$ becomes the bias when the reward rate is $-h_f^0$ (compare with the "bias" problem (12.35)). By using induction, it can be proven that (12.52) is equivalent to

$$h_f^k = A_f(-h_f^{k-1}). \tag{12.53}$$

Finally, from (12.51), we can easily deduce that

$$\mu_f(h_f^k) = 0, \quad \text{for all } f \in \mathbb{F}, k = 0, 1, \ldots. \tag{12.54}$$

**Poisson Equation and Average Reward HJB Equation.** The concept of $m$-discount optimality can be associated to the solution of a system of $m$ average reward HJB equations, defined in (12.58)–(12.60). In this section, we show the existence and uniqueness of a solution to this system of equations. Our approach is based on another auxiliary system of equations the so-called $-1$th, 0th, $\ldots$, $m$th Poisson equations, which are defined as follows.

**Definition 12.4.16.** Given $f \in \mathbb{F}$, and $m \geq -1$, we define the $-1$th, 0th,$\ldots$,$m$th Poisson equations by

$$g = r(x,i,f) + L^f h^0(x,i) \quad \text{for all } x \in \mathbb{R}^n \text{ and } i \in M. \tag{12.55}$$

$$h^0(x,i) = L^f h^1(x,i) \quad \text{for all } x \in \mathbb{R}^n \text{ and } i \in M. \tag{12.56}$$

$$\cdots \quad \cdots$$

$$h^m(x,i) = L^f h^{m+1}(x,i) \quad \text{for all } x \in \mathbb{R}^n \text{ and } i \in M. \tag{12.57}$$

The following result is concerned with the existence and uniqueness of a solution for (12.55)–(12.57). See [29, Theorem 5.7].

**Theorem 12.4.17.** *Fix $m \geq -1$. Then the constant $g \in \mathbb{R}$ and the functions $h^0$, $h^1$, ..., $h^m$ in $C^2(\mathbb{R}^n \times M) \cap B_w(\mathbb{R}^n \times M)$ are solutions to the Poisson equation (12.55)–(12.57) if and only if $g = \bar{r}(f)$, $h^k = h_f^k$, for $0 \leq k \leq m$, and $h^{m+1} = h_f^{m+1} + z$ for $z \in \mathbb{R}$.*

**Definition 12.4.18.** For all $x \in \mathbb{R}^n$, $i \in M$, and some fixed $m \geq -1$, we define the $-1$th, 0th,..., $m$th average reward HJB equation by

$$g = \max_{u \in U}[r(x,i,u) + L^u h^0(x,i)], \tag{12.58}$$

$$h^0(x,i) = \max_{u \in U^0(x,i)} [L^u h^1(x,i)], \tag{12.59}$$

$$\cdots \quad \cdots$$

$$h^m(x,i) = \max_{u \in U^m(x,i)} [L^u h^{m+1}(x,i)], \tag{12.60}$$

where, if we denote $U^{-1}(x,i) := U$ for all $x \in \mathbb{R}^n$ and $i \in M$, then the set $U^k(x,i)$, for $0 \leq k \leq m$, consists of the elements (controls) $u \in U^{k-1}(x,i)$ attaining the maximum in the $(k-1)$th average reward HJB equation.

*Remark 12.4.19.* From Assumptions 12.2.1 and 12.3.1, we can prove by induction that for each $m \geq 0$, the set $U^m(x,i)$ is compact for each $x \in \mathbb{R}^n$ and $i \in M$. Furthermore, for each $i \in M$, the mapping $x \longmapsto U^m(x,i)$ is upper semicontinuous and Borel measurable (see also Proposition 12.4.6).

**Definition 12.4.20.** For a fixed integer $m \geq -1$, let us denote by $\mathbb{F}_m$ the set of stationary policies $f \in \mathbb{F}$ such that $f(x,i) \in U^{m+1}(x,i)$ for each $x \in \mathbb{R}^n$, $i \in M$; that is, $f \in \mathbb{F}_m$ if and only if it attains the maximum in the $-1$th, 0th, ..., $m$th average reward HJB equations (12.58)–(12.60).

Our next result determines the *existence* and *uniqueness* of solutions to the average reward HJB equations defined above. In addition, it shows the existence of stationary policies $f \in \mathbb{F}$ that maximize the right-hand side of such equations. See [29, Theorem 5.11].

**Theorem 12.4.21.** *Fixed $m \geq -1$:*

(a) *There exists a unique solution $g \in \mathbb{R}$ $h^0$, ..., $h^{m+1}$ satisfying the average reward HJB equations (12.58)–(12.60), where $h^{m+1}$ is unique up to additive constants.*
(b) *The set $\mathbb{F}_m$ is nonempty.*

(c) If $f \in \mathbb{F}_m$, then the solution of (12.58)–(12.60) turns out $g = \bar{r}(f)$, $h^0 = h_f^0, \ldots,$
$h^{m+1} = h_f^{m+1} + z$, for some constant $z$.

**Laurent series.** In this section, we prove that the expected $\rho$-discounted $v$-reward defined in (12.45) can be expressed in terms of a Laurent series. This property will be a key point for the existence of $m$-discount optimal policies. Recall that $A_f^k$ is the $k$-composition of the operator $A_f$ in (12.48) by itself. We now establish our main result of this section with the constant $\delta$ given in (12.12). See [29, Theorem 5.12].

**Theorem 12.4.22.** *Fix $f \in \mathbb{F}$ and $v \in B_w(\mathbb{R}^n \times M)$. Then, for all $x \in \mathbb{R}^n$, and $i \in M$, the $\rho$-discounted $v$-reward (12.45) can be expressed through the following Laurent series:*

$$V_\rho(x, i, f, v) = \frac{1}{\rho}\bar{v}(f) + \sum_{k=0}^{\infty}(-\rho)^k A_f^{k+1} v(x, i). \tag{12.61}$$

*Moreover, this series converges in the $w$-norm for all $0 < \rho < \delta$.*

The following definition is associated with the residual terms of the Laurent series (12.61).

**Definition 12.4.23.** For each $v \in B_w(\mathbb{R}^m \times M \times U)$, $f \in \mathbb{F}$, and $k = 0, 1, \ldots$, we define the $k$-residual of the Laurent series (12.61) by

$$\Psi_k(f, v, \rho) := \sum_{j=k}^{\infty}(-\rho)^j A_f^{j+1} v. \tag{12.62}$$

Our next result shows that the above residual terms are *bounded* in the $w$-norm. See [29, Proposition 4.14].

**Proposition 12.4.24.** *Consider a positive constant $\gamma < \delta$, where $\delta$ is the constant in (12.12). Then, for all $\rho \leq \gamma$, and $k \geq 0$,*

$$\sup_{f \in \mathbb{F}} \|\Psi_k(f, v, \rho)\|_w \leq \frac{C_v \eta}{\delta^k(\delta - \gamma)} \rho^k. \tag{12.63}$$

A special case of (12.61) is $v \equiv r$. In this case, the $\rho$-discounted reward in (12.15) is expressed by

$$V_\rho(x, i, f) = \frac{1}{\rho}\bar{r}(f) + \sum_{k=0}^{\infty}\rho^k h_f^k(x, i) \tag{12.64}$$

for all $f \in \mathbb{F}$, $x \in \mathbb{R}^n$, $i \in M$, $h_f^k$ as in (12.53) and $\rho$ satisfying the condition of Theorem 12.4.22.

**Existence and Characterizations of $m$-Discounted Optimal Policies.** In this part, we show the existence of $m$-discount optimal policies for $m \geq -1$. To this end, we mainly use the results developed in this section, which lead us to state the characterizations of $m$-discount optimality. Before stating our main results, we shall introduce an important concept concerned with a *lexicographical order*.

**Definition 12.4.25.** For any two vectors $a$ and $b$ in $\mathbb{R}^n$, we say that $a$ is lexicographical greater than or equal to $b$, denoted by $a \succeq b$, if the first nonzero component of $a - b$ is positive. We also write $a \succ b$ when $a \succeq b$ and $a \neq b$.

The following theorem provides some characterizations of $m$-discount optimal policies. See [29, Theorem 5.16].

**Theorem 12.4.26.** *Fixed $m \geq -1$, the following statements are equivalents:*

(a) *$f$ lexicographically maximizes the terms $\bar{r}(f)$, $h_f^0$, ..., $h_f^m$ of the Laurent series* (12.64).
(b) *$f$ belongs to $\mathbb{F}_m$.*
(c) *$f$ is $m$-discount optimal.*

The next corollary is a straightforward consequence of Theorems 12.4.21 and 12.4.26.

**Corollary 12.4.27.** *For each $m \geq -1$, there exists an $m$-discount optimal policy.*

**Blackwell Optimality.** In the last sections, we analyzed $m$-discount optimality for any arbitrary $m \geq -1$. In this section, we focus on the study of Blackwell optimality, which is based on $m$-discount optimality. Blackwell optimality was introduced by [3]. Later on, the concept of Blackwell optimality was studied for different classes of control systems; see for instance, [8, 10, 11, 24, 25, 32] for discrete-time MDPs, [39] for continuous-time controlled Markov chains, [28, 42] for controlled diffusion processes, and [29] for switching diffusions. To proceed, the definition of Blackwell optimality is given next.

**Definition 12.4.28.** A policy $f \in \mathbb{M}$ is called *Blackwell optimal* if and only if, for any other policy $\hat{f} \in \mathbb{M}$, and any initial states $x \in \mathbb{R}^n$, $i \in M$, there exists a positive constant $\rho^* = \rho^*(x, i, \hat{f})$ such that

$$V_\rho(x, i, f) \geq V_\rho(x, i, \hat{f}), \quad \text{for all } 0 < \rho \leq \rho^*. \qquad (12.65)$$

We now introduce some preliminary results necessary for our analysis; see [29, Proposition 5.27].

**Proposition 12.4.29.** (a) *Consider the set $U^m(\cdot, \cdot)$ in Definition 12.4.18. Then, the sequence $\{U^m\}_m$ is decreasing. Furthermore, for each $x \in \mathbb{R}^n$ and $i \in M$,*

$$U^\infty(x, i) := \bigcap_{m=1}^\infty U^m(x, i) = \lim_{m \to \infty} U^m(x, i) \qquad (12.66)$$

*is a nonempty set.*

*(b)  Consider the set $\mathbb{F}_m$ in Definition 12.4.20. Hence, there exists a policy $f \in \mathbb{F}$
contained in*

$$\mathbb{F}_\infty := \bigcap_{m=1}^{\infty} \mathbb{F}_m = \lim_{m\to\infty} \mathbb{F}_m. \tag{12.67}$$

We now establish sufficient and necessary conditions for a policy $f \in \mathbb{F}$ to be
Blackwell optimal; see [29, Theorem 5.28].

**Theorem 12.4.30.** *A policy $f \in \mathbb{F}$ is Blackwell optimal if and only if it is contained
in $\mathbb{F}_\infty$ defined in* (12.67).

Theorem 12.4.30 and the definition of $\mathbb{F}_\infty$ in (12.67) lead to the existence of
Blackwell optimal policies, as established in the following corollary.

**Corollary 12.4.31.** *The set of Blackwell optimal policies is nonempty.*

## 12.5  Concluding Remarks

In this chapter, we have studied several classes of infinite-horizon optimal control
problems for a general class of hybrid switching diffusions. The analysis was done
by using different infinite-horizon criteria, namely, basic and advanced criteria. The
motivation for studying advanced criteria stems from the objective of improving the
basic criteria.

Our analysis provides a summary for the existence and characterizations of
optimal control policies related to a wide variety of infinite-horizon problems.
We can extend the results to the study for more advanced criteria, for instance,
the problem of finding average optimal policies that, in addition, minimize the
asymptotic variance. Currently, as an ongoing project, we are studying zero-sum
and non-zero-sum stochastic hybrid differential games. One of the disadvantages of
our approach is the assumption of uniform ellipticity hypothesis, which excludes
singular systems (e.g., piecewise deterministic systems). The degenerate cases and
singular systems deserve further thoughts and careful considerations.

**Acknowledgements**  The research of the first author was supported in part by the Air Force Office
of Scientific Research under FA9550-10-1-0210. The research of the second author was supported
in part by the National Science Foundation under DMS-0907753 and in part by the Air Force
Office of Scientific Research under FA9550-10-1-0210.

## References

1. Atsumi, H. (1965). Neoclassical growth and the efficient program of capital accumulation. *Rev. Econom. Studies* **32**, pp. 127–136.
2. Basak, G.K., Borkar, V.S., Ghosh, M.K. (1997). Ergodic control of degenerate diffusions. *Stoch. Anal. Appl.* **15**, pp. 1–17.

3. Blackwell D. (1962). Discrete dynamic programming. *Annals of mathematical statistics.* **33**, pp. 719–726.
4. Borkar, V.S. (2005). Controlled diffusion processes. *Probab. Surveys* **2**, pp. 213–244.
5. Borkar, V.S., Ghosh, M.K. (1988). Ergodic control of multidimensional diffusions I: The existence of results. *SIAM J. Control Optim* **26**, pp. 112–126.
6. Borkar, V.S., Ghosh, M.K. (1990). Ergodic control of multidimensional diffusions II: Adaptive control. *Appl. Math. Opt.* **21**, pp. 191–220.
7. Carlson, D.A., Haurie, D.B., Leizarowitz, A. (1991). *Infinite Horizon Optimal Control: Deterministic and Stochastic Systems.* Springer-Verlag, Berlin.
8. Carrasco, G., Hernández-Lerma, O. (2003). Blackwell optimality for Markov control processes in Borel spaces. Preliminary report.
9. Dana, R.A., Le Van, C. (1990). On the Bellman equation of the overtaking criterion. *J. Optim. Theory Appl.* **67**, pp. 587–600.
10. Dekker R., Hordijk A. (1988). Average, sensitive and Blackwell optimal policies in denumerable Markov decision chains with unbounded rewards. *Math. Oper. Res.* **13**, pp. 395–420.
11. Dekker R., Hordijk A. (1992). Recurrence conditions for averageand Blackwell optimality in denumerable state Markov decision chains. *Math. Oper. Res.* **17**, pp. 271–289.
12. Escobedo-Trujillo, B.A., Hernández-Lerma, O. (2011). Overtaking optimality for controlled Markov-modulated diffusions. Submitted.
13. Fort, G., Roberts, G.O. (2005). Subgeometric ergodicity of strong Markov processes. *Ann. Appl. Probab.* **15**, pp. 1565–1589.
14. Ghosh, M.K., Arapostathis, A., Marcus, S.I. (1993). Optimal control of switching diffusions to flexible manufacturing systems. *SIAM J. Control Optim.* **31**, pp. 1183–1204.
15. Ghosh, M.K., Arapostathis, A., Marcus, S.I. (1997). Ergodic control of switching diffusions. *SIAM J. Control Optim.* **35**, pp. 1952–1988.
16. Guo, X.P., Hernández-Lerma, O. (2003). Continuous-time controlled Markov chains with discounted rewards. *Acta Appl. Math.* **79**, pp. 195–216.
17. Guo, X.P., Hernández-Lerma, O. (2003). Drift and monotonicity conditions for continuous-time controlled Markov chains with average criterion. *IEEE Trans. Automatic Control.* **48**, pp. 236–244.
18. Hashemi, S.N., Heunis, A.J. (2005). On the Poisson equation for singular diffusions. *Stochastics* **77**, pp. 155–189.
19. Hernández-Lerma, O. (1994). *Lectures on Continuous-Time Markov Control Processes.* Sociedad Matemática Mexicana, Mexico City.
20. Hernández-Lerma, O., Lasserre, J.B. (1996). *Discrete-Time Markov Control Processes.* Springer, New York.
21. Hernández-Lerma, O., Lasserre, J.B. (1999). *Further Topics on Discrete-Time Markov Control Processes.* Springer, New York.
22. Hilgert, N., Hernández-Lerma, O. (2003). Bias optimality versus strong 0-discount optimality in Markov control processes with unbounded costs. *Acta Appl. Math.* **77**, 215–235.
23. Himmelberg, C.J., Parthasarathy, T., Van Vleck, F.S. (1976). Optimal plans for dynamic programming problems. *Math. Oper. Res.* **1**, pp. 390–394.
24. Hordijk, A., Yushkevich, A.A. (1999). Blackwell optimality in the class of stationary policies in Markov decision chains with a Borel state and unbounded rewards. *Math. Meth. Oper. Res.* **49**, pp. 1–39.
25. Hordijk, A., Yushkevich, A.A. (1999). Blackwell optimality in the class of all policies in Markov decision chains with a Borel state and unbounded rewards. *Math. Meth. Oper. Res.* **50**, pp. 421–448.
26. Jasso-Fuentes, H., Hernández-Lerma, O. (2008). Characterizations of overtaking optimality for controlled diffusion processes. *Appl. Math. Optim.* **57**, pp. 349–369.
27. Jasso-Fuentes, H., Hernández-Lerma, O. (2009). Ergodic control, bias and sensitive discount optimality for Markov diffusion processes. *Stoch. Anal. Appl.* **27**, pp. 363–385.
28. Jasso-Fuentes, H., Hernández-Lerma, O. (2009). Blackwell optimality for controlled diffusion processes. *J. Appl. Probab.* **46**, pp. 372–391.

29. Jasso-Fuentes, H., Yin, G. (2012). Advanced criteria for controlled Markov-modulated diffusions in an infinite horizon: overtaking, bias, and Blackwell optimality. Submitted.
30. Kushner, H.J. (1967). Optimal discounted stochastic control for diffusions processes. *SIAM J. Control* **5**, pp. 520–531.
31. Kushner, H.J. (1978). Optimality conditions for the average cost per unit time problem with a diffusion model. *SIAM J. Control Optim* **16**, pp. 330–346.
32. Lasserre J.B. (1988). Conditions for the existence of average and Blackwell optimal stationary policies in denumerable Markov decision processes. *J. Math. Analysis Appl.* **136**, pp. 479–490.
33. Le Van, C., Dana, R.-A. (2003). *Dynamic Programming in Economics*. Kluwer, Dordrecht.
34. Leizarowitz, A. (1988). Controlled diffusion process on infinite horizon with the overtaking criterion. *Appl. Math. Optim.* **17**, pp. 61–78.
35. Leizarowitz, A. (1990). Optimal controls for diffusion in $\mathbb{R}^d$—min-max max-min formula for the minimal cost growth rate. *J. Math. Anal. Appl.* **149**, pp. 180–209.
36. Mao, X., Yuan, C. (2006). *Stochastic differential equations with markovian switching*. Imperial College Press, London.
37. Meyn, S.P., Tweedie, R.L. (1993). Stability of Markovian processes III: Foster-Lyapunov criteria for continuous-time processes. *Adv. Appl. Prob.* **25**, pp. 518–548.
38. Prieto-Rumeau, T., Hernandez-Lerma, O. (2005). Bias and overtaking equilibria for zero-sum continuous-time Markov games. *Math. Meth. Oper. Res.* **61**, 437–454.
39. Prieto-Rumeau, T., Hernandez-Lerma, O. (2005). The Laurent series, sensitive discount and Blackwell optimality for continuous-time controlled Markov chains. *Math. Meth. Oper. Res.* **61**, 123–145.
40. Prieto-Rumeau, T., Hernández-Lerma, O. (2006). A unified approach to continuous-time discounted Markov control processes. *Morfismos* **10**. pp. 1–40.
41. Prieto-Rumeau, T., Hernández-Lerma, O. (2006). Bias optimality of continuous-time controlled Markov chains. *SIAM J. Control Optim.* **45**, pp. 51–73.
42. Puterman, M.L. (1994). Sensitive discount optimality in controlled one-dimensional diffusions. *Ann. Probab.* **2**, pp. 408–419.
43. Puterman, M.L. (1994). *Markov Decision Processes: Discrete Stochastic Dynamic Programming*, J. Wiley, New York.
44. Ramsey, F.P. (1928). A mathematical theory of savings. *Economics J.* **38**, pp. 543–559.
45. Ross, S. (1984). *Introduction to Stochastic Dynamic Programming*, Academic Press, New York.
46. Schäl, M. (1975). Conditions for optimality and for the limit of *n*-stage optimal policies to be optimal. *Z. Wahrs. verw. Gerb.* **32**, 179–196.
47. Tan, H., Rugh, W.J. (1998). Nonlinear overtaking optimal control: sufficiency, stability, and approximation. *IEEE Trans. Automatic Control* **43**, 1703–1718.
48. Tan, H., Rugh, W.J. (1998). On overtaking optimal tracking for linear systems. *Syst. Control Lett.* **33**, 63–72.
49. Tarres, R. (1985). Asymptotic evolution of a stochastic problem. *SIAM J. Control Optim.* **23**, pp. 614–631.
50. Veinott, A.F. Jr. (1969). Discrete dynamic programming with sensitive discount optimality criteria. *Ann. Math. Statist.* **40**, 1635–1660.
51. von Weizsäcker, C.C. (1965). Existence of optimal programs of accumulation for an infinite horizon. *Rev. Economic Stud.* **32**, pp. 85–104.
52. Yin, G., Zhang, Q. (1998). *Continuous-Time Markov Chains and Applications: A Singular Perturbation Approach*, Springer-Verlag, New York.
53. Yin, G., Zhang, Q. (2005), *Discrete-time Markov Chains: Two-time-scale Methods and Applications*, Springer, New York.
54. Yin, G., Zhu, C. (2010). *Hybrid switching diffusions*. Springer, New York.
55. Zaslavski, A.J. (2006). *Turnpike Properties in the Calculus of Variations and Optimal Control*. Springer, New York.

56. Zhu, C., Yin, G. (2007). Asymptotic properties of hybrid diffusion systems. *SIAM J. Control Optim.* **46**, pp. 1155–1179.
57. Zhu, Q., Guo, X.P. (2005). Another set of conditions for strong $n$ ($n = -1, 0$) discount optimality in Markov decision processes. *Stoch. Anal. Appl.* **23**, 953–974.

# Chapter 13
# Fluid Approximations to Markov Decision Processes with Local Transitions

Alexey Piunovskiy and Yi Zhang

## 13.1 Introduction

Markov decision processes (MDPs) model many practical problems that arise from queueing systems, telecommunication, inventories, and so on, see [6, 7, 15]. The fundamental results about an MDP model are the existence of an optimal policy and the sufficiency of the deterministic stationary policies out of the more general class of randomized history-dependent ones. On the other hand, from practical point of view, it is at least of equal importance to know how to obtain an optimal or nearly optimal policy. It is known that practically, the policy iteration and value iteration procedures fail to cope with MDP models with large state and action spaces. So for random walks, it is often the case that a deterministic continuous model is taken for analysis even when the underlying problem is in stochastic nature. This is called a fluid approximation.

Fluid approximations are widely used to solve practical problems; examples in the contexts of epidemiology and telecommunication can be found in [11, 16], respectively. In inventory control, the well-known (deterministic) economic-order quantity model can be viewed as a fluid approximation, too, cf. [14]. On the one hand, such fluid models can often be solved much more easily than the corresponding stochastic models. On the other hand, in most cases, they are applied without formal analytical justifications, or the justification focuses on the trajectory level, by showing the trajectory of a scaled stochastic model converges in some

A. Piunovskiy (✉) • Y. Zhang
Department of Mathematical Sciences,
University of Liverpool, Liverpool, L69 7ZL, UK
e-mail: piunov@liv.ac.uk; zy1985@liv.ac.uk

D. Hernández-Hernández and A. Minjárez-Sosa (eds.), *Optimization, Control, and Applications of Stochastic Systems*, Systems & Control: Foundations & Applications, DOI 10.1007/978-0-8176-8337-5_13, © Springer Science+Business Media, LLC 2012

sense to the one of the fluid model, and this is mainly considered for a continuous-time model, see [8–10] and the references therein. For the justification on the objective level, we refer the reader to [3, 4, 13] and the references therein.

In this chapter, we justify (at the level of objective functions) fluid approximations to a discrete-time MDP model with an undiscounted total cost criterion. This is done for an uncontrolled discrete-time model in [12] under more restrictive conditions. The argument is based on [1, 12, 13].

The rest of this chapter is organized as follows. We describe the concerned MDP model in Sect. 13.2 and formulate the main statements in Sect. 13.3, where two sections are devoted to the standard fluid approximation and the refined fluid approximation. We finish this chapter with a conclusion.

## 13.2 MDP Model

The MDP model under consideration is defined by the following elements:

- $X = \{0, 1, 2, \dots\}$ is the state space.
- $A$ is the action space, which can be an arbitrary non-empty Borel space, whose topology is omitted from the explicit presentation.
- $p(z|x,a)$ is the one-step transition probability, a stochastic kernel on $X$ given $X \times A$ and (Borel) measurable in $a \in A$.
- $r(x,a)$ is the one-step cost, which is (Borel) measurable in $a \in A$.

Assume that the real measurable functions $q^+(y,a), q^-(y,a)$ and $\rho(y,a)$ on $[0,\infty) \times A$ are given such that $q^+(0,a) = q^-(0,a) = \rho(0,a) = 0$, and on $(0,\infty) \times A$, $q^+(y,a) > 0$, $q^-(y,a) > 0$, and $q^+(y,a) + q^-(y,a) \leq 1$. Then we make the MDP model with the absorbing state zero and local transitions only by defining the one-step transition probability and cost via

$$p(z|x,a) = \begin{cases} q^+(x,a), & \text{if } z = x+1; \\ q^-(x,a), & \text{if } z = x-1; \\ 1 - q^+(x,a) - q^-(x,a), & \text{if } z = x; \\ 0 & \text{otherwise,} \end{cases}$$

$$r(x,a) = \rho(x,a).$$

Let $\varphi : X \to A$ be a deterministic stationary policy. For any fixed initial state $x$ and policy $\varphi$, the standard canonical construction gives a strategic measure $P_x^{\varphi}$ on the space of histories in the form of $x_0, a_0, x_1, a_1, \dots$, and the corresponding expectation is denoted by $E_x^{\varphi}$, see [7]. We denote the controlled process by $\{X_t, t = 0, 1, \dots\}$ and the action process by $\{A_t, t = 0, 1, \dots\}$. Then the MDP model is the following optimization problem, which is well defined after we impose some conditions below:

$$V^{\varphi}(x) := E_x^{\varphi}\left[\sum_{t=0}^{\infty} r(X_t, A_t)\right] \to \inf_{\varphi},$$

where the infimum is taken over the class of deterministic stationary policies only for simplicity and that under very general conditions, they suffice for the underlying optimization problem, see [1] for more details.

In this chapter, we shall actually scale the above described MDP model such that for any fixed scaling parameter $n = 1, 2, \ldots$, the elements of the $n$-MDP model are as follows:

- $X = \{0, 1, 2, \ldots\}$ and $A$ remain as the state and action spaces.
- 

$$
{}^{n}p(z|x,a) = \begin{cases} q^{+}(x/n,a), & \text{if } z = x+1; \\ q^{-}(x/n,a), & \text{if } z = x-1; \\ 1 - q^{+}(x/n,a) - q^{-}(x/n,a), & \text{if } z = x; \\ 0 & \text{otherwise} \end{cases}
$$

is the one-step transition probability.

- ${}^{n}r(x,a) = \frac{\rho(x/n,a)}{n}$ is the one-step cost, which is measurable in $a \in A$.

The $n$-MDP model reads

$$
{}^{n}V^{\varphi}(x) := E_{x}^{\varphi}\left[\sum_{t=0}^{\infty} {}^{n}r(X_{t}, A_{t})\right] \to \inf_{\varphi}.
$$

Below, we impose some conditions to guarantee that the $n$-MDP model under consideration is absorbing in the sense of [1]. To be exact, that means given any initial state $x$, $E_{x}^{\varphi}[T_{0}] < \infty$, where $T_{0} := \inf\{t > 0 : X_{t} = 0\}$. Here and below, the context always makes it clear what the scaling parameter is so that the controlled and action processes in the $n$-MDP model are still denoted by $\{X_{t}, t = 0, 1, \ldots\}$ and $\{A_{t}, t = 0, 1, \ldots\}$ for brevity.

The above scaling is called the fluid scaling. Its intuitive meaning together with its importance is now explained via the following example, where an uncontrolled situation is considered for simplicity. Accordingly, simpler denotations are employed. We remark that the example is better understood in the context of telecommunication, where fluid models are widely used to solve satisfactorily practical problems of stochastic nature, see [2, 16] and the reference therein.

*Example 13.2.1.* Suppose information packets, 1 kilobit (KB) each, arrive at a router (switch) at the (constant) rate $q^{+} > 0$ megabit/second (MB/s), and are served at the (constant) rate $q^{-} > q^{+}$ MB/s, where $q^{+} + q^{-} \leq 1$. We observe the process up to the moment when the router buffer is empty. Let the holding cost be $h$ per MB per second, so that $\rho(y) = hy$, where $y$ is the amount of information (MB). For simplicity, we consider the uncontrolled model so that the denotation of the policy $\varphi$ does not appear. One can consider batch arrivals and batch services of 1,000 packets every second; then $n = 1$ and $r(x) = hx = \rho(x)$.

On the other hand, it would be more accurate to consider particular packets; then probabilities $q^{+}$ and $q^{-}$ will be the same, but the time unit is $\frac{1}{1000}$ s, so that the arrival and service rates (MB/s) remain the same. Remembering that, we consider information up to the individual packets (cf. batches) and the time unit is $\frac{1}{1000}$ s, the cost function for the $n$-model will obviously change: $^nr(x) = \frac{hx}{1000}\frac{1}{1000} = \frac{\rho(x/n)}{n}$, where $n = 1,000$.                                                                                      □

The goal of this chapter is to estimate (from the above) the differences between

$$^nV^\varphi(^nX_0) := E^\varphi_{nX_0}\left[\sum_{t=0}^{\infty} {}^nr(X_t, A_t)\right]$$

and the objective functions of two related deterministic continuous models, namely, the standard fluid model and the refined fluid model, which are simpler to solve. So they are regarded and used as the fluid approximations to the original (stochastic) MDP model, see [11, 16] for examples. In greater detail, under some conditions, we provide explicit upper boundary estimates of the absolute differences between the objective functions of the stochastic and the corresponding fluid models in the initial data, which are understood as the level of accuracy of such fluid approximations. In a nutshell, under the imposed conditions, the absolute differences go to zero as fast as $\frac{1}{n}$, with $n$ being the scaling parameter.

## 13.3    Main Statements

### 13.3.1    Standard Fluid Approximation

Firstly, we motivate the standard fluid model by using Example 13.2.1. Then we give its formal definition and obtain its level of accuracy in approximating the $n$-MDP model.

*Example 13.2.1 continued.* Consider the situation in Example 13.2.1. The total holding cost of the $n$-model $^nV(x)$ up to the absorption at the state zero satisfies the equation (cf.[12, (10)])

$$\frac{\rho\left(\frac{x}{n}\right)}{n} + q^+ \, {}^nV(x+1) + q^- \, {}^nV(x-1) - (q^+ + q^-) \, {}^nV(x) = 0,$$

$$x = 1, 2, \ldots; {}^nV(0) = 0. \tag{13.1}$$

Since we measure information in MB, it is reasonable to introduce the function $\hat{v}(y)$ such that $^nV(x) = \hat{v}(x/n)$, where $\hat{v}(y)$ is the total holding cost up to the absorption given the initial queue was $y$ MB. After the obvious rearrangements of (13.1), we obtain that $\hat{v}(x/n)$ satisfies

$$\rho\left(\frac{x}{n}\right) + \frac{q^+\left\{\hat{v}\left(\frac{x}{n}+\frac{1}{n}\right) - \hat{v}\left(\frac{x}{n}\right)\right\}}{\frac{1}{n}} + \frac{q^-\left\{\hat{v}\left(\frac{x}{n}-\frac{1}{n}\right) - \hat{v}\left(\frac{x}{n}\right)\right\}}{\frac{1}{n}} = 0,$$

This is a version of the Euler method for solving the differential equation

$$\rho(y) + (q^+ - q^-)\frac{dv(y)}{dy} = 0. \tag{13.2}$$

Thus, we expect that $^nV(x) = \hat{v}(x/n) \approx v(x/n)$ at least for a big value of $n$.   □

The above example reveals that as far as the objective function is concerned, the $n$-MDP model can be approximated by a deterministic continuous model specified by a differential equation, at least for a big value of $n$. This gives the rise to the following standard fluid model.

*The standard fluid model:*

$$v^{\psi}(y_0) := \int_0^{\infty} \rho(y(\tau), \psi(y(\tau)))d\tau \to \inf_{\psi}$$

subject to $\dfrac{dy}{d\tau} = q^+(y, \psi(y)) - q^-(y, \psi(y))$, with the given initial state $y(0)$.

Here, $\psi$ is a measurable mapping from $[0, \infty)$ to $A$. Later, we often omit the argument $\tau$ from $y(\tau)$ for brevity. Under the conditions of Theorem 13.3.1 below, it can be seen that

$$v^{\psi}(y) = \int_0^y \frac{\rho(z, \psi(z))}{q^-(z, \psi(z)) - q^+(z, \psi(z))}dz, \tag{13.3}$$

cf. (13.2).

**Theorem 13.3.1 (cf. Theorem 1 in [12]).** *Let $n = 1, 2, \ldots$ and a policy $\psi$ for the fluid model be fixed, and $\hat{\phi}$ be given by $\hat{\phi}(x) := \psi(x/n)$. Suppose*

$$q^-(y, \psi(y)) > \underline{q} > 0, \ \inf_{y>0}\frac{q^-(y, \psi(y))}{q^+(y, \psi(y))} \geq \tilde{\eta} > 1, \ \sup_{y>0}\frac{|\rho(y, \psi(y))|}{\eta^y} < \infty,$$

*where $\underline{q} > 0$ is a constant, and $\eta \in (1, \tilde{\eta})$. Then:*

(a) $\sup_{x=1,2,\ldots}\frac{|^nV^{\hat{\phi}}(x)|}{\eta^{x/n}} < \infty$, *i.e.,* $^nV^{\hat{\phi}}(\cdot)$ *is* $^n\eta$-*bounded, where* $^n\eta(x) := \eta^{x/n}$.
(b) *If additionally the functions $q^+(y, \psi(y))$, $q^-(y, \psi(y))$, $\rho(y, \psi(y))$ are piecewise continuously differentiable, then for an arbitrarily fixed $\hat{y} \geq 0$*

$$\lim_{n\to\infty} \sup_{x\in\{0,1,\ldots,[n\hat{y}]\}} |^nV^{\hat{\phi}}(x) - v^{\psi}(x/n)| = 0,$$

*where the function $[\cdot]$ takes the integer part of its argument.*

(c) *If furthermore functions* $\rho(y, \psi(y))$, $q^-(y, \psi(y))$, $q^+(y, \psi(y))$ *are continuously differentiable with uniformly bounded derivatives so that* $\sup_{y>0} \left| \frac{d^2 v^\psi(y)}{dy^2} \right| :=$ $C < \infty$, *then for any arbitrarily fixed* $\hat{y} \geq 0$,

$$\sup_{x \in \{0,1,\ldots,[n\hat{y}]\}} \left| {}^n V^{\hat{\phi}}(x) - v^\psi(x/n) \right| \leq \frac{C(3\tilde{\eta}+1)}{2\gamma\tilde{\eta}} \frac{(\hat{y}+1)\tilde{\eta}^{\hat{y}}}{n}.$$

The detailed proof of the above theorem can be found in [12], see the proof of Theorem 1 therein.

The next example shows that the condition $\sup_{y>0} \frac{|\rho(y,\psi(y))|}{\eta^y} < \infty$ in the above theorem is important.

*Example 13.3.2 (cf. Example 3 in [12]).* Let $\mathbf{A} = [1,2]$, $q^+(y,a) = ad^+$, $q^-(y,a) = ad^-$ for $y > 0$, where $d^- > d^+ > 0$ are fixed numbers such that $2(d^+ + d^-) \leq 1$. Put $\rho(y,a) = a^2 \gamma^{y^2}$, where $\gamma > 1$ is a constant. So $\tilde{\eta} = \frac{d^-}{d^+} > 1$.

To solve the fluid model $v^\psi(y) \to \inf_\psi$, we use the dynamic programming approach. One can see that the Bellman function $v^*(y) := \inf_\psi v^\psi(y)$ has the form

$$v^*(y) = \int_0^y \inf_{a \in A} \left\{ \frac{\rho(u,a)}{q^-(u,a) - q^+(u,a)} \right\} du,$$

and satisfies the Bellman equation

$$\inf_{a \in \mathbf{A}} \left\{ \frac{dv^*(y)}{dy} [q^+(y,a) - q^-(y,a)] + \rho(y,a) \right\} = 0; \quad v^*(0) = 0,$$

cf. [13, Lemma 2] and the "incidental" statement in its proof. Here, we remark that the function $\inf_{a \in A} \frac{\rho(u,a)}{q^-(u,a) - q^+(u,a)}$ is universally measurable, see [5, Chap. 7] for more details. Hence, the function

$$v^*(y) = v^{\psi^*}(y) = \int_0^y \frac{\gamma^{u^2}}{d^- - d^+} du$$

is well defined, and $\psi^*(y) \equiv 1$ is the optimal policy.

We notice that the condition $\sup_{y>0} \frac{|\rho(y,\psi^*(y))|}{\eta^y} < \infty$ is not satisfied, while all the other requirements of Theorem 13.3.1 are met.

On the other hand, for the policy given by $\hat{\phi}(x) = \psi^*(\frac{x}{n}) \equiv 1$ and $n = 1, 2, \ldots$, ${}^n V^{\hat{\phi}}(x) = {}^n E^{\hat{\phi}} [\sum_{t=0}^\infty {}^n r(X_t, A_t)]$ satisfies the following equation

$$\frac{\gamma^{(\frac{x}{n})^2}}{n} + d^+ \, {}^n V^{\hat{\phi}}(x+1) + d^- \, {}^n V^{\hat{\phi}}(x-1) - (d^+ + d^-) {}^n V^{\hat{\phi}}(x) = 0; \quad {}^n V^{\hat{\phi}}(0) = 0,$$

cf. (13.1). But then, this equation does not admit non-negative finitely valued solutions, because if we put $^nV^{\hat{\phi}}(0) = 0, ^nV^{\hat{\phi}}(1) = b$, where $b \in [0,\infty)$ is a non-negative constant, then for any $x = 1,2,\dots,$

$$^nV^{\hat{\phi}}(x) = b\frac{\tilde{\eta}^x - 1}{\tilde{\eta} - 1} - \frac{1}{nd^+(\tilde{\eta}-1)}\sum_{j=1}^{x-1}\gamma^{(j/n)^2}(\tilde{\eta}^{x-j} - 1),$$

and thus, for a big enough value of $x$, we obtain $^nV^{\hat{\phi}}(x) < 0$. Therefore, Theorem 13.3.1 does not hold; $^nV^{\hat{\phi}}(x) = \infty$ for all $x = 1,2,\dots$.    □

## 13.3.2   Refined Fluid Approximation

Under the conditions of Theorem 13.3.1 except for $q^-(y,\psi(y)) > \underline{q}$, (13.3) may not hold, which in comparison with (13.2), suggests that the standard fluid approximation may fail to be accurate in this case; since $q^-(y)$ can now approach zero, and $q^+(y) < q^-(y)$, it could happen that the standard fluid model does not get absorbed at the state zero, while for any fixed $n = 1,2,\dots$, the stochastic process $\{^nX_t, t = 0,1,\dots\}$ indeed gets absorbed at the state zero, so that the standard fluid model and the $n$-MDP model could behave qualitatively differently. Example 13.3.4 below illustrates this situation. Nevertheless, in this case, the refined fluid model introduced below still approximates well the $n$-MDP model under the following condition.

Let $\psi(\cdot)$ be a measurable mapping from $[0,\infty)$ to $A$. We formulate the following condition.

**Condition A.** (a) $\inf_{y>0}\frac{q^-(y,\psi(y))}{q^+(y,\psi(y))} \geq \tilde{\eta} > 1$, $\sup_{y>0}\frac{|\rho(y,\psi(y))|}{\{q^+(y,\psi(y))+q^-(y,\psi(y))\}\eta^y} \leq C_1 <$ $\infty$, where $\eta \in (1,\tilde{\eta})$.
(b) For any $n$, there exists an ($n$-dependent) constant $K(n)$ such that

$$^nl_W(x) := K(n)q^-(x/n, \psi(x/n))\eta^{x/n} - 1 > 0, x = 1,2,\dots,$$

and

$$\sup_{x=1,2,\dots}\frac{|\rho(x/n, \psi(x/n))|}{^nl_W(x)} = \sup_{x=1,2,\dots}\frac{|\rho(x/n, \psi(x/n))|}{K(n)q^-(x/n, \psi(x/n))\eta^{x/n} - 1} < \infty,$$

where $\eta \in (1,\tilde{\eta})$ comes from part (a) of this condition.
(c) There exist points $y_1, y_2, \dots, y_l, \dots$ with $y_l \to \infty$ as $l \to \infty$ such that on each of the intervals $(0,y_1), (y_1,y_2), \dots$, the function $\frac{\rho(y,\psi(y))}{q^-(y,\psi(y)) - q^+(y,\psi(y))}$ is Lipschitz continuous.

Simple sufficient conditions for Condition A(b) are given below: see Proposition 13.3.1 and its proof.

*Refined fluid model:*

$$\int_0^\infty \frac{\rho(y, \psi(y))}{q^+(y, \psi(y)) + q^-(y, \psi(y))} du \to \inf_\psi$$

subject to $\dfrac{dy}{du} = \dfrac{q^+(y, \psi(y)) - q^-(y, \psi(y))}{q^+(y, \psi(y)) + q^-(y, \psi(y))}$, with a given initial state $y(0)$.

One can show under Condition A that

$$\tilde{v}^\psi(y_0) := \int_0^\infty \frac{\rho(y, \psi(y))}{q^+(y, \psi(y)) + q^-(y, \psi(y))} d\tau = \int_0^{y_0} \frac{\rho(z, \psi(z))}{q^-(z, \psi(z)) - q^+(z, \psi(z))} dz.$$

Note that $\frac{q^+(y, \psi(y)) - q^-(y, \psi(y))}{q^+(y, \psi(y)) + q^-(y, \psi(y))} \le \frac{1 - \tilde{\eta}}{1 + \tilde{\eta}} < 0$, so that $y(\cdot)$ in the fluid model is absorbed at the state zero in finite time.

**Theorem 13.3.2.** *Let $n = 1, 2, \ldots$ and $\hat{y} > 0$ be fixed, $\psi$ a measurable mapping from $[0, \infty)$ to $A$, and $\hat{\varphi}(\cdot)$ given by $\hat{\varphi}(x) := \psi(x/n)$. Suppose Condition A is satisfied for $\psi$ so that there exist an integer $L$ such that the function $\frac{\rho(y, \psi(y))}{q^-(y, \psi(y)) - q^+(y, \psi(y))}$ is Lipschitz continuous with the common Lipschitz constant $D$ on the intervals $(0, y_1)$, $(y_1, y_2), \ldots, (y_{L-1}, y_L), (y_l, y_{L+1})$ with $y_L < \hat{y} + 1 \le y_{L+1}$. Then*

$$\sup_{x \in \{0, 1, \ldots, [\hat{y}n]\}} \left| {}^n V^{\hat{\varphi}}(x) - \tilde{v}^\psi(x/n) \right| \le \frac{\varepsilon(n)}{2}.$$

*Here and below, we put*

$$\varepsilon(n) := \frac{2K_1}{n} + \frac{2K_2}{\tilde{\eta}^n} + 2K_3(\eta^{1/n} - 1),$$

$$K_1 := \frac{\tilde{\eta} + 1}{\tilde{\eta} - 1} [D(\hat{y} + 1) + 3C_1 L \eta^{\hat{y}+1}],$$

$$K_2 := \frac{\tilde{\eta} + 1}{\tilde{\eta} - 1} C_1 \left[ 1 + \frac{2(\tilde{\eta} + 1)}{(\tilde{\eta} - 1) \ln \eta} \right] \frac{\eta^{\hat{y}+1} \tilde{\eta}^2}{\tilde{\eta} - \eta},$$

$$K_3 := \left( \frac{\tilde{\eta} + 1}{\tilde{\eta} - 1} \right)^2 \frac{3C_1 L \eta^{\hat{y}+1}}{\ln \eta}.$$

*Proof.* Consider an $n$-birth-and-death process model whose birth rate is $nq^+$ $(x/n, \hat{\varphi}(x))$, death rate is $nq^-(x/n, \hat{\varphi}(x))$ and cost rate is $\rho(x/n, \hat{\varphi}(x))$. The underlying process is denoted by $\{Y_t, t \ge 0\}$. Then we have

$${}^n W^{\hat{\varphi}}(x) := E_x^{\hat{\varphi}} \left[ \int_0^\infty \rho(Y_t/n, \hat{\varphi}(Y_t)) dt \right] = {}^n V^{\hat{\varphi}}(x), \quad x = 0, 1, \ldots.$$

Indeed, both $V^{\hat{\varphi}}(x)$ and $^nW^{\hat{\varphi}}(x)$ are given by the unique $^n\eta$-bounded solution to the equation

$$0 = \rho(x/n, \hat{\varphi}(x)) + V(x+1)nq^+(x/n, \hat{\varphi}(x)) + nq^-(x/n, \hat{\varphi}(x))V(x-1)$$
$$-n(q^+(x/n, \hat{\varphi}(x)) + q^-(x/n, \hat{\varphi}(x)))V(x), x = 1, 2, \ldots;$$
$$0 = V(0)$$

by Remark 13.3.1, [12, Lem. 1, Lem. 2, Lem. 4] and [13, Lem. 1]. See also [1, Chap. 7].

It remains to apply [13, Thm.1].  □

For any $n = 1, 2, \ldots$, let us consider a subclass of deterministic stationary policies $^n\Pi$ whose elements are $\varphi$ such that the following are satisfied:

$$\inf_{x=1,2,\ldots} \frac{q^-(x/n, \varphi(x))}{q^+(x/n, \varphi(x))} \geq \tilde{\eta} > 1,$$
$$\sup_{x=1,2,\ldots} \frac{|\rho(x/n, \varphi(x))|}{\{q^+(x/n, \varphi(x)) + q^-(x/n, \varphi(x))\}\eta^x} \leq C_1 < \infty,$$

where $\eta \in (1, \tilde{\eta})$, and there exists an ($\varphi$, $n$-dependent) constant $K^{\varphi}(n)$ satisfying

$$K^{\varphi}(n)q^-(x/n, \varphi(x))\eta^{x/n} - 1 > 0, \ x = 1, 2, \ldots,$$

and

$$\sup_{x=1,2,\ldots} \frac{|\rho(x/n, \varphi(x))|}{K^{\varphi}(n)q^-(x/n, \varphi(x))\eta^{x/n} - 1} < \infty.$$

Note that if there exists a $\psi$ satisfying Condition A, then the set $^n\Pi$ is non-empty as $\hat{\varphi}(x) := \psi(x/n) \in {}^n\Pi$ for all $n = 1, 2, \ldots$. Under Condition B below, for any $n = 1, 2, \ldots$, $^n\Pi$ coincides with the whole class of deterministic stationary policies, see Proposition 13.3.1 below.

*Remark 13.3.1.* For any fixed $n = 1, 2, \ldots$, one can verify that under a fixed policy $\varphi \in {}^n\Pi$, the $n$-MDP model admits a Lyapunov function

$$^nl_L(x) := D\eta^{x/n}, x = 1, 2, \ldots, {}^nl_L(0) := 2,$$

where $D \geq 1$ is a big enough constant, and a weight function

$$^nl_W(x) := K^{\varphi}(n)q^-(x/n, \varphi(x))\eta^{x/n} - 1, x = 1, 2, \ldots, {}^nl_2(0) := 1,$$

cf. [1, Chap.7] and [12, Con.1].

**Condition B.** (a) $\inf_{y>0, a\in A} \frac{q^-(y,a)}{q^+(y,a)} \geq \tilde{\eta} > 1$, $\sup_{y>0, a\in A} \frac{|\rho(y,a)|}{(q^+(y,a)+q^-(y,a))\eta^y} \leq C_1 < \infty$,
where $\eta \in (1, \tilde{\eta})$.
(b) $\liminf_{y\to\infty} \inf_{a\in A} q^-(y,a) > 0$.

**Proposition 13.3.1.** *Suppose Condition B is satisfied. Then for any deterministic stationary policy $\varphi$, it holds that $\varphi \in {}^n\Pi$ for all $n = 1, 2, \ldots$.*

*Proof.* Let $n = 1, 2, \ldots$ be fixed and consider an arbitrarily fixed $\varphi$. Then under Condition B, there exists a $\zeta > 0$ such that $q^-(x/n, \varphi(x)) > \zeta$, $x = 1, 2, \ldots$. So there exists a constant $K^\varphi(n) > 0$ such that $q^-(x/n, \varphi(x))\eta^{x/n} - \frac{1}{K^\varphi(n)} > \tilde{\zeta}$, $x = 1, 2, \ldots$ where $\tilde{\zeta} > 0$. Thus,

$$K^\varphi(n)q^-(x/n, \varphi(x))\eta^{x/n} - 1 > 0,$$

and

$$\sup_{x=1,2,\ldots} \frac{|\rho(\frac{x}{n}, \varphi(x))|}{K^\varphi(n)q^-(\frac{x}{n}, \varphi(x))\eta^{\frac{x}{n}} - 1} = \frac{1}{K^\varphi(n)} \sup_{x=1,2,\ldots} \frac{|\rho(\frac{x}{n}, \varphi(x))|}{q^-(\frac{x}{n}, \varphi(x))\eta^{\frac{x}{n}} - \frac{1}{K^\varphi(n)}}$$

$$\leq \frac{1}{K^\varphi(n)} \sup_{x=1,2,\ldots} \frac{|\rho(x/n, \varphi(x))|}{(q^+(x/n, \varphi(x)) + q^-(x/n, \varphi(x)))\eta^{x/n}}$$

$$\times \sup_{x=1,2,\ldots} \frac{2q^-(x/n, \varphi(x))\eta^{x/n}}{q^-(x/n, \varphi(x))\eta^{x/n} - \frac{1}{K^\varphi(n)}}$$

$$\leq \frac{2C_1}{K^\varphi(n)} \sup_{x=1,2,\ldots} \frac{q^-(x/n, \varphi(x))\eta^{x/n}}{q^-(x/n, \varphi(x))\eta^{x/n} - \frac{1}{K^\varphi(n)}} < \infty.$$

The other requirements for a policy to be in ${}^m\Pi$ are satisfied by $\varphi$ is evident. $\square$

**Theorem 13.3.3.** *Suppose the policy $\psi^*$ solves the refined fluid model and satisfies Condition A. Then for any fixed $n = 1, 2, \ldots$,*

$$\sup_{x\in\{0,1,\ldots,[n\hat{y}]\}} \left| {}^nV^{\varphi^*}(x) - \inf_{\varphi\in{}^n\Pi} {}^nV^\varphi(x) \right| \leq \varepsilon(n),$$

*where $\varphi^*(x) := \psi^*(x/n), x = 0, 1, 2, \ldots$.*

*Proof.* It can be shown that $\forall \varphi \in {}^n\Pi$, ${}^nV^\varphi(x) \geq \tilde{v}^{\psi^*}(x/n) - \frac{\varepsilon(n)}{2}$. The proof is similar to the one of Theorem 13.3.2. On the other hand, from Theorem 13.3.2, we have $\inf_{\varphi\in{}^n\Pi} {}^nV^\varphi(x) \leq {}^nV^{\varphi^*} \leq \tilde{v}^{\psi^*}(x/n) + \frac{\varepsilon(n)}{2} \leq \inf_{\varphi\in{}^n\Pi} {}^nV^\varphi(x) + \varepsilon(n)$. $\square$

The above theorem asserts that if one solves the refined fluid model and obtains the optimal policy $\psi^*$, then the policy given by $\varphi^*(x) = \psi^*(x/n)$ is $\varepsilon(n)$-optimal in the underlying $n$-MDP, and $\varepsilon(n)$ goes to zero as $n$ grows large.

The next proposition comes from [13, Lem.3].

**Proposition 13.3.2.** *Under Condition B, assume that there exist finite intervals* $(0, y_1')$, $(y_1', y_2'), \ldots$, *with* $\lim_{j \to \infty} y_j' = \infty$, *such that on each of them, the function* $\frac{\rho(y,a)}{q^-(y,a) - q^+(y,a)}$ *is Lipschitz continuous with respect to* $y$ *for each* $a \in A$, *and the Lipschitz constants are* $a$-*independent. Then for any fixed* $\hat{y} > 0$, *there exists an* $\psi^*$ *satisfying Condition A and solving the refined fluid model on* $[0, \hat{y}]$.

Now, we give an example where the main results (Theorems 13.3.2 and 13.3.3) of this work are applicable. In fact, by Propositions 13.3.1 and 13.3.2, it suffices to verify Condition B and the other condition of Proposition 13.3.2.

*Example 13.3.3.* Consider a discrete-time single-server queueing system, where during each time step, the probability of having an arrival is given by the function $q^+(y,a) = \frac{a}{2y+2+a}$, and the probability of having a service completion (given there is at least one job) is given by the function $q^-(y,a) = \frac{2y+2}{2y+2+a}$, where $y > 0$ and $a \in A := [\frac{1}{2}, 1]$. Suppose the cost function is given by $\rho(y,a) = 2y - a$, which means that we aim at minimizing the holding cost, which is incurred at a rate of $2\pounds$ per time step, while admitting more jobs is encouraged. The state zero is taken as the absorbing state, so that $q^-(0,0) = q^+(0,0) = \rho(0,0) = 0$.

It is easy to see that

$$\inf_{y>0, a \in [\frac{1}{2},1]} \frac{q^-(y,a)}{q^+(y,a)} = \inf_{y>0, a \in [\frac{1}{2},1]} \frac{2+2y}{a} = 2 =: \tilde{\eta} > 1,$$

$$\sup_{y>0, a \in [\frac{1}{2},1]} \frac{|\rho(y,a)|}{(q^+(y,a) + q^-(y,a))(1.5)^y} = \frac{|\rho(y,a)|}{(1.5)^y} < \infty,$$

$$1 < \eta := 1.5 < \tilde{\eta},$$

$$\liminf_{y \to \infty} \inf_{a \in [\frac{1}{2},1]} q^-(y,a) > 0,$$

and the function given by

$$\frac{\rho(y,a)}{q^-(y,a) - q^+(y,a)} = \frac{(2y-a)(2y+2+a)}{2y+2-a} = 2y + a - \frac{4a}{2y+2-a}$$

is obviously Lipschitz in $y > 0$ for any $a \in [\frac{1}{2}, 1]$. Hence, Condition B and the other condition of Proposition 13.3.2 are satisfied by this example. □

The next example indicates that the condition $q^-(y, \psi(y)) > \underline{q} > 0$ is important for the standard fluid model to approximate the underlying MDP model accurately. It also illustrates that the standard and the refined fluid models can behave qualitatively differently.

**Fig. 13.1** The graph of $q^-(y)$

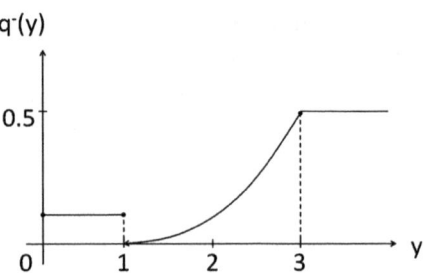

*Example 13.3.4 (cf. Example 1 in [13]).* For brevity, we deal with an uncontrolled model, i.e., $A$ is taken as a singleton, so that denotations such as $q^-(y), q^+(y)\rho(y)$ are used for brevity. We put

$$q^-(y) = 0.1I\{y \in (0,1]\} + 0.125(y-1)^2I\{y \in (1,3]\} + 0.5I\{y > 3\},$$
$$q^+(y) = 0.2q^-(y), \ \rho(y) = 8q^-(y).$$

Clearly, $q^-(y)$ is not separated from zero, see Fig. 13.1, while Condition A is satisfied.

For the original fluid model, we have $\frac{dy}{d\tau} = -0.1\,(y-1)^2$, and, if the initial state $y_0 = 2$, then $y(\tau) = 1 + \frac{10}{\tau+10}$, so that $\lim_{\tau \to \infty} y(\tau) = 1$.

On the other hand, since $q^-(y), q^+(y) > 0$ for $y > 0$, and there is a negative trend, the state process $X_t$ in the $n$-stochastic model starting from $^nX_0/n = y_0 = 2$ will be absorbed at zero, see [1, Lem. 7.2, Def. 7.4], while the moment of the absorbtion is postponed for later and later as $n \to \infty$ because the process spends more and more time in the neighborhood of 1, see Figs. 13.2 and 13.3.

When using the original fluid model, we have

$$v(2) = \int_0^\infty \rho(y(\tau))d\tau = 10 = \lim_{T \to \infty} \lim_{n \to \infty} E_{2n}\left[\sum_{t=1}^\infty I\{t/n \le T\}\,^nr(X_{t-1}, A_t)\right].$$

When using the refined fluid model, we can calculate $\frac{\rho(y)}{q^+(y)+q^-(y)} = \frac{8}{1.2}$ for $y > 0$ and $y(u) = 2 - \frac{2}{3}u$, so that the process in the refined fluid model is absorbed at the state zero at the time moment $u = 3$. Therefore,

$$\lim_{n \to \infty}\,^nV(2n) = \bar{v}(2) = \int_0^\infty \hat{\rho}(y(u))du = \int_0^3 \frac{8}{1.2}du = 20$$

$$= \lim_{n \to \infty} \lim_{T \to \infty} E_{2n}\left[\sum_{t=1}^\infty I\{t/n \le T\}\,^nr(X_{t-1}, A_t)\right] \ne v(2).$$

So the standard fluid model fails to be accurate in this example.                    □

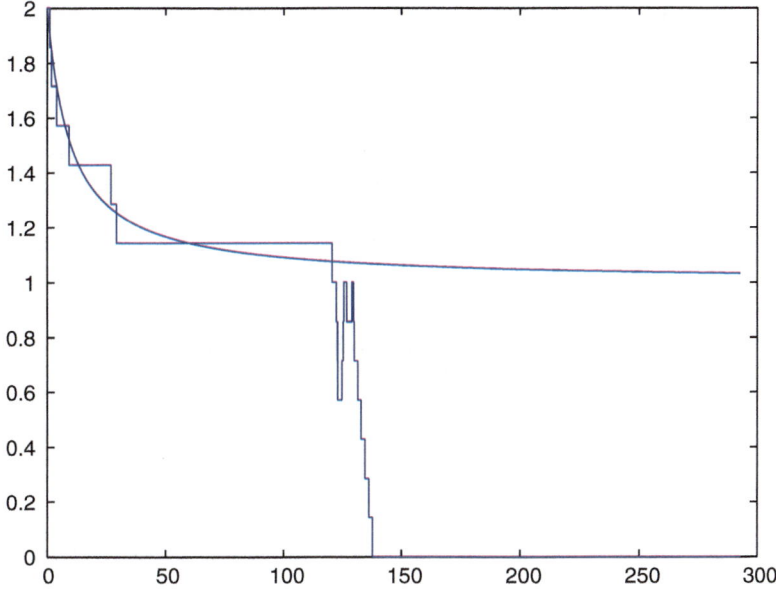

**Fig. 13.2** The state process in the $n$-stochastic model and its fluid approximation, $n = 7$

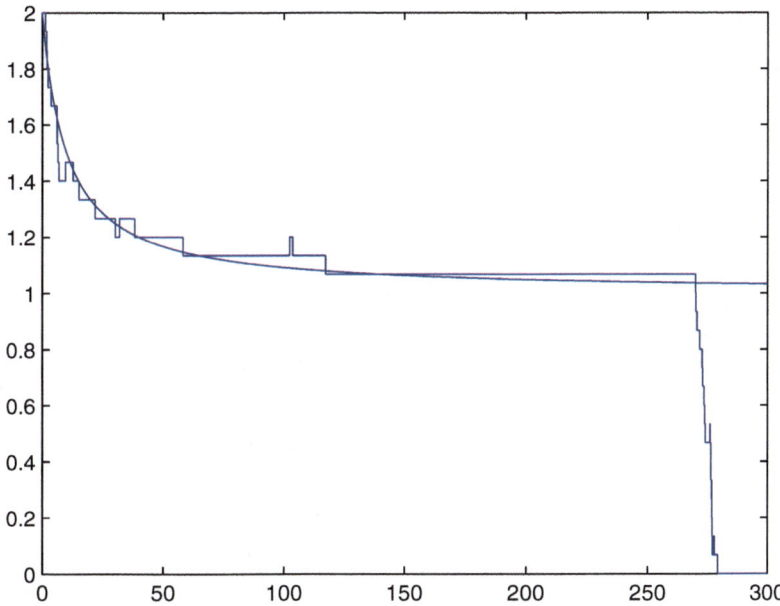

**Fig. 13.3** The state process in the $n$-stochastic model and its fluid approximation, $n = 15$

## 13.4  Conclusion

In this chapter, the convergence of the objective function of a scaled absorbing MDP model, with a total undiscounted cost, to the one of the (standard and refined) fluid model is shown. The upper boundary estimate of the rate of convergence is presented based on the initial data, which is of order $\frac{1}{n}$, where $n$ is the scaling parameter. Hence, the level of accuracy of the fluid approximation is obtained. By examples, we also show that the standard fluid model may fail to approximate the $n$-MDP model, while the refined fluid model is still accurate.

**Acknowledgements**  Mr. Mantas Vykertas kindly helped us improve the English presentation of this chapter. We thank the referee for valuable comments, too.

## References

1. Altman, E.: Constrained Markov Decision Processes. Chapman and Hall/CRC, Boca Raton (1999)
2. Avrachenkov, K., Ayesta, U., Piunovskiy, A.: Convergence of trajectories and optimal buffer sizing for AIMD congestion control. Perform. Evaluation. **67**, 501–527 (2010)
3. Avrachenkov, K., Piunovskiy, A., Zhang, Y.: Asymptotic fluid optimality and efficiency of tracking policy for bandwidth-sharing networks. J. Appl. Probab. **48**, 90–113 (2011)
4. Bäuerle, N.: Optimal control of queueing networks: an approach via fluid models. Adv. Appl. Prob. **34**, 313–328 (2002)
5. Bertsekas, D., Shreve, S.: Stochastic Optimal Control. Academic Press, NY (1978)
6. Jacko, P., Sansó, B.: Optimal anticipative congestion control of flows with time-varying input stream. Perform. Evaluation. **69**, 86–101 (2012)
7. Hernández-Lerma, O., Lasserre, J.: Discrete-time Markov Control Processes. Springer-Verlag, NY (1996)
8. Dai, J.: On positive Harris recurrence of multiclass queueing networks: a unified approach via fluid limit models. Ann. Appl. Prob. **5**, 49–77 (1995)
9. Foss, S., Kovalevskii, A.: A stability criterion via fluid limits and its application to a Polling system. Queueing. Syst. **32**, 131–168 (1999)
10. Mandelbaum, A., Pats, G.: State-dependent queues: approximations and applications. In Kelly, F., Williams, R. (eds.) Stochastic Networks, pp. 239–282. Springer, NY (1995)
11. Piunovskiy, A., Clancy, D.: An explicit optimal intervention policy for a deterministic epidemic model. Optim. Contr. Appl. Met. **29**, 413–428 (2008)
12. Piunovskiy, A.: Random walk, birth-and-death process and their fluid approximations: absorbing case. Math. Meth. Oper. Res. **70**, 285–312 (2009)
13. Piunovskiy, A., Zhang, Y.: Accuracy of fluid approximations to controlled Birth-and-Death processes: absorbing case. Math. Meth. Oper. Res. **73**, 159–187 (2011)
14. Piunovskiy, A., Zhang, Y.: On the fluid approximations of a class of general inventory level-dependent EOQ and EPQ models. Adv. Oper. Res. (2011) doi: 10.1155/2011/301205
15. Puterman, M.: Markov Decision Processes: Discrete Stochastic Dynamic Programming. Wiley, NY (1994)
16. Zhang, Y., Piunovskiy, A., Ayesta, U., Avrachenkov, K.: Convergence of trajectories and optimal buffer sizing for MIMD congestion control. Com. Com. **33**, 149–159 (2010)

# Chapter 14
# Minimizing Ruin Probabilities by Reinsurance and Investment: A Markovian Decision Approach

**Rosario Romera and Wolfgang Runggaldier**

## 14.1 Introduction

Consider a classical Cramér-Lundberg model

$$X_t = x - ct - \sum_{i=1}^{N_t} Y_i, \tag{14.1}$$

where the claim number process $\{N_t\}$ is a Poisson process with intensity $\lambda$, and the claim payment $\{Y_t\}$ is a sequence of independent and identically distributed (i.i.d.) positive random variables independent of $\{N_t\}$ and with support on the positive half line. Let $c$ be the constant premium rate (income) paid by the insurer. We assume the *safety loading* condition $c > \lambda E[Y]$.

One of the key quantities in the classical risk model is the *ruin probability*, as a function of the initial reserve $x$,

$$\psi(x) = \Pr\{X_t < 0 \text{ for some } t > 0\}.$$

In general, it is very difficult to derive explicit and closed expressions for the ruin probability. The pioneering works on approximations to the ruin probability were achieved by Cramér and Lundberg as early as the 1930s under Cramér-Lundberg condition. This condition is to assume that there exists a constant $\kappa > 0$ called *adjustment coefficient*, satisfying the following Lundberg equation:

$$\int_0^\infty e^{\kappa y} \bar{F}(y) dy = \frac{c}{\lambda},$$

R. Romera (✉)
University Carlos III, Madrid 28903, Spain
e-mail: mrromera@est-econ.uc3m.es

W. Runggaldier
Dipartimento di Matematica Pura ed Applicata, University of Padova, Padova 35121, Italy
e-mail: runggal@math.unipd.it

D. Hernández-Hernández and A. Minjárez-Sosa (eds.), *Optimization, Control, and Applications of Stochastic Systems*, Systems & Control: Foundations & Applications, DOI 10.1007/978-0-8176-8337-5_14, © Springer Science+Business Media, LLC 2012

with $F(y) = \Pr\{Y \le y\}$. Under this condition, the Cramér-Lundberg asymptotic formula states that if

$$\int_0^\infty y e^{\kappa y} dP(y) < \infty,$$

where $P(y) = \frac{1}{E[Y]} \int_0^y \bar{F}(s) ds$ is the equilibrium distribution of $F$, then

$$\psi(x) \le e^{-\kappa x}, x \ge 0. \tag{14.2}$$

The insurer has now the possibility to reinsure the claims. In the case of proportional reinsurance Schmidli [22] showed that there exists an optimal strategy that can be derived from the corresponding Hamilton-Jacobi-Bellman equation. Hipp and Vogt [15] derived by similar methods the same result for excess of loss reinsurance. In Schäl [21], the control problem of controlling ruin by investment in a financial market is studied. The insurance business is described by the usual Cramér-Lundberg-type model, and the risk driver of the financial market is a compound Poisson process. Conditions for investment to be profitable are derived by means of discrete-time dynamic programming. Moreover, Lundberg bounds are established for the controlled model. Diasparra and Romera [3, 4] consider a discrete-time process driven by proportional reinsurance and an interest rate process which behaves as a Markov chain. To reduce the risk of ruin, the insurer may reinsure a part or even all of the reserve. Recursive and integral equations for the ruin probabilities are given, and generalized Lundberg inequalities for the ruin probabilities are derived.

We consider a discrete-time insurance risk/reserve process which can be controlled by reinsurance and investment in the financial market, and we study the ruin probability problem in the finite-horizon case. Although controlling a risk/reserve process is a very active area of research (see [2, 6, 16, 24, 26], and references therein), obtaining explicit optimal solutions minimizing the ruin probability is in general a difficult task even for the classical Cramér-Lundberg risk process. Thus, an alternative method commonly used in ruin theory is to derive inequalities for ruin probabilities. The inequalities can be used to obtain upper bounds for the ruin probabilities [8, 23, 27], and this is the approach followed in this chapter. The basis of this approach is the well-known fact that in the classical Cramér-Lundberg model, if the claim sizes have finite exponential moments, then the ruin probability decays exponentially as the initial surplus increases (see for instance the book by Asmussen [1]). For the heavy-tailed claims' case, it is also shown to decay with a rate depending on the distribution of the claim size (see, e.g., [7]). Paulsen [18] reviews general processes for the ruin problem when the insurance company invests in a risky asset. Xiong and Yang [25] give conditions for the ruin probability to be equal to 1 for any initial endowment and without any assumption on the distribution of the claim size as long as it is not identically zero.

Control problems for risk/reserve processes are commonly formulated in continuous time. Schäl [20] introduces a formulation of the problem where events (arrivals of claims and asset price changes) occur at discrete points in time that may be deterministic or random, but their total number is fixed. Diasparra and Romera [3] consider a similar formulation in discrete time. Having a fixed total number of events implies that in the case of random time points the horizon is random as well.

In this chapter, we follow an approach inspired by Edoli and Runggaldier [5] who claim that a more natural way to formulate the problem in case of random time points is to consider a given fixed time horizon so that also the number of event times becomes random, and this makes the problem nonstandard. Accordingly, it is reasonable to assume that also the control decisions (level of reinsurance and amount invested) correspond to these random time points. Notice that this formulation can be seen equivalently in discrete or continuous time. The stochastic elements that affect the evolution of the risk/reserve process are thus the timing and size of the claims, as well as the dynamics of the prices of the assets in which the insurer is investing. This evolution is controlled by the sequential choice of the reinsurance and investment levels. Claims occur at random points in time and also their sizes are random, while asset price evolutions are usually modeled as continuous-time processes. On small time scales, prices actually change at discrete random time points and vary by tick size. In the proposed model, we let also asset prices change only at discrete random time points with their sizes being random as well. This will allow us to consider the timing of the events, namely, the arrivals of claims and the changes of the asset prices, to be triggered by a same continuous-time semi-Markov process (SMP), that is, a stochastic process where the embedded jump chain (the discrete process registering what values the process takes) is a Markov chain, and where the holding times (time between jumps) are random variables, whose distribution function may depend on the two states between which the move is made. Since between event times the situation for the insurer does not change, we shall consider controls only at event times.

Under this setting, we construct a methodology to achieve an optimal solution that minimizes the bounds of the ruin probability previously derived. Admissible strategies ranging in a compact set are in fact reinsurance levels and investment positions. From a general perspective, and due to the Markovian feature of the risk process, it seems quite natural to look at the minimization of the ruin probability as a Markov decision problem (MDP) for which suitable MDP techniques may hold. Although this is not a standard approach in actuarial risk models, we present in this chapter a connection between our optimization problem and the use of MDP techniques. Many of the most relevant contributions in the literature related to MDP techniques have been developed by Onesimo Hernández-Lerma and his coauthors, and some of them have inspired the optimization part of this chapter [9–14].

The rest of the chapter is organized as follows: Section 14.2 describes the elements of the considered model and introduces the formulation of the risk process. Section 14.3 presents the previous recursive relations on ruin probabilities needed to derived our main contributions on the exponential bounds for the ruin probabilities. In Sect. 14.4, the derivation of the reinsurance and investment policy that minimizes an exponential bound is described in connection with MDP techniques, namely, policy improvement and value iteration.

## 14.2   The Risk Process

We start this section by fixing the elements of the model studied in this chapter. According to the model proposed in Romera and Runggaldier [19], we consider a finite time horizon $T > 0$. More precisely, to model the timing of the events (arrival of claims and asset price changes), inspired by Schäl [21], we introduce the process $\{K_t\}_{t>0}$ for $t \leq T$, a continuous-time SMP on $\{0, 1\}$, where $K_t = 0$ holds for the arrival of a claim, and $K_t = 1$ for a change in the asset price. The embedded Markov chain, that is, the jump chain associated to the SMP $\{K_t\}_{t>0}$, evolves according to a transition probability matrix $P = \|p_{ij}\|_{i,j \in \{0,1\}}$ that is supposed to be given, and the holding times (time between jumps) are random variables whose probability distribution function may depend on the two states between which the move is made.

Let $T_n$ be the random time of the $n-th$ event, $n \geq 1$, and let the counting process $N_t$ denote the number of events having occurred up to time $t$ defined as follows:

$$N_t = \sum_{j=1}^{\infty} 1_{\{T_j \leq t\}} \tag{14.3}$$

and so

$$T_n = \min\{t \geq 0 | N_t = n\}. \tag{14.4}$$

We introduce now the dynamics of the controlled risk process $X_t$ for $t \in [0, T]$ with $T$ a given fixed horizon. For this purpose, let $Y_n$ be the $n-th$ $(n \geq 1)$ claim payment represented by a sequence of i.i.d. random variables with common probability distribution function (p.d.f.) $F(y)$. Let $Z_n$ be the random variable denoting the time between the occurrence of the $n-1st$ and $nth$ $(n \geq 1)$ jumps of the SMP $\{K_t\}_{t>0}$. We assume that $\{Z_n\}$ is a sequence of i.i.d. random variables with p.d.f. $G(.)$. From this, we may consider that the transition probabilities of the SMP $\{K_t\}_{t>0}$ are

$$P\{K_{T_{n+1}} = j, Z_{n+1} \leq s | K_{T_n} = i\} = p_{ij} G(s).$$

Notice that for a full SMP model, the distribution function $G(.)$ depends also on $i$ and $j$. While the results derived below go through in the same way also for a $G_{ij}(.)$ depending on $i, j$, we restrict ourselves to independent $G(.)$. A specific form of SMP arises, for example, as follows: let $N_t^1$ and $N_t^2$ be independent Poisson processes with intensities $\lambda^1$ and $\lambda^2$, respectively. We may think of $N_t^1$ as counting the number of claims and $N_t^2$ that of price changes and we have that $N_t = N_t^1 + N_t^2$ is again a Poisson process with intensity $\lambda = \lambda^1 + \lambda^2$. It then follows easily that

$$\begin{cases} p_{ij} = \frac{\lambda^j}{\lambda} := p_j, \quad \forall i, \\ G(s) = \left[1 - e^{-\lambda s}\right]. \end{cases}$$

The risk process is controlled by reinsurance and investment. In general, this means that we may choose adaptively at the event times $T_{N_t}$ (they correspond to the jump times of $N_t$) the retention level (or proportionality factor or risk exposure) $b_{N_t}$ of a reinsurance contract as well as the amount $\delta_{N_t}$ to be invested in the risky asset, namely, in $S_{N_t}$ with $S_t$ denoting discounted prices. For the values $b$ that the various $b_{N_t}$ may take, we assume that $b \in (b_{\min}, 1] \subset (0, 1]$, where $b_{\min}$ will be introduced below, and for the values of $\delta$ for the various $\delta_{N_t}$, we assume $\delta \in [\underline{\delta}, \overline{\delta}]$ with $\underline{\delta} \leq 0$ and $\overline{\delta} > 0$ exogenously given. Notice that this condition allows also for negative values of $\delta$ meaning that, see also [22], short selling of stocks is allowed. On the other hand, with an exogenously given upper bound $\overline{\delta}$, it might occasionally happen that $\delta_{N_t} > X_{N_t}$, implying a temporary debt of the agent beyond his/her current wealth in order to invest optimally in the financial market. By choosing a policy that minimizes the ruin probability, this debt is however only instantaneous and with high probability leads to a positive wealth already at the next event time.

Assume that prices change only according to

$$\frac{S_{N_t+1} - S_{N_t}}{S_{N_t}} = \left(e^{W_{N_t+1}} - 1\right) K_{T_{N_t+1}}, \tag{14.5}$$

where $W_n$ is a sequence of i.i.d. random variables taking values in $[\underline{w}, \overline{w}]$ with $\underline{w} < 0 < \overline{w}$, where one may also have $\underline{w} = -\infty$, $\overline{w} = +\infty$ and with p.d.f. $H(w)$. For simplicity and without loss of generality, we consider only one asset to invest in. An immediate generalization would be to allow for investment also in the money market account.

Let $c$ be the premium rate (income) paid by the customer to the company, fixed in the contract. Since the insurer pays to the reinsurer a premium rate, which depends on the retention level $b_{N_t}$ chosen at the various event times $T_{N_t}$, we denote by $C(b_{N_t})$ the net income rate of the insurer at time $t \in [0, T]$. For $b \in (b_{\min}, 1]$, we let $h(b, Y)$ represent the part of the generic claim $Y$ paid by the insurer, and in what follows, we take the function $h(b, Y)$ to be of the form $h(b, Y) = b \cdot Y$ (proportional reinsurance). We shall call *policy* a sequence $\pi = (b_n, \delta_n)$ of *control actions*. Control actions over a single period will be denoted by $\phi_n = (b_n, \delta_n)$. According to the expected value principle with safety loading $\theta$ of the reinsurer, for a given starting time $t < T$, the function $C(b)$ can be chosen as follows:

$$C(b) := c - (1 + \theta) \frac{E\{Y_1 - h(b, Y_1)\}}{E\{Z_1 \wedge (T - t)\}}, \quad 0 < t < T, \tag{14.6}$$

We use $Z_1$ and $Y_1$ in the above formula since, by our i.i.d. assumption, the various $Z_n$ and $Y_n$ are all independent copies of $Z_1$ and $Y_1$. Notice also that, in order to keep formula (14.6) simple and possibly similar to standard usage, in the denominator of the right-hand side, we have considered the random time $Z_1$ between to successive events, while more correctly, we should have taken the random time between two successive claims, which is larger. For this, we can however play with the safety loading factor. As explained in Romera and Runggaldier [19], we can now define

$b_{\min} := \min\{b \in (0,1] \mid c \geq C(b) \geq c^*\}$, where $c^* \geq 0$ denotes the minimal value of the premium considered by the insurer. We have to consider the following technical restrictions:

**Assumption 14.2.1.** *Let*

(i) *The random variables* $(Z_n, Y_n, W_n)_{n \geq 1}$ *are, conditionally on* $K_t$, *mutually independent.*

(ii) $E\{e^{rY_1}\} < +\infty$ *for* $r \in (0, \bar{r})$ *with* $\bar{r} \in (0, \infty)$.

(iii) $c - (1 + \theta)\frac{E\{Y_1\}}{E\{Z_1 \wedge T\}} \geq 0$.

Notice that, since the support of $Y_1$ is the positive half line, we have $\lim_{r \uparrow \bar{r}}\{E\{e^{rY_1}\}\} = \infty$ ($\bar{r}$ may be equal to $+\infty$, e.g., if the support of $Y_1$ is bounded).

In the given setting, we obtain for the insurance risk process (surplus) $X$ the following one-step transition dynamics between the generic random times $T_n$ and $T_{n+1}$ when at $T_n$ a control action $\phi = (b, \delta)$ is taken for a certain $b \in (b_{\min}, 1] \subset (0, 1]$, and $\delta \in [\underline{\delta}, \bar{\delta}]$,

$$X_{T_{n+1}} = X_{T_n} + C(b)Z_{n+1} - (1 - K_{T_{n+1}})h(b, Y_{n+1}) + K_{T_{n+1}}\delta(e^{W_{n+1}} - 1). \quad (14.7)$$

**Definition 14.2.1.** Letting $U := [b_{\min}, 1] \times [\underline{\delta}, \bar{\delta}]$, we shall say that a control action $\phi = (b, \delta)$ is *admissible* if $(b, \delta) \in U$. Notice that $U$ is compact.

We want now to express the one-step dynamics in (14.7) when starting from a generic time instant $t < T$ with a capital $x$. For this purpose, note that if, for a given $t < T$ one has $N_t = n$, the time $T_{N_t}$ is the random time of the $n - th$ event and $T_n \leq t \leq T_{n+1}$. Since, when standing at time $t$, we observe the time that has elapsed since the last event in $T_{N_t}$, it is not restrictive to assume that $t = T_{N_t}$ [see the comment below after (14.8)]. Furthermore, since $Z_n, Y_n, W_n$ are i.i.d., in the one-step random dynamics for the risk process $X_t$, we may replace the generic $(Z_{n+1}, Y_{n+1}, W_{n+1})$ by $(Z_1, Y_1, W_1)$. We may thus write

$$X_{N_t+1} = x + C(b)Z_1 - (1 - K_{T_{N_t+1}})h(b, Y_1) + K_{T_{N_t+1}}\delta(e^{W_1} - 1) \quad (14.8)$$

for $0 < t < T$, $T > 0$ and with $X_t = x \geq 0$ (recall that we assumed $t = T_{N_t}$). Notice that, if we had $t \neq T_{N_t}$ and therefore $t > T_{N_t}$, the second term on the right in (14.8) would become $C(b)[Z_1 - (t - T_{N_t})]$, and (14.8) could then be rewritten as

$$X_{N_t+1} = [x - C(b)(t - T_{N_t})] + C(b)Z_1 - (1 - K_{T_{N_t+1}})h(b, Y_1) + K_{T_{N_t+1}}\delta(e^{W_1} - 1)$$

with the quantity $[x - C(b)(t - T_{N_t})]$, which is known at time $t$, replacing $x$. This is the sense in which above, we mentioned that it is not restrictive to assume that $t = T_{N_t}$. In what follows, we shall work with the risk process $X_t$, (or $X_{N_t}$) as defined by (14.8). For convenience, we shall denote by $(b_n, \delta_n)$ the values of $\phi = (b, \delta)$ at $t = T_{N_t}$. Accordingly, we shall also write $(b_{N_t}, \delta_{N_t})$ for $(b_{T_{N_t}}, \delta_{T_{N_t}})$.

Following [24], we shall also introduce an absorbing (cemetery) state $\kappa$, such that if $X_{N_t} < 0$ or $X_{N_t} = \kappa$, then $X_{N_t+1} = \kappa$, $\forall t \leq T$. The state space is then $\mathbb{R} \cup \{\kappa\}$.

## 14.3  Ruin Probabilities

We present first the general expression of the ruin probability corresponding to the risk model (14.8). Thus, using the policy $\pi$, given the initial surplus $x$ at time $t$ and initial event $k \in \{0,1\}$ for the Markov chain $K_t$ at time $t$, the ruin probability is given by

$$\psi^\pi(t,x;k) := P^\pi \left( \bigcup_{s=N_t+1}^{N_T} \{X_s < 0 \,|\, X_{N_t} = x, K_t = k\} \right). \tag{14.9}$$

Note that the finite-horizon character of the considered model imposes a specific definition for the ruin probabilities. In order to obtain recursive relations for the ruin probability, we specify some notation and introduce the basic definitions concerning the finite-horizon ruin probabilities when one or $n$ intra-event periods are considered.

Given a policy $\pi$, namely. a predictable process pair $\pi_t := (b_t, \delta_t)$ with $(b_t, \delta_t)$ in $U$ [of which in the definitions below we need just to consider the generic individual control action $\phi = (b, \delta)$], we introduce the following functions:

$$u^\pi(y,z,w,k) : = (1-k)by - C(b)z - k\delta(e^w - 1),$$

$$\tau^\pi(y,w,k,x) : = \frac{(1-k)by - k\delta(e^w - 1) - x}{C(b)}, \tag{14.10}$$

so that $u^\pi(y,z,w,k) < x \iff z > \tau^\pi(y,w,k,x)$.

The ruin probability over one intra-event period (namely, the period between to successive event times) when using the control action $\phi = (b, \delta)$ is, for a given initial surplus $x$ at time $t \in (0,T)$ and initial event $K_{T_{N_t}} = k \in \{0,1\}$,

$$\psi_1^\pi(t,x;k) := \sum_{h=0}^{1} p_{k,h} \int_{\underline{w}}^{\bar{w}} \int_0^\infty G(\tau^\pi(y,w,h,x) \wedge (T-t)) dF(y) dH(w). \tag{14.11}$$

We want to obtain a recursive relation for

$$\psi_n^\pi(t,x;k) : = P^\pi \left\{ \bigcup_{k=N_t+1}^{(N_t+n)\wedge N_T} \{X_k < 0\} \,|\, X_{N_t} = x, K_{T_{N_t}} = k \right\}$$

$$: = P_{x,k}^\pi \left\{ \bigcup_{k=N_t+1}^{(N_t+n)\wedge N_T} \{X_k < 0\} \right\}, \tag{14.12}$$

namely, for the ruin probability when at most $n$ events are considered in the interval $[t,T]$ and a policy $\pi$ is adopted.

In Romera and Runggaldier [19], the following recursive relation is obtained:

**Proposition 14.3.1.** *For an initial surplus x at a given time $t \in [0,T]$, as well as an initial event $K_{T_{N_t}} = k$ and a given policy $\pi$, one has*

$$\psi_n^\pi(t,x,k)$$
$$= P\{N_T - N_t > 0\} \Sigma_{h=0}^1 p_{k,h} \int_{\underline{w}}^{\bar{w}} \int_0^\infty G(\tau^\pi(y,w,h,x) \wedge (T-t)) dF(y) dH(w)$$
$$+ P\{N_T - N_t > 1\} \Sigma_{h=0}^1 p_{k,h}$$
$$\cdot \int_{\underline{w}}^{\bar{w}} \int_0^\infty \int_{\tau^\pi(y,w,h,x)}^{T-t} \psi_{n-1}^\pi(t+z, x-u^\pi(y,z,w,h),h) dG(z) dF(y) dH(w)$$

$$(14.13)$$

*from which it immediately also follows that*

$$\psi_1^\pi(t,x,k) = P\{N_T - N_t = 1\} \sum_{h=0}^1 p_{k,h} \int_{\underline{w}}^{\bar{w}} \int_0^\infty G(\tau^\pi(y,w,h,x) \wedge (T-t)) dF(y) dH(w)$$

$$(14.14)$$

*since in the case of at most one jump, one has that $P\{N_T - N_t > 0\} = P\{N_T - N_t = 1\}$ and $P\{N_T - N_t > 1\} = 0$.*

## 14.4 Minimizing the Bounds

In the following, we base ourselves on results in Diasparra and Romera [3,4] that are here extended to the general setup of this chapter to obtain the exponential bounds and then to minimize them.

To stress the fact that the process $X$ defined in (14.7) corresponds to the choice of a specific policy $\pi$, in what follows, we shall use the notation $X^\pi$.

Given a policy $\pi_t = (b_t, \delta_t)$ and defining for $t \in [0,T]$, the random variable

$$V_t^\pi := C(b)(Z_1 \wedge (T-t)) - \mathbf{1}_{\{Z_1 \leq T-t\}} \left[ (1 - K_{T_{N_t}+1})bY_1 + K_{T_{N_t}+1}\delta \left( e^{W_1} - 1 \right) \right],$$

$$(14.15)$$

where $b = b_t$ and $\delta = \delta_t$ let, for $r \in (0,\bar{r})$ and $k \in \{0,1\}$,

$$l_r^\pi(t,k) := E_{t,k}\{e^{-rV_t^\pi}\} - 1,$$

$$(14.16)$$

where, for reasons that should become clear below, we distinguish the dependence of $l^\pi$ on $r$ from that on $(t,k)$.

*Remark 14.4.1.* Notice that, by its definition, $l_r^\pi(t,k)$ is, for given $\pi$ and $r \in (0,\bar{r})$, a bounded function of $(t,k) \in [0,T] \times \{0,1\}$. Given its continuity in $r$, it is uniformly bounded on any compact subset of $(0,\bar{r})$, for example, on $[\eta, \bar{r}-\eta]$ for $\eta \in (0,\bar{r})$. Having fixed $\eta > 0$, denote this bound by $L$, that is,

$$\sup_{(t,k)\in[0,T]\times\{0,1\}, \, r\in[\eta,\bar{r}-\eta]} |l_r^\pi(t,k)| \leq L.$$

$$(14.17)$$

**Definition 14.4.2.** We shall call a policy $\pi$ *admissible* and denote their set by $A$ if at each $t \in [0,T]$, the corresponding control action $(b_t, \delta_t) \in U$, and for any $t \in [0,T]$ and $k \in \{0,1\}$, $E_{t,k}\{V_t^\pi\} > 0 \; \forall \pi \in A$.

Notice that $A$ is nonempty since (see Assumption 14.2.1, (iii) it contains at least the stationary policy $(b_{N_t}, \delta_{N_t}) \equiv (b_{\min}, 0)$.

According to Romera and Runggaldier [19], we obtain the following result:

**Proposition 14.4.2.** *For each $(t,k)$ and each $\pi \in A$, we have that:*

*(a) As a function of $r \in (0, \bar{r})$, $l_r^\pi(t,k)$ is convex with a negative slope at $r = 0$.*
*(b) The equation $l_r^\pi(t,k) = 0$ has a unique positive root in $(0, \bar{r})$ that we simply denote by $R^\pi$ so that the defining relation for $R^\pi$ is*

$$l_{R^\pi}^\pi(t,k) = 0. \tag{14.18}$$

Notice that $R^\pi$ actually depends also on $(t,k)$, but for simplicity of notation, we denote it just by $R^\pi$.

In view of the main result of this section, namely, Theorem 14.4.1 below, we first obtain [19]:

**Lemma 14.4.1.** *Given a surplus $x > 0$ at a given initial time $t \in [0,T]$ and an initial event $k \in \{0,1\}$, we have*

$$\psi_1^\pi(t,x,k) \leq e^{-R^\pi x} \tag{14.19}$$

*for each $\pi \in A$, where $R^\pi$ is the unique positive root of (14.18) that depends on $t$ and $k$ but is independent of $x$.*

**Lemma 14.4.2.** *For given $(t,x,k)$, we have*

$$\psi_n^\pi(t,x,k) \leq \gamma_n e^{-R^\pi x} \tag{14.20}$$

*for all $n \in N, \pi \in A$, where $R^\pi$ is the unique positive solution with respect to $r$ of $l_r^\pi(t,k) = 0$ (see (14.18)), and $\gamma_n$ is defined recursively b*

$$\gamma_1 = 1,$$
$$\gamma_n = \gamma_{n-1} P\{N_T - N_t > 1\} + P\{N_T - N_t = 1\}. \tag{14.21}$$

*Remark 14.4.2.* Due to the defining relations (14.21), it follows immediately that $\gamma_n \leq 1$ for all $n \in N$. In fact, using forward induction, we see that the inequality is true for $n = 1$, and assuming it true for $n - 1$, we have

$$\gamma_n = \gamma_{n-1} P\{N_T - N_t > 1\} + P\{N_T - N_t = 1\} \leq P\{N_T - N_t > 0\} \leq 1. \tag{14.22}$$

We come now to our main result in this section, namely, Theorem 14.4.1 whose proof follows immediately from Lemma 14.4.2 noticing that, see Remark 14.4.2, one has $\gamma_n \leq 1$.

**Theorem 14.4.1.** *Given an initial surplus $x > 0$ at a given time $t \in [0,T]$, we have, for all $n \in N$ and any initial event $k \in \{0,1\}$ and for all $\pi \in A$,*

$$\psi_n^\pi(t,x,k) \leq e^{-R^\pi x}$$

*with $R^\pi$ that may depend on $(t,k)$ in $[0,T] \times \{0,1\}$.*

### 14.4.1  Minimizing the Bounds by a Policy Improvement Procedure

As mentioned previously, it is in general a difficult task to obtain an explicit solution to the given reinsurance-investment problem in order to minimize the ruin probability and this even for a classical risk process. We shall thus choose the reinsurance level and the investment in order to minimize the bounds that we have derived. By Theorem 14.4.1, this amounts to choosing a strategy $\pi \in A$ such that, for each pair $(t,k)$, the value of $R^\pi$ is as large as possible. This strategy will thus depend in general also on $t$ and $k$ but not on the level $x$ of wealth. By Proposition 14.4.2, this $R^\pi$ is, for each $\pi \in A$, the unique positive solution of the equation $l_r^\pi(t,k)) = 0$, where $l_r^\pi(t,k)$ is, as a function of $r \in [0,\bar{r}]$ (and for every fixed $(t,k)$), convex with a negative slope at $r = 0$. To obtain, for a given $(t,k)$, the largest value of $R^\pi$, it thus suffices to choose $\pi \in A$ that minimizes $l_r^\pi(t,k)$ just at $r = R^\pi$. This, in fact, appeals also to intuition since, according to the definition in (14.16), minimizing $l_r^\pi(t,k)$ amounts to penalizing negative values of $X_t^\pi = x + V_t^\pi$, thereby minimizing the possibility of ruin.

Concerning the minimization of $l_r^\pi(t,k)$ at $r = R^\pi$, notice that decisions concerning the control actions $\phi = (b,\delta)$ have to be made only at the event times $T_n$. The minimization of $l_r^\pi(t,k)$ with respect to $\pi \in A$ has thus to be performed only for pairs $(t,k)$ corresponding to event times, namely, those of the form $(T_n, K_{T_n})$, thus leading to a policy $\pi$ with individual control actions $\phi_{T_n} = (b_{T_n}, \delta_{T_n})$.

Our problem to determine an investment and insurance policy to minimize the bounds on the ruin probability may thus be solved by solving the following subproblems:

1. For a given policy, $\bar{\pi} \in A$ determine $l_r^{\bar{\pi}}(t,k)$ for pairs $(t,k)$ of the form $(T_n, K_{T_n})$.
2. Determine $R^{\bar{\pi}}(T_n, K_{T_n})$ that is solution with respect to $r$ of $l_r^{\bar{\pi}}(T_n, K_{T_n}) = 0$.
3. Improve the policy by minimizing $l_{R^{\bar{\pi}}}^\pi(T_n, K_{T_n})$ with respect to $\pi \in A$.

This leads to a policy improvement-type approach, more precisely, one can proceed as follows:

- Start from a given policy $\pi^0$ (e.g., the one requiring minimal reinsurance and no investment in the financial market).
- Determine $R^{\pi^0}$ corresponding to $\pi^0$ for the various $(T_n, K_{T_n})$.

- For $r = R^{\pi^0}$, determine $\pi^1$ that minimizes $l_{R^{\pi^0}}^{\pi}(T_n, K_{T_n})$.
- Repeat the procedure until a stopping criterion is met (notice that by the above procedure $R^{\pi^n} > R^{\pi^{n-1}}$).

One crucial step in this procedure is determining the function $l_r^{\pi}(t,k)$ corresponding to a given $\pi \in A$, and this will be discussed in the next section.

## 14.4.2  Computing the Value Function in the Policy Improvement Procedure

Recall again that the decisions have to be made only at the event times over a given finite horizon, and consequently, the function $l_r^{\pi}(t,k)$ has to be computed only for pairs of the form $(T_n, K_{T_n})$. The number of these events is however random and may be arbitrarily large; furthermore, the timing of these events is random as well. On the other hand, notice that if we can represent the function $l_r^{\pi}(t,k)$ to be minimized as the fixed point of a contraction operator involving expectations of functions of a Markov process, then the computation can be performed iteratively as in value iteration algorithms.

For this purpose, recalling that $Z_n$ are i.i.d. random variables with probability distribution function $G(.)$ and that, for given $\pi \in A$ and $r \in [\eta, \bar{r} - \eta]$, the functions $l_r^{\pi}(t,k)$ are bounded by some $L$ (see Remark 14.4.1), we start with the following:

**Definition 14.4.3.** For given $\pi \in A$, define $T^{\pi}$ as the operator acting on bounded functions $v(t,k)$ of $(t,k)$ in the following way:

$$T^{\pi}(v(t,k))$$

$$= 1_{\{t \leq T\}} E_{t,k}^{\pi} \left\{ 1_{\{t+Z_1 \leq T\}} v(t+Z_1, K_{t+Z_1}) + 1_{\{t \leq T \leq t+Z_1\}} \left[ e^{-rC(b)(T-t)} - 1 \right] \right\}$$

$$= \sum_{h=0}^{1} p_{k,h} \left\{ \int_0^{T-t} v(t+z, h) dG(z) + \bar{G}(T-t) \left[ e^{-rC(b)(T-t)} - 1 \right] \right\}$$

with $\bar{G}(z) = 1 - G(z)$ and where, given $\pi_t = (b_t, \delta_t)$, the value of $b$ is $b = b_t$.

The following lemma is now rather straightforward:

**Lemma 14.4.3.** For a given $\pi \in A$ and any value of the parameter $r \in [\eta, \bar{r} - \eta]$, the function $l_r^{\pi}(.)$ is a fixed point of $T^{\pi}$, that is,

$$l_r^{\pi}(t,k) = T^{\pi}(l_r^{\pi})(t,k). \tag{14.23}$$

On the basis of the above definitions and results, we may now consider the following recursive relations:

$$l_r^{\pi,0}(T_n, k) = \bar{G}(T - T_n)[e^{-rC(b_n)(T-T_n)} - 1],$$

$$l_r^{\pi,m}(T_n, k) = T^{\pi}(l_r^{\pi,m-1})(T_n, k) \quad for\ m = 1, 2, .. \tag{14.24}$$

that we may view as a *value iteration* algorithm. Since between any event time $T_n$ and the terminal time $T$ there may be any number of events occurring, to obtain $l_r^\pi(.)$, the recursions in (14.24) would have to be iterated an infinite number of times. If however the mappings $T^\pi$ are contracting in the sense that

$$\|T^\pi(v_1) - T^\pi(v_2)\| \le \gamma \|v_1 - v_2\| \tag{14.25}$$

for bounded functions $v_1(.)$ and $v_2(.)$ and with $\gamma < 1$, then $l_r^{\pi,m}(T_n,k)$ approximates $l_r^\pi(T_n,k)$ arbitrarily well in the sup-norm, provided $m$ is sufficiently large.

The above assumption can be seen to be satisfied in various cases of practical interest [19].

### 14.4.2.1  Reduction of Dimensionality and Particular Cases

For the policy improvement and value iteration-type procedure in the previous section, the "Markovian state" was seen to be the tuple $(T_n, K_{T_n})$, which makes the problem two dimensional. It is shown in Romera and Runggaldier [19] that in the particular case when the inter-event time and the claim size distributions are (negative) exponential, a case that has been most discussed in the literature under different settings, then the state space can be further reduced to only the time variable $t$ (the sequence of event times $T_n$ is then in fact a Markov process by itself), and so, the optimal policy becomes dependent only on the event time. This particular case can also be shown [19] to be a concrete example where the mappings $T^\pi$ are contracting as assumed in (14.25). Always for this particular case, it can furthermore be shown (see again [19]) that the fixed point $l_r^\pi$ of the mapping $T^\pi$ which, as discussed above, depends here only on $t$, can be computed as a semianalytic solution involving a Volterra integral equation.

We conclude this section by pointing out that, although by our procedure, we minimize only an upper bound on the ruin probability, the optimal upper bound can also be used as a benchmark with respect to which other standard policies may be evaluated.

Finally, as explained in Romera and Runggaldier [19], our procedure allows also to obtain some qualitative insight into the impact that investment in the financial market may have on the ruin probability. It turns in fact out that, in line with some of the findings in the more recent literature (see e.g., [17]), the choice of investing also in the financial market has little impact on the ruin probability unless, as we do here, one allows also for reinsurance.

**Acknowledgements**  Rosario Romera was supported by Spanish MEC grant SEJ2007-64500 and Comunidad de Madrid grant S2007/HUM-0413. Part of the contribution by Wolfgang Runggaldier was obtained while he was visiting professor in 2009 for the chair in Quantitative Finance and Insurance at the LMU University in Munich funded by LMU Excellent. Hospitality and financial support are gratefully acknowledged.

# References

1. Asmussen, S., Ruin Probabilities. *World Scientific, River Edge, NJ.* 2000.
2. Chen, S., Gerber, H. and Shiu, E., Discounted probabilities of ruin in the compound binomial model. *Insurance: Mathematics and Economics*, 2000, 26, 239–250.
3. Diasparra, M.A. and Romera, R., Bounds for the ruin probability of a discrete-time risk process. *Journal of Applied Probability*, 2009, 46(1), 99–112.
4. Diasparra, M.A. and Romera, R., Inequalities for the ruin probability in a controlled discrete-time risk process. *European Journal of Operational Research*, 2010, 204(3), 496–504.
5. Edoli, E. and Runggaldier, W.J., On Optimal Investment in a Reinsurance Context with a Point Process Market Model. *Insurance: Mathematics and Economics*, 2010, 47, 315–326.
6. Eisenberg, J. and Schmidli, H., Minimising expected discounted capital injections by reinsurance in a classical risk model. *Scandinavian Actuarial Journal*, 2010, 3, 1–22.
7. Gaier, J.,Grandits, P. and Schachermayer, W., Asymptotic ruin probabilities and optimal investment. *Annals of Applied Probability*, 2003, 13, 1054–1076.
8. Grandell, J., Aspects of Risk Theory, *Springer*, New York. 1991.
9. Guo, X. P. and Hernandez-Lerma, O., Continuous-Time Markov Decision Processes: Theory and Applications. *Springer-Verlag Berlin Heidelberg*, 2009.
10. Hernandez-Lerma, O. and Lasserre, J. B., Discrete-Time Markov Control Processes: Basic Optimality Criteria. *Springer-Verlag, New York*, 1996.
11. Hernandez-Lerma, O. and Lasserre, J. B., Further Topics on Discrete-Time Markov Control Processes. *Springer-Verlag, New York*, 1999.
12. Hernandez-Lerma, O. and Romera, R., Limiting discounted-cost control of partially observable stochastic systems. *Siam Journal on Control and Optimization* 40 (2), 2001, 348–369.
13. Hernandez-Lerma, O. and Romera, R., The scalarization approach to multiobjective Markov control problems: Why does it work? *Applied Mathematics and Optimization* 50 (3), 2004, 279–293.
14. Hernandez-Lerma, O. and Prieto-Romeau, T., Selected Topics on Continuous-Time Controlled Markov Chains and Markov Games, *Imperial College Press*, 2012.
15. Hipp, C. and Vogt, M., Optimal dynamic XL insurance, *ASTIN Bulletin* 33 (2), 2003, 93–207.
16. Huang, T., Zhao, R. and Tang, W., Risk model with fuzzy random individual claim amount. *European Journal of Operational Research*, 2009, 192, 879–890.
17. Hult, H. and Lindskog, F., Ruin probabilities under general investments and heavy-tailed claims. *Finance and Stochastics*, 2011, 15(2), 243–265.
18. Paulsen, J., Sharp conditions for certain ruin in a risk process with stochastic return on investment. *Stochastic Processes and Their Applications*, 1998, 75, 135–148.
19. Romera, R. and Runggaldier, W., Ruin probabilities in a finite-horizon risk model with investment and reinsurance. UC3M Working papers. *Statistics and Econometrics*, 2010, 10–21.
20. Schäl, M., On Discrete-Time Dynamic Programming in Insurance: Exponential Utility and Minimizing the Ruin Probability. *Scandinavian Actuarial Journal*, 2004, 3, 189–210.
21. Schäl, M., Control of ruin probabilities by discrete-time investments. *Mathematical Methods of Operational Research*, 2005, 62, 141–158.
22. Schmidli, H., Optimal proportional reinsurance policies in a dynamic setting. *Scandinavian Actuarial Journal*, 2001, 12, 55–68.
23. Schmidli, H., On minimizing the ruin probability by investment and reinsurance. *Annals of Applied Probability*, 2002, 12, 890–907.
24. Schmidli, H., Stochastic Control in Insurance. *Springer, London.* 2008.
25. Xiong, S. and Yang, W.S., Ruin probability in the Cramér-Lundberg model with risky investments. *Stochastic Processes and their Applications*, 2011, 121, 1125–1137.

26. Wang, R.,Yang, H. and Wang, H., On the distribution of surplus immediately after ruin under interest force and subexponential claims. *Insurance: Mathematics and Economics*, 2004, 34, 703–714.
27. Willmot, G. and Lin, X., Lundberg Approximations for Compound Distributions with Insurance Applications *Lectures Notes in Statistics* 156, Springer, New York, 2001.

# Chapter 15
# Estimation of the Optimality Deviation in Discounted Semi-Markov Control Models

**Luz del Carmen Rosas-Rosas**

## 15.1 Introduction

Semi-Markov control models (SMCMs) is a class of continuous-time stochastic control models where the distribution of the times between consecutive decision epochs (holding or sojourn times) is arbitrary and the actions or controls are selected at the transition times. Its evolution is as follows: At time of the $n$th decision epoch $T_n$, the system is in the state $x_n = x$ and the controller chooses a control $a_n = a$. Then, the system remains in the state $x$ during a nonnegative random time $\delta_{n+1}$ with distribution $F\left(\cdot \mid x,a\right)$, and the following happens: (1) an immediate cost $D\left(x,a\right)$ is incurred; (2) the system jumps to a new state $x_{n+1} = y$ according to a transition law $Q\left(\cdot \mid x,a\right)$; and (3) a cost rate $d\left(x,a\right)$ is imposed until the transition occurs. Once the transition to state $y$ occurs, the process is repeated. The actions are selected according to rules $\pi$ known as control policies, and the costs are accumulated throughout the evolution of the system in an infinite horizon using a discounted cost criterion, where total cost under the policy $\pi$ and initial stated $x$ is denoted by $V_\alpha\left(x, \pi\right)$. Thus, the control problem is to find a policy $\pi^*$ such that

$$V_\alpha\left(x, \pi^*\right) = \inf_{\pi \in \Pi} V_\alpha\left(x, \pi\right).$$

However, there exist situations where the holding time distribution $F$ is unknown by the controller; and therefore, it is necessary to implement approximation procedures together with control schemes to obtain nearly optimal policies.

Let us assume that it is possible to get an approximate distribution $\tilde{F}\left(\cdot \mid x,a\right)$. Under this setting, considering the control problem for $\tilde{F}$ and the corresponding

---

L.C. Rosas-Rosas (✉)

Departamento de Matemáticas, Universidad de Sonora, Hermosillo, Sonora 83000, México

e-mail: lcrosas@gauss.mat.uson.mx

D. Hernández-Hernández and A. Minjárez-Sosa (eds.), *Optimization, Control, and Applications of Stochastic Systems*, Systems & Control: Foundations & Applications, DOI 10.1007/978-0-8176-8337-5_15, © Springer Science+Business Media, LLC 2012

total cost $\tilde{V}_\alpha$, we can obtain a policy $\tilde{\pi}$ minimizing $\tilde{V}_\alpha$. Then, if the controller uses $\tilde{\pi}$ to control the original semi-Markov process, our main objective is to estimate the *optimality deviation* of the control policy $\tilde{\pi}$ defined as

$$\mathscr{D}(x,\tilde{\pi}) := V_\alpha(x,\tilde{\pi}) - \inf_{\pi \in \Pi} V_\alpha(x,\pi).$$

Specifically, our main result states that

$$\mathscr{D}(x,\tilde{\pi}) \leq \psi\left(d\left(F,\tilde{F}\right)\right),$$

where $\psi : [0,\infty) \to [0,\infty)$ is a function such that $\psi(s) \to 0$ as $s \to 0$, and $d$ is some "distance" between $F$ and $\tilde{F}$.

Moreover, our result can be used to construct asymptotically discounted optimal (ADO) policies as is shown in [7]. Indeed, if we consider the particular case when $F$ has an unknown density $g$ independent of the state-actions pairs, and the holding times are observable, it is possible to apply some statistical density estimation method to obtain a consistent estimator $g_n$. Then, using $g_n$ as the approximate holding time distribution, the resulting policy $\tilde{\pi} = \{f_n\}$ will be ADO for the original SMCM (see Sect. 15.3).

Similar estimation problems of the deviation of optimality have been studied by several authors but for Markov control processes and/or controlled diffusion processes, and under different names, namely, robustness, perturbation, or stability index, (see, e.g., [1, 2, 4, 10–12]). In addition, asymptotic optimality implementing estimation and control procedures has been analyzed, for instance in [3, 6, 9].

The chapter is organized as follows: Section 15.2 contains the SMCM and the required assumptions, whereas in Sect. 15.2, the main result and a special case of asymptotic optimality are introduced. Finally, the proofs are presented in Sect. 15.4.

## 15.2   Semi-Markov Control Models

A SMCM is defined by the collection

$$\mathscr{M} = (\mathbb{X}, \mathbb{A}, \{A(x) : x \in \mathbb{X}\}, Q, F, D, d),$$

where $\mathbb{X}$ is the *state space* and $\mathbb{A}$ is the *control (or action) space*; both are assumed to be Borel spaces. For each $x \in \mathbb{X}$, we associate a nonempty measurable subset $A(x)$ of $\mathbb{A}$, whose elements are the *admissible controls* for the controller when the system is in state $x$. The set $\mathbb{K} = \{(x,a) : x \in \mathbb{X}, a \in A(x)\}$ of admissible state-action pairs is assumed to be a Borel subset of $\mathbb{X} \times \mathbb{A}$. The *transition law* $Q(\cdot \mid \cdot)$ is a stochastic kernel on $\mathbb{X}$ given $\mathbb{K}$, and $F(\cdot \mid x,a)$ is the *distribution function of the holding time* at state $x \in \mathbb{X}$ when the control $a \in A(x)$ is chosen. Finally, the *cost functions* $D$ and $d$ are possibly unbounded and measurable real-valued functions on $\mathbb{K}$.

On the other hand, we assume that the costs are continuously discounted, that is, for a discount factor $\alpha > 0$, a cost $K$ incurred at time $t$ is equivalent to a cost $K\exp(-\alpha t)$ at time 0. In this sense, according to points (1) and (3) of the interpretation of the SMCM given in the introduction, the *one-stage cost* takes the form

$$c(x,a) := D(x,a) + d(x,a) \int_0^\infty \int_0^t e^{-\alpha s} ds F (dt \mid x,a), \quad (x,a) \in \mathbb{K},$$

which it is not difficult to see that may be rewritten as

$$c(x,a) = D(x,a) + \tau_\alpha(x,a) d(x,a), \quad (x,a) \in \mathbb{K}, \qquad (15.1)$$

where,

$$\tau_\alpha(x,a) := \frac{1 - \beta_\alpha(x,a)}{\alpha} \quad \text{and} \quad \beta_\alpha(x,a) := \int_0^\infty e^{-\alpha t} F(dt \mid x,a), \quad (x,a) \in \mathbb{K}.$$

$$(15.2)$$

**Optimality Criterion.** We denote by $\Pi$ the set of all admissible control policies, and by $\mathbb{F} \subset \Pi$ the subset of stationary policies (see, for instance, [5] for definitions). As usual, each stationary policy $\pi \in \mathbb{F}$ is identified with a measurable function $f : \mathbb{X} \to A$ such that $f(x) \in A(x), x \in \mathbb{X}$, so that $\pi$ is of the form $\pi = \{f, f, \ldots\}$. In this case, we denote $\pi$ by $f$.

For each $x \in \mathbb{X}$ and $\pi \in \Pi$, we define the *total expected discounted cost* as

$$V_\alpha(x,\pi) := E_x^\pi \left[ \sum_{n=0}^\infty e^{-\alpha T_n} c(x_n, a_n) \right].$$

Hence, a policy $\pi^* \in \Pi$ is said to be *optimal* if for all $x \in \mathbb{X}$,

$$V_\alpha(x) := V_\alpha(x, \pi^*) = \inf_{\pi \in \Pi} V_\alpha(x, \pi),$$

and $V_\alpha(x)$ is called the *optimal value function* for the model $\mathcal{M}$.

**Assumptions.** We shall require two sets of assumptions. The first one ensures that in a bounded time interval, there are at most a finite number of transitions of the process. Assumption 15.2.2 contains standard continuity and compactness conditions to guarantee the existence of minimizers.

**Assumption 15.2.1** *There exists positive constants $\zeta$ and $\varepsilon$, such that for every* $(x,a) \in \mathbb{K}$,

$$F(\zeta \mid x,a) \leq 1 - \varepsilon.$$

*Remark 15.2.1.* As was noted in [7], Assumption 15.2.1 implies that

$$\rho_\alpha := \sup_{(x,a) \in \mathbb{K}} \beta_\alpha(x,a) < 1.$$

**Assumption 15.2.2** *(a) The function $F(t \mid x,a)$ is continuous on $a \in A(x)$ for every $x \in \mathbb{X}$ and $t \in \mathbb{R}$.*

*(b) For each $x \in \mathbb{X}$, $A(x)$ is a compact set, and the cost functions $D(x,a)$ and $d(x,a)$ are lower semicontinuous (l.s.c.) on $A(x)$. Moreover, there exist a measurable function $W : \mathbb{X} \to [1,\infty)$ and positive constants $b$, $\bar{c}_1$, and $\bar{c}_2$ such that $1 \leq b < (\rho_\alpha)^{-1}$ and*

$$\sup_{a \in A(x)} |D(x,a)| \leq \bar{c}_1 W(x), \quad \sup_{a \in A(x)} |d(x,a)| \leq \bar{c}_2 W(x) \quad \forall x \in \mathbb{X}, \quad (15.3)$$

$$\int_{\mathbb{X}} W(y) Q(dy \mid x,a) \leq bW(x), \quad x \in \mathbb{X}, \ a \in A(x). \quad (15.4)$$

*(c) For each $x \in \mathbb{X}$,*

$$a \mapsto \int_{\mathbb{X}} W(y) Q(dy \mid x,a)$$

*is a continuous function on $A(x)$.*

*(d) For every bounded and continuous function $u$ on $\mathbb{X}$,*

$$a \mapsto \int_{\mathbb{X}} u(y) Q(dy \mid x,a)$$

*is a bounded and continuous function on $A(x)$.*

We denote by $\mathbb{B}_W$ the normed linear space of all measurable functions $u : \mathbb{X} \to \mathbb{R}$ with norm

$$\|u\|_W := \sup_{x \in \mathbb{X}} \frac{|u(x)|}{W(x)} < \infty. \quad (15.5)$$

## 15.3 Optimality Deviation

Let $\tilde{F}(\cdot \mid x,a)$ be a distribution function that approximates the holding time distribution $F(\cdot \mid x,a)$, $(x,a) \in \mathbb{K}$. We consider the SMCM

$$\tilde{\mathcal{M}} = (\mathbb{X}, \mathbb{A}, \{A(x) : x \in \mathbb{X}\}, Q, \tilde{F}, D, d),$$

where $\mathbb{X}$, $\mathbb{A}$, $A(x)$, $Q$, $D$ and $d$ are as the model $\mathcal{M}$. The functions $\tilde{c}$, $\tilde{\tau}_\alpha$, $\tilde{\beta}_\alpha$, $\tilde{V}_\alpha(x,\pi)$, $\tilde{V}_\alpha(x)$, and $\tilde{\rho}_\alpha$ are defined accordingly. Furthermore, we assume that the Assumptions 15.2.1 and 15.2.2 are satisfied for the model $\tilde{\mathcal{M}}$. Hence, as a consequence, we have the following result (see, e.g., [7,8]):

**Proposition 15.3.1.** *Suppose that Assumptions 15.2.1 and 15.2.2 hold. Then:*

*(a) There exist stationary optimal policies $f^*$ and $\tilde{f}$ for the models $\mathcal{M}$ and $\tilde{\mathcal{M}}$, respectively.*

(b) *The value functions $V_\alpha$ and $\tilde{V}_\alpha$ satisfy the corresponding optimality equation, that is,*

$$V_\alpha(x) = \min_{a \in A(x)} \left\{ c(x,a) + \beta_\alpha(x,a) \int_{\mathbb{X}} V_\alpha(y) Q(dy \mid x,a) \right\}, \quad x \in \mathbb{X},$$

*and*

$$\tilde{V}_\alpha(x) = \min_{a \in A(x)} \left\{ \tilde{c}(x,a) + \tilde{\beta}_\alpha(x,a) \int_{\mathbb{X}} \tilde{V}_\alpha(y) Q(dy \mid x,a) \right\}, \quad x \in \mathbb{X}.$$

According to the Proposition 15.3.1, we have that for $x \in \mathbb{X}$,

$$V_\alpha(x) = \inf_{\pi \in \Pi} V_\alpha(x,\pi) = V_\alpha(x,f^*)$$

and

$$\tilde{V}_\alpha(x) = \inf_{\pi \in \Pi} \tilde{V}_\alpha(x,\pi) = \tilde{V}_\alpha(x,\tilde{f}).$$

Hence, our objective is to estimate the optimality deviation of the stationary policy $\tilde{f}$. Then, we can state our main result as follows:

**Theorem 15.3.1.** *Under Assumptions 15.2.1 and 15.2.2,*

$$\mathscr{D}(x,\tilde{f}) := V_\alpha(x,\tilde{f}) - V_\alpha(x) \leq M^* \lVert \tilde{F} - F \rVert W(x) \quad \forall x \in \mathbb{X}, \tag{15.6}$$

*for some positive constant $M^*$, where*

$$\lVert \tilde{F} - F \rVert := \sup_{(x,a) \in \mathbb{K}} \int_0^\infty e^{-\alpha t} \lvert \mu_{\tilde{F}}(\cdot \mid x,a) - \mu_F(\cdot \mid x,a) \rvert (dt),$$

*whereas $\lvert \mu_{\tilde{F}}(\cdot \mid x,a) - \mu_F(\cdot \mid x,a) \rvert$ represents the total variation of the signed measure $\mu_{\tilde{F}}(\cdot \mid x,a) - \mu_F(\cdot \mid x,a)$, being $\mu_{\tilde{F}}$ and $\mu_F$ the measures induced, respectively, by the distribution functions $\tilde{F}$ and $F$ for every fixed pair $(x,a) \in \mathbb{K}$.*

### 15.3.1  Asymptotic Optimality

As in [7], we consider the special case where the distributions $F$ and $\tilde{F}$ have densities $g$ and $\tilde{g}$, respectively, which are independent of the state-action pairs $(x,a)$. That is,

$$F(t \mid x,a) = \int_0^t g(s)\,ds, \quad \text{and} \quad \tilde{F}(t \mid x,a) = \int_0^t \tilde{g}(s)\,ds.$$

Then, clearly, $\beta_\alpha$ and $\tau_\alpha$, as well as $\tilde{\beta}_\alpha$ and $\tilde{\tau}_\alpha$ are independent of $(x, a)$. That is,

$$\beta_\alpha(x, a) := \beta_\alpha = \int_0^\infty e^{-\alpha s} g(s) \, ds,$$

$$\tau_\alpha(x, a) := \tau_\alpha = \frac{1 - \beta_\alpha}{\alpha},$$

and similarly for $\tilde{\beta}_\alpha$ and $\tilde{\tau}_\alpha$.

Furthermore, note that

$$\tilde{\beta}_\alpha - \beta_\alpha = \int_0^\infty e^{-\alpha s} \{\tilde{g}(s) - g(s)\} \, ds,$$

and in such case, (15.6) becomes

$$\mathscr{D}(x, \tilde{f}) \leq M^* \|\tilde{g} - g\| W(x) \quad \forall x \in \mathbb{X}, \tag{15.7}$$

where

$$\|\tilde{g} - g\| := \int_0^\infty e^{-\alpha s} |\tilde{g}(s) - g(s)| \, ds.$$

In particular, assuming that the holding times are observable, let $\delta_1, \ldots, \delta_n$ be independent realizations (observed up to the moment of the $n$th decision epoch) of random variables with the unknown density $g$, and $g_n(s) := g_n(s; \delta_1, \ldots, \delta_n), s \in \mathbb{R}_+$ be an arbitrary estimator of $g$ such that

$$E\left[\int_0^\infty |g_n(s) - g(s)| \, ds\right] \to 0 \quad \text{as} \quad n \to \infty \tag{15.8}$$

being $g_n$ a density.

Letting

$$f_n := f_n^{g_n} \tag{15.9}$$

the optimal stationary policy corresponding to the density $g_n$, where the minimization is done almost sure (a.s.) with respect to the probability measure induced by the common distribution of the random variables $\delta_1, \ldots, \delta_n$, and applying (15.7) with $g_n$ instead of $\tilde{g}$, but observing that $g_n$ is a random variable, we obtain

$$E[\mathscr{D}(x, f_n)] \leq M^* E[\|g - g_n\|] W(x) \to 0 \quad \text{as} \quad n \to \infty, x \in \mathbb{X}.$$

Moreover, it is possible to prove that this fact yields the following result:

**Theorem 15.3.2.** *The policy* $\tilde{\pi} = \{f_n\}$ *is (pointwise) asymptotically discounted optimal, which means that for each* $x \in \mathbb{X}$,

$$E\left[\Phi\left(x, f_n\right)\right] \to 0 \quad \text{as} \quad n \to \infty,$$

*where for each* $(x, a) \in \mathbb{K}$,

$$\Phi\left(x, a\right) := c\left(x, a\right) + \beta_{\alpha} \int_{\mathbb{X}} V_{\alpha}\left(y\right) Q\left(\mathrm{d}y \mid x, a\right) - V_{\alpha}\left(x\right), \qquad (15.10)$$

*is the well-known discrepancy function.*

## 15.4  Proofs

We will use repeatedly the following inequalities: For $u \in \mathbb{B}_W$ and $(x, a) \in \mathbb{K}$,

$$|u\left(x\right)| \leq \|u\|_W \, W\left(x\right) \qquad (15.11)$$

and

$$\int_{\mathbb{X}} u\left(y\right) Q\left(\mathrm{d}y \mid x, a\right) \leq b \|u\|_W \, W\left(x\right). \qquad (15.12)$$

The relation (15.11) is a consequence of the definition of $\|u\|_W$ in (15.5), whereas (15.12) follows from (15.4). Furthermore, from Assumption 15.2.2(b), and following a straightforward calculation (see, e.g., [3, 5]), we can prove that, for all $x \in \mathbb{X}$ and $\pi \in \Pi$,

$$\sup_{n \geq 0} E_x^{\pi} \left[W\left(x_n\right)\right] < \infty,$$

which in turn implies that there exist positive constants $M$ and $\tilde{M}$, such that, for all $\pi \in \Pi$,

$$V_{\alpha}\left(x\right) \leq V_{\alpha}\left(x, \pi\right) \leq MW\left(x\right) \qquad (15.13)$$

and, similarly,

$$\tilde{V}_{\alpha}\left(x\right) \leq \tilde{V}_{\alpha}\left(x, \pi\right) \leq \tilde{M}W\left(x\right).$$

On the other hand, observe that for all $(x, a) \in \mathbb{K}$,

$$\left|\tilde{\beta}_{\alpha}\left(x, a\right) - \beta_{\alpha}\left(x, a\right)\right| \leq \left\|\tilde{F} - F\right\|.$$

Hence, from (15.3) and the definition of the costs $c$ and $\tilde{c}$ [see (15.1) and (15.2)], we obtain

$$|c\left(x, a\right) - \tilde{c}\left(x, a\right)| \leq \frac{\bar{c}_2}{\alpha} \left\|\tilde{F} - F\right\| W\left(x\right), \quad \forall \left(x, a\right) \in \mathbb{K}. \qquad (15.14)$$

In addition, for any function $u \in \mathbb{B}_W$, we have

$$\left| \tilde{B}_\alpha (x,a) - B_\alpha (x,a) \right| \int_{\mathbb{X}} u(y) Q(dy \mid x,a) \leq b \|u\|_W \|\tilde{F} - F\| W(x) \qquad (15.15)$$

and

$$B_\alpha (x,a) \int_{\mathbb{X}} u(y) Q(dy \mid x,a) \leq b \|u\|_W \rho_\alpha W(x) \qquad (15.16)$$

for all $(x,a) \in \mathbb{K}$, where $\rho_\alpha$ is defined in Remark 15.2.1 (similarly, we obtain (15.15) and (15.16) substituting $B_\alpha$ and $\rho_\alpha$ by $\tilde{B}_\alpha$ and $\tilde{\rho}_\alpha$, respectively).

Thus, denoting $u_1(x) := V_\alpha(x, \tilde{f})$, where $\tilde{f}$ is the optimal policy for the model $\tilde{\mathcal{M}}$, from (15.11), we obtain

$$\tilde{B}_\alpha (x, \tilde{f}) \int_{\mathbb{X}} \left| V_\alpha (y, \tilde{f}) - \tilde{V}_\alpha (y, \tilde{f}) \right| Q(dy \mid x, \tilde{f})$$

$$= \tilde{B}_\alpha (x, \tilde{f}) \int_{\mathbb{X}} \left| u_1(y) - \tilde{V}_\alpha(y) \right| Q(dy \mid x, \tilde{f})$$

$$\leq b \tilde{\rho}_\alpha \left\| u_1 - \tilde{V}_\alpha \right\|_W W(x), \quad x \in \mathbb{X}. \qquad (15.17)$$

**Lemma 15.4.1.** *For all $x \in \mathbb{X}$, there exist positive constants $M_1^*$ and $M_2^*$, such that*

$$\left\| u_1 - \tilde{V}_\alpha \right\|_W \leq M_1^* \left\| \tilde{F} - F \right\| \qquad (15.18)$$

*and*

$$\left\| \tilde{V}_\alpha - V_\alpha \right\|_W \leq M_2^* \left\| \tilde{F} - F \right\|. \qquad (15.19)$$

*Proof.* First observe that from the Markov property, for any stationary policy $f \in \mathbb{F}$ and for each $x \in \mathbb{X}$

$$V_\alpha (x,f) = c(x,f) + B_\alpha (x,f) \int_{\mathbb{X}} V_\alpha (y,f) Q(dy \mid x,f).$$

Similarly for $\tilde{V}_\alpha$. Then,

$$\left| u_1(x) - \tilde{V}_\alpha(x) \right| = \left| V_\alpha (x, \tilde{f}) - \tilde{V}_\alpha (x, \tilde{f}) \right|$$

$$\leq \left| \left\{ c(x, \tilde{f}) + B_\alpha (x, \tilde{f}) \int_{\mathbb{X}} V_\alpha (y, \tilde{f}) Q(dy \mid x, \tilde{f}) \right\} \right.$$

$$\left. - \left\{ \tilde{c}(x, \tilde{f}) + \tilde{B}_\alpha (x, \tilde{f}) \int_{\mathbb{X}} \tilde{V}_\alpha (y, \tilde{f}) Q(dy \mid x, \tilde{f}) \right\} \right|.$$

Adding and subtracting the term

$$\tilde{B}_\alpha (x, \tilde{f}), \int_{\mathbb{X}} V_\alpha (y, \tilde{f}) Q(dy \mid x, \tilde{f}),$$

we obtain, for all $x \in \mathbb{X}$,

$$
\begin{aligned}
\left| u_1(x) - \tilde{V}_\alpha(x) \right| &\leq \left| c(x, \tilde{f}) - \tilde{c}(x, \tilde{f}) \right| \\
&+ \left| \beta_\alpha(x, \tilde{f}) - \tilde{\beta}_\alpha(x, \tilde{f}) \right| \int_{\mathbb{X}} \left| V_\alpha(y, \tilde{f}) \right| Q(dy \mid x, \tilde{f}) \\
&+ \tilde{\beta}_\alpha(x, \tilde{f}) \int_{\mathbb{X}} \left| V_\alpha(y, \tilde{f}) - \tilde{V}_\alpha(y, \tilde{f}) \right| Q(dy \mid x, \tilde{f}).
\end{aligned}
$$

Hence, from (15.14), (15.15), and (15.17), we have (recall $W(\cdot) \geq 1$),

$$
\left\| u_1 - \tilde{V}_\alpha \right\|_W \leq \frac{\bar{c}_2}{\alpha} \left\| \tilde{F} - F \right\| + b \left\| u_1 \right\|_W \left\| \tilde{F} - F \right\| + b \tilde{\rho}_\alpha \left\| u_1 - \tilde{V}_\alpha \right\|_W.
$$

Now, since $b\tilde{\rho}_\alpha < 1$ (see Assumption 15.2.2(b)), we get

$$
\left\| u_1 - \tilde{V}_\alpha \right\|_W \leq M_1^* \left\| \tilde{F} - F \right\|,
$$

where

$$
M_1^* := \frac{\frac{\bar{c}_2}{\alpha} + b \left\| u_1 \right\|_W}{1 - b \tilde{\rho}_\alpha},
$$

which proves (15.18).

On the other hand, to prove (15.19), we have from Proposition 15.3.1(b),

$$
\begin{aligned}
\left| \tilde{V}_\alpha(x) - V_\alpha(x) \right| &= \left| \min_{a \in A(x)} \left\{ \tilde{c}(x, a) + \tilde{\beta}_\alpha(x, a) \int_{\mathbb{X}} \tilde{V}_\alpha(y) Q(dy \mid x, a) \right\} \right. \\
&\quad \left. - \min_{a \in A(x)} \left\{ c(x, a) + \beta_\alpha(x, a) \int_{\mathbb{X}} V_\alpha(y) Q(dy \mid x, a) \right\} \right| \\
&\leq \sup_{a \in A(x)} \left\{ \left| \tilde{c}(x, a) - c(x, a) \right| + \left| \tilde{\beta}_\alpha(x, a) \int_{\mathbb{X}} \tilde{V}_\alpha(y) Q(dy \mid x, a) \right. \right. \\
&\quad \left. \left. - \beta_\alpha(x, a) \int_{\mathbb{X}} V_\alpha(y) Q(dy \mid x, a) \right| \right\}.
\end{aligned}
$$

Now, it is easy to see that, by adding and subtracting the term

$$
\beta_\alpha(x, a) \int_{\mathbb{X}} \tilde{V}_\alpha(y) Q(dy \mid x, a)
$$

and by applying similar arguments as the proof of (15.18), we get

$$
\begin{aligned}
\left| \tilde{V}_\alpha(x) - V_\alpha(x) \right| &\leq \frac{\bar{c}_2}{\alpha} \left\| \tilde{F} - F \right\| W(x) + b \left\| \tilde{V}_\alpha \right\|_W \left\| \tilde{F} - F \right\| W(x) \\
&+ b \rho_\alpha \left\| \tilde{V}_\alpha - V_\alpha \right\|_W W(x).
\end{aligned}
$$

Since $W(\cdot) \geq 1$, this relation implies

$$\left\|\tilde{V}_\alpha - V_\alpha\right\|_W \leq \frac{\bar{c}_2}{\alpha}\left\|\tilde{F} - F\right\| + b\left\|\tilde{V}_\alpha\right\|_W\left\|\tilde{F} - F\right\| + b\rho_\alpha\left\|\tilde{V}_\alpha - V_\alpha\right\|_W,$$

which yields

$$\left\|\tilde{V}_\alpha - V_\alpha\right\|_W \leq M_2^*\left\|\tilde{F} - F\right\|,$$

where

$$M_2^* := \frac{\frac{\bar{c}_2}{\alpha} + b\left\|\tilde{V}_\alpha\right\|_W}{1 - b\rho_\alpha}.$$

$\square$

### 15.4.1 Proof of Theorem 15.3.1

In order to prove the main result, first observe that

$$\tilde{V}_\alpha(x, \tilde{f}) = \tilde{V}_\alpha(x), \quad x \in \mathbb{X}.$$

Then,

$$\mathscr{D}(x, \tilde{f}) = V_\alpha(x, \tilde{f}) - V_\alpha(x)$$
$$= \left\{V_\alpha(x, \tilde{f}) - \tilde{V}_\alpha(x, \tilde{f})\right\} + \left\{\tilde{V}_\alpha(x) - V_\alpha(x)\right\}, \quad x \in \mathbb{X}.$$

Hence, from (15.11), Lemma 15.4.1 yields

$$\mathscr{D}(x, \tilde{f}) \leq \left|V_\alpha(x, \tilde{f}) - \tilde{V}_\alpha(x, \tilde{f})\right| + \left|\tilde{V}_\alpha(x) - V_\alpha(x)\right|$$
$$\leq M^*\left\|\tilde{F} - F\right\|W(x) \quad \forall x \in \mathbb{X},$$

where

$$M^* := M_1^* + M_2^*.$$

$\square$

### 15.4.2 Proof of Theorem 15.3.2

First, for every $n \in \mathbb{N}$, we define (see (15.10)) the approximate discrepancy function

$$\Phi_n(x, a) := c_n(x, a) + \beta_n \int_\mathbb{X} V_n(y)\,Q(\mathrm{d}y \mid x, a) - V_n(x), \quad (x, a) \in \mathbb{K}, \quad (15.20)$$

where, for each $n \in \mathbb{N}$,

$$c_n(x,a) := D(x,a) + \tau_n d(x,a), \quad (x,a) \in \mathbb{K},$$

$$\tau_n := \frac{1 - \beta_n}{\alpha},$$

$$\beta_n := \int_0^\infty e^{-\alpha s} g_n(s)\, ds,$$

$$V_n(x) := \inf_{\pi \in \Pi} V_n(x, \pi),$$

$$V_n(x, \pi) := E_x^\pi \left[ \sum_{k=0}^\infty e^{-\alpha T_k} c_n(x_k, a_k) \right].$$

Note that, for each $n \in \mathbb{N}$, (15.9), (15.20), and Proposition 15.3.1 imply that

$$\Phi_n(x, f_n) = 0, \quad a.s., \ x \in \mathbb{X}.$$

Hence, and since $\Phi_n$ is a nonnegative function, for each $n \in \mathbb{N}$, we can write the following:

$$\Phi(x, f_n) = \big| \Phi(x, f_n) - \Phi_n(x, f_n) \big|$$

$$\leq \sup_{a \in A(x)} \big| \Phi(x, a) - \Phi_n(x, a) \big| \quad a.s. \ \forall x \in \mathbb{X}. \qquad (15.21)$$

In particular observe that from (15.10) and (15.20), for each $n \in \mathbb{N}$, we have

$$\big| \Phi(x, a) - \Phi_n(x, a) \big| \leq \big| c(x, a) - c_n(x, a) \big| + \big| V_\alpha(x) - V_n(x) \big|$$

$$+ \left| \beta_\alpha \int_{\mathbb{X}} V_\alpha(y) Q(dy|\, x, a) - \beta_n \int_{\mathbb{X}} V_n(y) Q(dy|\, x, a) \right| \quad a.s.$$

Then, by adding and subtracting the term

$$\beta_\alpha \int_{\mathbb{X}} V_n(y) Q(dy|\, x, a),$$

we get

$$\big| \Phi(x, a) - \Phi_n(x, a) \big| \leq \big| c(x, a) - c_n(x, a) \big| + \big| V_\alpha(x) - V_n(x) \big|$$

$$+ \beta_\alpha \int_{\mathbb{X}} |V_\alpha(y) - V_n(y)|\, Q(dy|\, x, a)$$

$$+ |\beta_\alpha - \beta_n| \int_{\mathbb{X}} |V_\alpha(y)|\, Q(dy|\, x, a) \quad a.s.$$

Therefore, using (15.11), (15.14), (15.15), (15.17), and (15.19), it follows that for every $n \in \mathbb{N}$,

$$\left| \Phi(x,a) - \Phi_n(x,a) \right| \leq \frac{\bar{c}_2}{\alpha} \left\| g_n - g \right\| W(x) + M_2^* \left\| g_n - g \right\| W(x)$$

$$+ b\rho_n M_2^* \left\| g_n - g \right\| W(x) + b \left\| V_\alpha \right\|_W \left\| g_n - g \right\| W(x) \quad a.s.$$

Hence, considering that $b\rho_n < 1$ a.s., and from (15.21), for each $x \in \mathbb{X}$, we obtain the following,

$$E\left[ \Phi(x, f_n) \right] \leq M_0^* W(x) E\left[ \left\| g_n - g \right\| \right] \quad \forall n \in \mathbb{N},$$

where

$$M_0^* := \frac{\bar{c}_2}{\alpha} + 2M_2^* + b \left\| V_n \right\|_W.$$

Consequently, since $\Phi(\cdot, \cdot) \geq 0$, from (15.8), we obtain, for every $x \in \mathbb{X}$,

$$E\left[ \Phi(x, f_n) \right] \to 0 \quad as \ n \to \infty. \qquad \square$$

# References

1. Abbad, M.; Filar, J.A. Perturbation and stability theory for Markov control processes. IEEE Trans. Automat. Control. **1992**, 37: 1415–1420.
2. Gordienko, E.; Lemus-Rodríguez, E. Estimation of robustness for controlled diffusion processes. Stochastics. **1999**, 17: 421–441.
3. Gordienko, E.; Minjárez-Sosa, J.A. Adaptive control for discrete-time Markov processes with unbounded costs: discounted criterion. Kybernetika. **1998**, 34: 217–234.
4. Gordienko, E.; Salem-Silva, F. Robustness inequality for Markov control processes with unbounded costs. Systems Control Lett. **1998**, 33, 125–130.
5. Hernández-Lerma, O.; Lasserre, J.B. *Further Topics on Discrete-Time Markov Control Processes*; Springer, New York, 1999.
6. Hilgert, N.; Minjárez-Sosa, J.A. Adaptive policies for time-varying stochastic systems under discounted criterion. Math. Meth. Oper. Res. **2001**, 54: 491–505.
7. Luque-Vásquez, F.; Minjárez-Sosa, J.A. Semi-Markov control processes with unknown holding times distribution under a discounted criterion. Math. Meth. Oper. Res. **2005**, 61: 455–468.
8. Luque-Vásquez, F.; Robles-Alcaraz, M.T. Controlled semi-Markov models with discounted unbounded costs. Bol. Soc. Mat. Mexicana. **1994**, 39: 51–68.
9. Minjárez-Sosa, J.A. Approximation and estimation in Markov control processes under a discounted criterion. Kybernetika. **2004**, 40,6: 681–690.
10. Montes-de-Oca, R.; Sakhanenko, A.I.; Salem-Silva, F. Estimates for perturbations of general discounted Markov control chains. Appl. Math. **2003**, 30,3 pp.287–304.
11. Van Dijk, N.M. Perturbation theory for unbounded Markov reward processes with applications to queueing. Adv. Appl. Probab. **1988**, 20: 99–111.
12. Van Dijk, N.M.; Puterman, M.L. Perturbation theory for Markov reward processes with applications to queueing systems. Adv. Appl. Probab. **1988**, 20: 79–98.

# Chapter 16
# Discrete Time Approximations of Continuous Time Finite Horizon Stopping Problems

Lukasz Stettner

## 16.1 Introduction

We assume that on a locally compact metric space $E$, we are given a right continuous standard Markov process $(x(t))$ with weakly Feller transition operator $(P_t)$, i.e., with transition operator $P_t$ that for $t \geq 0$ transforms the class $\mathscr{C}_0(E)$ of continuous functions on $E$ vanishing at infinity into itself (see [3] for the properties of such Markov processes). Let $C([0,T] \times E)$ denote the class of continuous bounded functions on $[0,T] \times E$. For a given functions $f, g, h \in C([0,T] \times E)$ and a discount factor $\alpha > 0$, consider the value function the of following continuous time cost functional:

$$
w(s,x) = \sup_{\tau \leq T-s} E_{sx} \left\{ \int_0^\tau e^{-\alpha u} f(s+u, x(u)) du + \chi_{\tau < T-s} e^{-\alpha \tau} g(s+\tau, x(\tau)) \right.
$$

$$
\left. + \chi_{\tau = T-s} e^{-\alpha(T-s)} h(T, x(T-s)) \right\}, \tag{16.1}
$$

where supremum is taken over all stopping times which do not exceed the final horizon $T - s$. As is pointed out in [14], in the case of finite horizon stopping problems with bounded functionals, by suitable redefinition of the functions in the functional, we can neglect discount factor $\alpha > 0$. Therefore, the assumption $\alpha > 0$ is made for simplicity of the presentation only. In the chapter, we consider two methods to approximate the value function $w$. The first one will be based on a solution to a suitable penalty equation. For a given $\beta > 0$, we shall find a solution $w^\beta$ to the following equation,

L. Stettner (✉)
Institute of Mathematics Polish Academy of Sciences, 00-956 Warsaw, Poland
e-mail: stettner@impan.gov.pl

D. Hernández-Hernández and A. Minjárez-Sosa (eds.), *Optimization, Control, and Applications of Stochastic Systems*, Systems & Control: Foundations & Applications, DOI 10.1007/978-0-8176-8337-5_16, © Springer Science+Business Media, LLC 2012

$$w^\beta(s,x) = E_{sx}\left\{\int_0^{T-s} \mathrm{e}^{-\alpha u}\left[f(s+u,x(u)) + \beta(g(s+u,x(u))\right.\right.$$

$$\left.\left. - w^\beta(s+u,x(u)))^+\right]\mathrm{d}u + \mathrm{e}^{-\alpha(T-s)}h(T,x(T-s))\right\}, \quad (16.2)$$

and then shall approximate its intensity version using discretized intensities. The second method consists in direct discretization of stopping times. For a given small $\Delta$, we consider the family of stopping times $\mathcal{T}_\Delta$ of the discretized Markov process $(x(n\Delta))$ taking values in $\{0,\Delta,2\Delta,\ldots,n\Delta,\ldots\}$. Assuming that $s$ is a multiplicity of $\Delta$, we define a discrete time optimal stopping problem

$$w_\Delta(s,x) = \sup_{\mathcal{T}_\Delta \ni \tau \leq T-s} E_{sx}\left\{\sum_{i=0}^{\frac{\tau}{\Delta}-1} \mathrm{e}^{-i\Delta}\bar{f}_\Delta(s+i\Delta,x(i\Delta)) + \chi_{\tau<T-s}\mathrm{e}^{-\alpha\tau}g(s+\tau,x(\tau))\right.$$

$$\left. + \chi_{\tau=T-s}\mathrm{e}^{-\alpha(T-s)}h(T,x(T-s))\right\} \qquad (16.3)$$

with $\bar{f}_\Delta(s,x) = E_{sx}\{\int_0^\Delta \mathrm{e}^{-\alpha u}f(s+u,x(u))\mathrm{d}u\}$.

The value function $w_\Delta$ of the above of discrete time stopping problem can be approximated using either Bellman's equation or by a discrete time penalty equation of the form

$$w_\Delta^b(s,x) = \bar{f}_\Delta(s,x) + \frac{b}{1-b}(g(s,x) - w_\Delta^b(s,x))^+ + \mathrm{e}^{-\alpha\Delta}\int_E w_\Delta^b(s+\Delta,y)P_\Delta(x,\mathrm{d}y) \qquad (16.4)$$

or by a deterministic pathwise stopping which is in particular convenient for Monte Carlo methods. Discrete time penalty equation seems to be new. Continuous time penalty equation was studied in the case of diffusion processes in [1, 6] and for Feller Markov processes in [9, 10, 13, 14]. Time discretization is a natural way to approximate optimal stopping problems of general Markov processes. In this chapter, we approximate this way, both value functions and optimal stopping times. We provide also errors of such approximation.

## 16.2 Discrete Time Optimal Stopping: Bellman Equation, Penalty Equation and Deterministic Approach

We assume now that $(x(n))$ is a discrete time Markov process with transition operator $P(x,\mathrm{d}y)$ on a given Borel measurable state space $E$. In this section, we consider a deterministic terminal time $T$, which is a positive integer and solve optimal stopping problem for discrete time Markov process $(x(n))$ over finite time horizon $T$. We shall also assume that time parameter $s$ takes only nonnegative integer values. Consider the following value function of discrete time optimal stopping problem:

$$w(s,x) = \sup_{\tau \le T-s} E_{sx} \left\{ \sum_{i=0}^{\tau-1} \gamma^i f(s+i,x(i)) + \chi_{\tau < T-s} \gamma^\tau g(s+\tau,x(\tau)) \right.$$

$$\left. + \chi_{\tau = T-s} \gamma^{T-s} h(T,x(T-s)) \right\}. \tag{16.5}$$

We assume that $\gamma \in (0,1)$. Notice that by [4]

**Lemma 16.2.1.** *w is a solution to the following system of Bellman equations* $w(T,x) = h(T,x)$ *and for $s < T$*

$$w(s,x) = \max \left\{ f(s,x) + \gamma \int_E w(s+1,y)P(x,dy), g(s,x) \right\} \tag{16.6}$$

*and the optimal stopping time $\hat{\tau}(s)$ for (16.5) is of the form*

$$\hat{\tau}(s) = \inf\{i \ge 0 : w(s+i,x(i)) = g(s+i,x(i))\} \wedge (T-s).$$

Let $b$ be a parameter taking values from the interval $(0,1)$. We consider the following equation for nonnegative integer $s < T$:

$$w^b(s,x) = f(s,x) + \frac{b}{1-b}(g(s,x) - w^b(s,x))^+ + \gamma \int_E w^b(s+1,y)P(x,dy) \tag{16.7}$$

with $w^b(T,x) = h(T,x)$.

**Theorem 16.2.1.** *For $b \in (0,1)$ and bounded measurable functions $f,g,h$, there is a unique bounded solution $w^b$ to (16.7). Moreover, $w^b$ for $s < T$ is also a solution to*

$$w^b(s,x) = \max_{0 \le b(s,x) \le b} (1-b(s,x))f(s,x) + b(s,x)g(s,x)$$

$$+ (1-b(s,x))\gamma \int_E w^b(s+1,y)P(x,dy), \tag{16.8}$$

*where $b(s,x)$ takes values from the interval $[0,b]$ and maximum is attained for $b(s,x)$ with values from $\{0,b\}$. Furthermore, we also have that*

$$w^b(s,x) = \sup_{\tau \le T-s} E_{sx} \left\{ \sum_{i=0}^{\tau-1} \gamma^i f(s+i,x(i)) + \chi_{\tau < T-s} \gamma^\tau (g(s+\tau,x(\tau)) \right.$$

$$\left. -(g(s+\tau,x(\tau)) - w^b(s+\tau,x(\tau)))^+) + \chi_{\tau = T-s} \gamma^{T-s} h(T,x(T-s)) \right\}.$$

$$\tag{16.9}$$

*Moreover, $w^b$ converges uniformly to $w$ as $b \to 1$.*

*Proof.* Multiplying (16.7) by $(1-b)$ and rearranging the terms, we obtain

$$w^b(s,x) = (1-b)f(s,x) + b\left((g(s,x) - w^b(s,x))^+ + w^b(s,x)\right)$$
$$+\gamma(1-b)\int_E w^b(s+1,y)P(x,dy). \qquad (16.10)$$

Denote by $F_b w(s,x)$ the right-hand side of (16.10). For two bounded functions $w_1$ and $w_2$, we have

$$|F_b w_1(s,x) - F_b w_2(s,x)| \le b|w_1(s,x) - w_2(s,x)| + \gamma(1-b)\|w_1 - w_2\|$$
$$\le (1-(1-b)(1-\gamma))\|w_1 - w_2\| \qquad (16.11)$$

with $\|\cdot\|$ standing for the supremum norm. Therefore, $F_b$ is a contraction in the class of bounded measurable functions, and there is a unique fixed point $w^b$, which is the unique solution to (16.10) and (16.7). Multiplying (16.7) by $(1-b(s,x))$ after small rearrangement, we obtain

$$w^b(s,x) = (1-b(s,x))f(s,x) + \frac{b(1-b(s,x))}{1-b}(g(s,x) - w^b(s,x))^+$$
$$+b(s,x)w^b(s,x) + (1-b(s,x))\gamma\int_E w^b(s+1,y)P(x,dy). \qquad (16.12)$$

Note that for $0 \le b(s,x) \le b$,

$$\frac{b(1-b(s,x))}{1-b}(g(s,x) - w^b(s,x))^+ + b(s,x)w^b(s,x) \ge b(s,x)g(s,x) \qquad (16.13)$$

with equality for $b(s,x) = 0$ when $w^b(s,x) \ge g(s,x)$ and $b(s,x) = b$ otherwise. Consequently, we have (16.8). Now, iterating (16.7) for any stopping time $\tau \le T - s$, we obtain

$$w^b(s,x) = E_{sx}\left\{\sum_{i=0}^{\tau-1}\gamma^i\left(f(s+i,x(i)) + \frac{b}{1-b}(g(s+i,x(i)) - w^b(s+i,x(i)))^+\right)\right.$$
$$\left. +\gamma^\tau w^b(s+\tau,x(\tau))\right\}. \qquad (16.14)$$

Note that for $\tau < T - s$,

$$w^b(s+\tau,x(\tau)) \ge g(s+\tau,x(\tau)) - (g(s+\tau,x(\tau)) - w^b(s+\tau,x(\tau)))^+ \qquad (16.15)$$

so that

$$w^b(s,x) \geq E_{sx}\left\{\sum_{i=0}^{\tau-1}\gamma^i f(s+i,x(i)) + \chi_{\tau<T-s}\gamma^\tau\,(g(s+\tau,x(\tau)))\right.$$

$$\left. -(g(s+\tau,x(\tau))-w^b(s+\tau,x(\tau)))^+) + \chi_{\tau=T-s}\gamma^{T-s}h(T,x(T-s))\right\}$$

(16.16)

with equality for $\tau = \inf\{i \geq s : g(s+i,x(i)) \geq w^b(s+i,x(i))\} \wedge (T-s)$, which completes the proof of (16.9). It remains to show the convergence of $w^b$ to $w$ when $b \to 1$. For this purpose, note first that

$$g(s,x) = E_{sx}\left\{\sum_{i=0}^{T-s-1}\gamma^i(g(s+i,x(i))-\gamma Pg(s+i+1,x(i))) + \gamma^{T-s}g(T,x(T-s))\right\}$$

(16.17)

with $Pg(s,x) = \int_E g(s+1,y)P(x,dy)$. For $\hat{w}^b(s,x) = w^\beta(s,x) - g(s,x)$ from (16.7), we therefore have

$$\hat{w}^b(s,x) = E_{sx}\left\{\sum_{i=0}^{T-s-1}\gamma^i\left(f(s+i,x(i)) - g(s+i,x(i)) + \gamma Pg(s+i+1,x(i))\right.\right.$$

$$\left.\left. +\frac{b}{1-b}(0-\hat{w}^b(s+i,x(i)))^+\right) + \gamma^{T-s}(h(T,x(T-s))-g(T,x(T-s)))\right\},$$

(16.18)

which means that $\hat{w}^b$ is a solution to the penalty equation (16.7) with $f(s,x)$ replaced by $f(s,x) - g(s,x) + \gamma Pg(s+1,x)$, $g$ by $0$, and $h$ by $h - g$. Consequently, a suitable version of (16.8) holds and

$$\hat{w}^b(s,x) \geq (1-b)\,(f(s,x) - g(s,x) + \gamma Pg(s+1,x))) + (1-b)\gamma P\hat{w}^b(s+1,x)$$

(16.19)

from which by iteration we obtain

$$\hat{w}^b(s,x) \geq E_{sx}\left\{\sum_{i=0}^{T-s-1}\gamma^i(1-b)^{i+1}\,(f(s+i,x(i)) - g(s+i,x(i))\right.$$

$$\left. +\gamma Pg(s+i+1,x(i))) + \gamma^{T-s}(1-b)^{T-s}(h(T,x(T-s))-g(T,x(T-s)))\right\}.$$

Since in (16.17) without loss of generality, we could choose $g(T,x) = h(T,x)$ from (16.9), we finally obtain for $s < T$

$$w(s,x) - g(s,x) \geq - \sum_{i=0}^{T-s-1} \gamma^i (1-b)^{i+1} \|f - g + \gamma Pg\| \geq -\frac{(1-b)\|f - g + \gamma Pg\|}{1 - \gamma(1-b)},$$

(16.20)

and the right-hand side converges uniformly in $s$ and $x$ to 0, as $b$ increases to 1. From (16.9), we now immediately obtain that $w^b$ converges uniformly to $w$ as $b \to 1$. □

An alternative method to approximate the discrete time optimal stopping value function is to solve the deterministic optimal stopping problem following [2, 5], which in particular is convenient for the use of Monte Carlo methods.

**Proposition 16.2.1** *For $n = 0, 1, \ldots, T - s$, we have*

$$w(s+n,x(n)) = \sup_{0 \leq i \leq T-(s+n)} \left\{ \sum_{j=0}^{i-1} \gamma^j f(s+n+j,x(n+j)) + \gamma^i g(s+n+i,x(n+i)) \right.$$

$$\left. - \sum_{j=0}^{i-1} \gamma^{j+1} \left( w(s+n+j,x(n+j)) - Pw(s+n+j+1,x(n+j)) \right) \right\}$$

(16.21)

*and consequently*

$$w(s+n,x(n)) = \frac{1}{1-\gamma} \sup_{0 \leq i \leq T-(s+n)} \left\{ \sum_{j=0}^{i-1} \gamma^j f(s+n+j,x(n+j)) \right.$$

$$+ \gamma^i g(s+n+i,x(n+i)) - \sum_{j=1}^{i-1} \gamma^{j+1} \left( w(s+n+j,x(n+j)) \right.$$

$$\left. - Pw(s+n+j+1,x(n+j)) \right) + \gamma Pw(s+n+1,x(n)) \right\}.$$ (16.22)

*Proof.* Note first that $w(s+n+j,x(j)) - f(s+n+j,x(j)) - \gamma Pw(s+n+j+1,x(j)) \geq 0$ and

$$\gamma^i w(s+n+i,x(i)) + \sum_{j=0}^{i-1} \gamma^j f(s+n+j,x(j)) - \sum_{j=0}^{i-1} \gamma^{j+1} \left( w(s+n+j+1,x(j+1)) \right.$$

$$\left. - Pw(s+n+j+1,x(j)) \right) = w(s+n,x(0)) - \sum_{j=0}^{i-1} \gamma^j \left( w(s+n+j,x(j)) \right.$$

$$\left. - f(s+n+j,x(j)) - \gamma Pw(s+n+j+1,x(j)) \right)$$

(16.23)

and $\sum_{j=0}^{i-1} \gamma^{j+1} \left( w(s+n+j+1,x(j+1)) - Pw(s+n+j+1,(x(j))) \right)$ is a martingale. Therefore,

$$
w(s+n,x(n)) = \sup_{0 \le \tau \le T-s-n} E_{s+n,x(n)} \left\{ \sum_{j=0}^{\tau-1} \gamma^j f(s+n+j,x(j)) \right.
$$

$$
+ \gamma^\tau g(s+n+\tau,x(\tau)) - \sum_{j=0}^{\tau-1} \gamma^{j+1} \left( w(s+n+j+1,x(j+1)) \right)
$$

$$
\left. - Pw(s+n+j+1,(x(j))) \right\} \le E_{s+n,x(n)} \left\{ \sup_{0 \le i \le T-s-n} \left( \sum_{j=0}^{i-1} \gamma^j f(s+n+j,x(j)) \right) \right.
$$

$$
+ \gamma^i g(s+n+i,x(i)) - \sum_{j=0}^{i-1} \gamma^{j+1} \left( w(s+n+j+1,x(j+1)) \right)
$$

$$
\left. - Pw(s+n+j+1,(x(j))) \right) \right\} \le E_{s+n,x(n)} \left\{ \sup_{0 \le i \le T-s-n} \left( \sum_{j=0}^{i-1} \gamma^j f(s+n+j,x(j)) \right) \right.
$$

$$
+ \gamma^i w(s+n+i,x(i)) - \sum_{j=0}^{i-1} \gamma^{j+1} \left( w(s+n+j+1,x(j+1)) \right)
$$

$$
\left. - Pw(s+n+j+1,(x(j))) \right) \right\} \le E_{s+n,x(n)} \left\{ \sup_{0 \le i \le T-s-n} \left( w(s+n,x(0)) \right) \right.
$$

$$
\left. - \sum_{j=0}^{i-1} \gamma^j \left( w(s+n+j,x(j)) - f(s+n+j,x(j)) - \gamma Pw(s+n+j+1,x(j)) \right) \right) \right\}
$$

$$
\le E_{s+n,x(n)} \left\{ w(s+n,x(0)) \right\} = w(s+n,x(n)).
$$

Since by (16.23) and $w(s+n+j,x(n+j)) - f(s+n+j,x(n+j)) - \gamma Pw(s+n+j+1,x(n+j)) \ge 0$ again

$$
\sup_{0 \le i \le T-(s+n)} \left\{ \sum_{j=0}^{i-1} \gamma^j f(s+n+j,x(n+j)) + \gamma^i g(s+n+i,x(n+i)) \right.
$$

$$
\left. - \sum_{j=0}^{i-1} \gamma^{j+1} \left( w(s+n+j,x(n+j)) - Pw(s+n+j+1,x(n+j)) \right) \right\}
$$

$$
\le \sup_{0 \le i \le T-(s+n)} \left\{ \sum_{j=0}^{i-1} \gamma^j f(s+n+j,x(n+j)) + \gamma^i w(s+n+i,x(n+i)) \right.
$$

$$
\left. - \sum_{j=0}^{i-1} \gamma^{j+1} \left( w(s+n+j,x(n+j)) - Pw(s+n+j+1,x(n+j)) \right) \right\}
$$

$$\leq \sup_{0\leq i\leq T-(s+n)} \left\{ w(s+n,x(n)) + \sum_{j=0}^{i-1} \gamma^j \left( w(s+n+j,x(n+j)) \right) \right.$$

$$\left. -f(s+n+j,x(n+j)) - \gamma P w(s+n+j+1,x(n+j))) \right\}$$

$$\leq w(s+n,x(n)), \tag{16.24}$$

we obtain (16.21). The formula (16.22) follows directly from (16.21).                    □

**Remark 16.2.2** *Notice that the formula (16.21) follows from the dual representation of the value function taking into account an explicit form of the martingale in the* **Doob-Meyer** *decomposition of the discrete time Snell envelope (for details, see formula (1.1) of [11] and formula (2.4) of [12]).*

## 16.3   Continuous Time Penalty Equation

In this section, we assume that $(x_t)$ is a right continuous standard Markov process introduced in Sect. 16.1. Assuming that the functions $f$, $g$, and $h$ are continuous bounded, we summarize below the results concerning the use of penalty method for continuous time optimal stopping problems [9, 14].

**Theorem 16.3.2.** *For $\beta > 0$, there is a unique solution $w^\beta$ to (16.2), which is a continuous bounded function and also has the form*

$$w^\beta(s,x) = \sup_{(b(t))\in M_\beta} E_{sx} \left\{ \int_0^{T-s} e^{-\int_0^u (\alpha+b(s+r))\mathrm{d}r} (f(s+u,x(u))) \right.$$

$$\left. +b(s+u)g(s+u,x(u))\mathrm{d}u + e^{-\int_0^{T-s}(\alpha+b(s+u))\mathrm{d}u} h(T,x(T-s)) \right\},$$

$$\tag{16.25}$$

*where $M_\beta$ is the set of progressively measurable processes $(b(t))$ with values from the interval $[0,\beta]$. The optimal value $\hat{b}^\beta(s+t)$ in (16.25) is of the form $\hat{b}^\beta(s+t)=\beta$ whenever $g(s+t,x(t)) > w^\beta(s+t,x(t))$ and $\hat{b}^\beta(s+t) = 0$ otherwise. Furthermore, $w^\beta(s,x)$ converges increasing to $w(s,x)$ (defined in (16.1)) uniformly on compact subsets of $[0,T] \times E$. Moreover, whenever the function $g$ is of the form*

$$g(s,x) = E_{sx} \left\{ \int_0^\infty e^{-\alpha t} \phi(s+t,x(t))\mathrm{d}t \right\} \tag{16.26}$$

*with $\phi \in C([0,\infty) \times E)$, then we have the uniform estimate*

$$w(s,x) - w^\beta(s,x) \leq \frac{\|f-\phi\|}{\alpha+\beta} + e^{-\alpha T}\|(h-g)^+\| \tag{16.27}$$

with $(f - \phi)^+$ standing for a positive part of $f - \phi$. Finally, the stopping time $\tau^\beta(s)$ defined as

$$\tau^\beta(s) := \inf\left\{t \geq 0 : w^\beta(s+t, x(t)) \leq g(s+t, x(t))\right\} \wedge (T - s) \qquad (16.28)$$

increases , as $\beta \to \infty$ to the stopping time $\hat{\tau}(s)$

$$\hat{\tau}(s) = \inf\{t \geq 0 : w(s+t, x(t)) = g(s+t, x(t))\} \wedge (T - s), \qquad (16.29)$$

which in the case when $g \leq h$ is an optimal stopping time for (16.1).


## 16.4  Discrete Time Semigroup Approximation

In this section, for $s$ and $T$ being **multiplicities** of $\Delta$, we shall approximate stopping times $\tau \leq T - s$ by stopping times of the family $\mathcal{T}_\Delta$ defined in Sect. 16.1. Namely, let $\tau_\Delta = k\Delta$, whenever $\tau \in ((k-1)\Delta, k\Delta]$, for $k = 1, 2, \ldots$. We have

**Lemma 16.4.2.** *For a stopping time $\tau$ and its discrete approximation $\tau_\Delta$, we have the following estimation;*

$$\left| E_{sx}\left\{\chi_{\tau < T - s}\left(g(s + \tau, x(\tau)) - g(s + \tau_\Delta, x(\tau_\Delta))\right)\right\}\right|$$

$$\leq E_{sx}\left\{\sup_{u \leq \Delta}|\mathscr{P}_u g(s + \tau, x(\tau)) - g(s + \tau, x(\tau))|\right\} \qquad (16.30)$$

*with $\mathscr{P}_u g(s, x) = \int_E g(s + u, y) P_u(x, dy)$.*

*Proof.* We use strong Markov property of $(x(t))$. We have

$$E_{sx}\left\{\chi_{\tau < T - s} g(s + \tau, x(\tau)) - g(s + \tau_\Delta, x(\tau_\Delta))\right\}$$
$$= E_{sx}\left\{\chi_{\tau < T - s} E_{s + \tau, x(\tau)}\left\{g(s + \tau, x) - g(s + \tau, x(\tau_\Delta - \tau))\right\}\right\}$$
$$\leq E_{sx}\left\{\chi_{\tau < T - s} \sup_{u \leq \Delta}\left(g(s + \tau, x(\tau)) - \mathscr{P}_u g(s + \tau, x(\tau))\right)\right\}$$

from which (16.30) immediately follows.                                    □

Let

$$r_\Delta(s, x) = \sup_{u \leq \Delta}|\mathscr{P}_u g(s, x) - g(s, x)|. \qquad (16.31)$$

We have the following:

**Lemma 16.4.3.** *$r_\Delta$ converges uniformly on compact subsets of $[0,T] \times E$ to 0 as $\Delta \to 0$. When $g \in C_0([0,T] \times E)$, the convergence is uniform.*

*Proof.* By Theorem T1 Chap. XIII in [7] for $g \in C_0([0,T] \times E)$, the function $r_\Delta$ converges uniformly to 0 as $\Delta \to 0$. Using Proposition 2.1 of [8] for a given $\varepsilon > 0$ and a compact set $K \subset E$, we have for a sufficiently large $R$

$$\sup_K P_x \left\{ \exists_{s \in [0,T]} \rho(x(s),x) \geq R \right\} \leq \varepsilon, \tag{16.32}$$

where $\rho$ is a metric compatible with the topology of $E$. We approximate function $g$ outside of the ball with radius $R$ and center in the set $K$ by a function $g' \in C([0,T] \times E)$. Clearly

$$\sup_{x \in K} |\mathscr{P}_u g(s,x) - g(s,x)| \leq \varepsilon(\|g\| + \|g'\|) + |\mathscr{P}_u g'(s,x) - g'(s,x)|,$$

and the last term converges to 0 as $\Delta \to 0$, which completes the proof. $\square$

Summarizing Lemmas 16.4.2 and 16.4.3, we obtain

**Theorem 16.4.3.** *For $s$ being a multiplicity of $\Delta$ assuming that $g \leq h$, we have*

$$0 \leq w(s,x) - w_\Delta(s,x) \leq \Delta \|f\| + \sup_\tau E_{sx} \{ r_\Delta(s + \tau, x(\tau)) \}. \tag{16.33}$$

*Whenever $g \in C_0([0,T] \times E)$ the convergence $w_\Delta$ to $w$, as $\Delta \to 0$, is uniform, and stopping time*

$$\hat{\tau}_\Delta(s) = \inf \{ i\Delta : w_\Delta(s + i\Delta, x(i\Delta)) = g(s + i\Delta, x(i\Delta) \} \wedge (T - s) \tag{16.34}$$

*is $\left( \Delta \|f\| + \sup_{u \leq \Delta} \|\mathscr{P}_u g - g\| \right)$ optimal.*

*Proof.* By the definition of (16.1) and (16.3), it is clear that $w(s,x) \geq w_\Delta(s,x)$. For any stopping time $\tau$, taking into account that $g \leq h$, we have

$$E_{sx} \left\{ \int_0^\tau e^{-\alpha u} f(s+u,x(u)) du + \chi_{\tau < T-s} e^{-\alpha \tau} g(s+\tau,x(\tau)) \right.$$

$$\left. + \chi_{\tau = T-s} e^{-\alpha(T-s)} h(T,x(T-s)) \right\} \leq E_{sx} \left\{ \int_0^{\tau_\Delta} e^{-\alpha u} f(s+u,x(u)) du + \Delta \|f\| \right.$$

$$+ \chi_{\tau < T-s} e^{-\alpha \tau_\Delta} g(s + \tau_\Delta, x(\tau_\Delta)) + r_\Delta(s + \tau, x(\tau))$$

$$\left. + \chi_{\tau = T-s} e^{-\alpha(T-s)} h(T,x(T-s)) \right\} \leq E_{sx} \left\{ \int_0^{\tau_\Delta} e^{-\alpha u} f(s+u,x(u)) du \right.$$

$$+ \chi_{\tau_\Delta < T-s} e^{-\alpha \tau_\Delta} g(s + \tau_\Delta, x(\tau_\Delta)) + \chi_{\tau_\Delta = T-s} e^{-\alpha(T-s)} h(T,x(T-s)) \right\}$$

$$+ \Delta \|f\| + E_{sx} \{ r_\Delta(s + \tau, x(\tau)) \}.$$

Now by Lemma 16.2.1, an optimal stopping for $\Delta$ discretized problem is within the class of $\mathcal{T}_\Delta$. Consequently (16.33) follows. When $g \in C_0([0,T] \times E)$ using Lemmas 16.4.2 and 16.4.3 and (16.33), we have that

$$\|w - w_\Delta\| \leq \Delta \|f\| + \sup_{u \leq \Delta} \|\mathscr{P}_u g - g\|.$$

Therefore, since by Lemma 16.2.1, $\hat{\tau}_\Delta(s)$ is optimal for the stopping problem (16.3), we obtain

$$w(s,x) \leq w_\Delta(s,x) + \|w - w_\Delta\| = E_{sx}\left\{ \int_0^{\hat{\tau}_\Delta(s)} e^{-\alpha u} f(s+u, x(u)) du \right.$$

$$+ \chi_{\hat{\tau}_\Delta(s) < T-s} e^{-\alpha \hat{\tau}_\Delta(s)} g(s + \hat{\tau}_\Delta(s), x(\hat{\tau}_\Delta(s)))$$

$$\left. + \chi_{\hat{\tau}_\Delta(s) = T-s} e^{-\alpha(T-s)} h(T, x(T-s)) \right\} + \|w - w_\Delta\|,$$

which means that $\hat{\tau}_\Delta(s)$ is $\|w - w_\Delta\|$ optimal.                $\square$

## 16.5   Discrete Time Intensity of Stopping Approximation

Assume now that $T$ is a multiplicity of $\Delta > 0$. Let $P_t g(s,x) = E_{sx}\{g(s, x(t))\}$, for $t \geq 0$ and a bounded Borel measurable function $g$.

Define a sequence of equations for the function $\tilde{w}_\Delta^\beta$ defined successively in the intervals $[(n-1)\Delta, n\Delta]$ with $n = 0, 1, \ldots, \frac{T}{\Delta}$. For $s \in [T - \Delta, T]$, let

$$\tilde{w}_\Delta^\beta(s,x) = \int_0^{T-s} e^{-\alpha u} \left( P_u f(s+u,x) + \beta \left( P_u g - P_u \tilde{w}_\Delta^\beta \right)^+ (s+u,x) \right) du$$

$$+ e^{-\alpha(T-s)} P_{T-s} h(T,x) \qquad (16.35)$$

and inductively for $s \in [(n-1)\Delta, n\Delta]$

$$\tilde{w}_\Delta^\beta(s,x) = \int_0^{n\Delta-s} e^{-\alpha u} \left( P_u f(s+u,x) + \beta \left( P_u g - P_u \tilde{w}_\Delta^\beta \right)^+ (s+u,x) \right) du$$

$$+ e^{-\alpha(n\Delta-s)} P_{n\Delta-s} \tilde{w}_\Delta^\beta(n\Delta, x). \qquad (16.36)$$

**Proposition 16.5.1** *There is a unique continuous bounded function $\tilde{w}_\Delta^\beta$ satisfying (16.35) and (16.36). Moreover, this solution is of the form for $s \in [T - \Delta, T]$*

$$\tilde{w}_\Delta^\beta(s,x) = \sup_{b \in D_\beta^\Delta} \left[ \int_0^{T-s} e^{-\int_0^u (\alpha + b(s+r)) dr} \left( P_u f(s+u,x) \right. \right.$$

$$\left. + \beta(s+u) P_u g(s+u,x) \right) du + e^{-\int_0^{T-s}(\alpha + b(s+r))dr} P_{T-s} h(T,x) \Bigg]$$

$$(16.37)$$

*and inductively for* $s \in [(n-1)\Delta, n\Delta]$

$$\tilde{w}_\Delta^\beta(s,x) = \sup_{b \in D_\beta^\Delta} \left[ \int_0^{n\Delta-s} e^{-\int_0^u (\alpha+b(s+r))dr} \left(P_u f(s+u,x) \right. \right.$$

$$\left. \left. + \beta(s+u)P_u g(s+u,x)\right) du + e^{-\int_0^{n\Delta-s}(\alpha+b(s+r))dr} P_{n\Delta-s} \tilde{w}_\Delta^\beta(n\Delta,x) \right]$$

(16.38)

*with* $D_\beta^\Delta$ *the class of processes that are deterministic in the intervals* $[T-\Delta,T]$ *and* $[(n-1)\Delta, n\Delta]$ *respectively and take values from the interval* $[0,\beta]$.

*Proof.* Using Lemma 1 of [14], we obtain the following equivalent formulae for (16.35) and (16.36)):

$$\tilde{w}_\Delta^\beta(s,x) = \int_0^{T-s} e^{-(\alpha+\beta)u} \left( P_u f(s+u,x) + \beta \left( P_u g - P_u \tilde{w}_\Delta^\beta \right)^+ (s+u,x) \right.$$

$$\left. + \beta P_u \tilde{w}_\Delta^\beta(s+u,x) \right) du + e^{-(\alpha+\beta)(T-s)} P_{T-s} h(T,x) \qquad (16.39)$$

and inductively for $s \in [(n-1)\Delta, n\Delta]$

$$\tilde{w}_\Delta^\beta(s,x) = \int_0^{n\Delta-s} e^{-(\alpha+\beta)u} \left( P_u f(s+u,x) + \beta \left( P_u g - P_u \tilde{w}_\Delta^\beta \right)^+ (s+u,x) \right.$$

$$\left. + \beta P_u \tilde{w}_\Delta^\beta(s+u,x) \right) du + e^{-(\alpha+\beta)(n\Delta-s)} P_{n\Delta-s} \tilde{w}_\Delta^\beta(n\Delta,x). \qquad (16.40)$$

Since $\left( P_u g - P_u \tilde{w}_\Delta^\beta \right)^+ (s+u,x) + P_u \tilde{w}_\Delta^\beta(s+u,x) = \max \left\{ P_u g(s+u,x), P_u \tilde{w}_\Delta^\beta(s+u,x) \right\}$, the right-hand sides of (16.39) and (16.40) define contractive operators in the class of continuous bounded functions defined on the time intervals $[T-\Delta,T]$ or $[(n-1)\Delta, n\Delta]$, respectively. Consequently, we have the existence of unique continuous bounded solutions to (16.35) and (16.36). To show the formulae (16.36) and (16.37), we use Lemma 1 of [14] again. For $((b(s)) \in D_\beta^\Delta$, we obtain from (16.35) and (16.36) for $s \in [T-\Delta,T]$

$$\tilde{w}_\Delta^\beta(s,x) = \int_0^{T-s} e^{-\int_0^{T-s}(\alpha+b(s+r))dr} \left( P_u f(s+u,x) + \beta \left( P_u g - P_u \tilde{w}_\Delta^\beta \right)^+ (s+u,x) \right.$$

$$\left. + \beta(s+u)P_u \tilde{w}_\Delta^\beta(s+u,x) \right) du + e^{-\int_0^{T-s}(\alpha+b(s+r))dr} P_{T-s} \tilde{w}_\Delta^\beta(T,x)$$

(16.41)

and inductively for $s \in [(n-1)\Delta, n\Delta]$

$$\tilde{w}_\Delta^\beta(s,x) = \int_0^{n\Delta-s} e^{-\int_0^{n\Delta-s}(\alpha+b(s+r))dr} \left( P_u f(s+u,x) \right.$$

$$+ \beta \left( P_u g - P_u \tilde{w}_\Delta^\beta \right)^+ (s+u,x) + \beta(s+u) P_u \tilde{w}_\Delta^\beta(s+u,x) \right) du$$

$$+ e^{-\int_0^{n\Delta-s}(\alpha+b(s+r))dr} P_{n\Delta-s} \tilde{w}_\Delta^\beta(n\Delta,x). \tag{16.42}$$

Note now that

$$\int_0^{T-s} e^{-\int_0^{T-s}(\alpha+b(s+r))dr} \left( P_u f(s+u,x) + \beta \left( P_u g - P_u \tilde{w}_\Delta^\beta \right)^+ (s+u,x) \right.$$

$$+ \beta(s+u) P_u \tilde{w}_\Delta^\beta(s+u,x) \right) du \geq \int_0^{T-s} e^{-\int_0^{T-s}(\alpha+b(s+r))dr} \left( P_u f(s+u,x) \right.$$

$$+ \beta(s+u) P_u g(s+u,x)) du$$

with equality for $b(u) = 0$, whenever $P_u \tilde{w}_\Delta^\beta((s+u,x) \geq T_u g(s+u,x)$ and $b(u) = \beta$ for $P_u \tilde{w}_\Delta^\beta(s+u,x) < T_u g(s+u,x)$. Therefore we obtain (16.37), and in a similar way, we also have (16.38).  $\square$

Let

$$\hat{w}_\Delta^\beta(s,x) = \sup_{b \in D_\beta^{c,\Delta}} \left[ \int_0^{T-s} e^{-\int_0^u(\alpha+b(s+r))dr} \left( P_u f(s+u,x) \right. \right.$$

$$\left. + b(s+u) P_u g(s+u,x)) du + e^{-\int_0^{T-s}(\alpha+b(s+r))dr} P_{T-s} h(T,x) \right] \tag{16.43}$$

and inductively for $s \in [(n-1)\Delta, n\Delta]$

$$\hat{w}_\Delta^\beta(s,x) = \sup_{b \in D_\beta^{c,\Delta}} \left[ \int_0^{n\Delta-s} e^{-\int_0^u(\alpha+b(s+r))dr} \left( P_u f(s+u,x) \right. \right.$$

$$\left. + b(s+u) P_u g(s+u,x)) du + e^{-\int_0^{n\Delta-s}(\alpha+b(s+r))dr} P_{n\Delta-s} \hat{w}_\Delta^\beta(n\Delta,x) \right]$$

$$\tag{16.44}$$

with $D_\beta^{c,\Delta}$ the class of processes that are constant in the intervals $[T-\Delta, T]$ and $[(n-1)\Delta, n\Delta]$ respectively and take values from the interval $[0, \beta]$. Clearly, $\hat{w}_\Delta^\beta(T,x) = h(T,x)$, and for $s \leq T - \Delta$ being multiplicity of $\Delta$, we have

$$\hat{w}_\Delta^\beta(s,x) = \sup_{b \in [0,\beta]} \left[ \int_0^{n\Delta} e^{-(\alpha+b)u} \left( P_u f(s+u,x) \right. \right.$$

$$\left. + b P_u g(s+u,x)) du + e^{-(\alpha+b)\Delta} P_\Delta \hat{w}_\Delta^\beta(\Delta,x) \right]. \tag{16.45}$$

**Lemma 16.5.4.** *For s being a multiplicity of $\Delta$, we have that $w(s,x) \geq \tilde{w}_\Delta^\beta(s,x) \geq \hat{w}_\Delta^\beta(s,x)$ and*

$$\lim_{\beta \to \infty} \hat{w}_\Delta^\beta(s,x) \geq w_\Delta(s,x) \tag{16.46}$$

*with $w_\Delta(s,x)$ defined in (16.3).*

*Proof.* The fact that $w(s,x) \geq \tilde{w}_\Delta^\beta(s,x)$ follows directly from (16.25), (16.37), and (16.38). Note that for $s \in [(n-1)\Delta, n\Delta]$,

$$\hat{w}_\Delta^\beta(s,x) \geq \max\left[ \int_0^{n\Delta-s} e^{-\alpha u} P_u f(s+u,x) du + e^{-\alpha(n\Delta-s)} P_{n\Delta-s} \hat{w}_\Delta^\beta(n\Delta,x), \right.$$

$$\int_0^{n\Delta-s} e^{-(\alpha+\beta)u} \left( P_u f(s+u,x) + \beta P_u g(s+u,x) \right) du$$

$$\left. + e^{-(\alpha+\beta)(n\Delta-s)} P_{n\Delta-s} \hat{w}_\Delta^\beta(n\Delta,x) \right] \tag{16.47}$$

and letting $\beta \to \infty$, we obtain for $\hat{w}_\Delta^\infty$

$$\hat{w}_\Delta^\infty(s,x) \geq \max\left[ \int_0^{n\Delta-s} e^{-\alpha u} P_u f(s+u,x) du + e^{-\alpha(n\Delta-s)} P_{n\Delta-s} \hat{w}_\Delta^\infty(n\Delta,x), g(s,x) \right] \tag{16.48}$$

from which we easily obtain that $\hat{w}_\Delta^\infty(s,x) \geq w_\Delta(s,x)$. □

In what follows, we are going to show that $\hat{w}_\Delta^\beta(s,x)$ approximate $w^\beta(s,x)$, provided that $\Delta$ is sufficiently small. For this purpose, we first prove the following Lemma

**Lemma 16.5.5.** *Let $\hat{b}_\Delta^\beta(t) = \beta$ whenever $n\Delta \leq t < (n+1)\Delta$ and $\hat{b}^\beta(t) = \beta$, and $\hat{b}_\Delta^\beta(t) = 0$ otherwise, with $\hat{b}^\beta$ defined in Theorem 16.3.2. For fixed $\beta > 0$, we have that*

$$\int_0^{T-s} |\hat{b}_\Delta^\beta(s+u) - \hat{b}^\beta(s+u)| du \to 0 \tag{16.49}$$

*$P_{sx}$ a.e. as $\Delta \to 0$.*

*Proof.* Notice first that because of the continuity of $w^\beta$ and $g$ and right continuity of $(x(t))$ when $\hat{b}^\beta(t) = \beta$ then also $\hat{b}^\beta(t+u) = \beta$ for a sufficiently small positive $u$. Consequently, the set $\{t : \hat{b}^\beta(t) = \beta\}$ consists of at most countable disjoint intervals. If to each interval of the set $\{t : \hat{b}^\beta(t) = \beta\}$ we add adjacent intervals in which $\hat{b}^\beta = 0$, the sum of such intervals shall form a countable covering of the interval $[0,T]$. For any $\varepsilon > 0$, there is a finite family of disjoint intervals $I_k$, $k = 1, 2, \ldots, N$ from the above-mentioned family such that $\mathscr{L}([0,T] \setminus \cup_{i=1}^N I_k) \leq \frac{\varepsilon}{2\beta}$, where $\mathscr{L}$ stands for Lebesgue measure. Note that on each set $I_k$ the value of $|\hat{b}_\Delta^\beta(s+u) - \hat{b}^\beta(s+u)| = 0$ outside of at most two small subintervals of the length $\Delta$ on the right- and left-hand sides of the suitable interval where $\hat{b}^\beta = \beta$. Let now $\Delta < \frac{\varepsilon}{4N\beta}$. Then

$$\int_0^{T-s} |\hat{b}_\Delta^\beta(s+u) - \hat{b}^\beta(s+u)|du \le \beta\frac{\varepsilon}{2\beta} + \sum_{k=1}^N \int_{I_k} |\hat{b}_\Delta^\beta(s+u) - \hat{b}^\beta(s+u)|du \le \varepsilon$$

from which (16.49) easily follows.                                                  □

**Corollary 16.5.2** *For each $\beta > 0$, we have*

$$\hat{w}_\Delta^\beta(s,x) \to w^\beta(s,x) \tag{16.50}$$

*as $\Delta \to 0$ uniformly in $(s,x)$ from compact subsets of $[0,T] \times E$.*

*Proof.* Comparing (16.43) and (16.44) with (16.25), we clearly have that $\hat{w}_\Delta^\beta(s,x) \le w^\beta(s,x)$. Since

$$w^\beta(s,x) = E_{sx}\left\{ \int_0^{T-s} e^{-\int_0^u (\alpha+\hat{b}^\beta(s+r))dr}(f(s+u,x(u))\right.$$

$$\left. + \hat{b}^\beta(s+u)g(s+u,x(u)))du + e^{-\int_0^{T-s}(\alpha+\hat{b}^\beta(s+u))du}h(T,x(T-s)) \right\}$$

by Lemma 16.5.5 and the dominated convergence theorem, we obtain that

$$E_{sx}\left\{ \int_0^{T-s} e^{-\int_0^u(\alpha+\hat{b}_\Delta^\beta(s+r))dr}(f(s+u,x(u)) + \hat{b}_\Delta^\beta(s+u)g(s+u,x(u)))du\right.$$

$$\left. + e^{-\int_0^{T-s}(\alpha+\hat{b}_\Delta^\beta(s+u))du}h(T,x(T-s)) \right\} \to w^\beta(s,x)$$

as $\Delta \to 0$. Consequently, we obtain that $\hat{w}_{2^{-n}\Delta}^\beta(s,x) \to w^\beta(s,x)$ as $n \to \infty$, and since both functions $\hat{w}_\Delta^\beta$ and $w^\beta$ are continuous and the convergence is monotonic by Dini lemma, we obtain uniform convergence on compact sets.                     □

For $s$ being a multiplicity of $\Delta$, let

$$\bar{w}_\Delta^\beta(s,x) = \sup_{b\in[0,\beta]} \left[ e^{-b\Delta} \int_0^\Delta e^{-\alpha u} P_u f(s+u,x)du \right.$$

$$\left. + (1-e^{-b\Delta})g(s,x) + e^{-(\alpha+b)\Delta}P_\Delta \bar{w}_\Delta^\beta(s+\Delta,x) \right] \tag{16.51}$$

with $\bar{w}_\Delta^\beta(T,x) = h(T,x)$.

**Proposition 16.5.3** *For $n < \frac{T}{\Delta}$, we have*

$$\left|\hat{w}_\Delta^\beta(n\Delta,x)-\bar{w}_\Delta^\beta(n\Delta,x)\right| \leq (1-e^{-\beta\Delta})\left(\int_{n\Delta}^T e^{-\alpha(u-n\Delta)}\|P_uf\|du+G_\Delta\frac{e^{-n\Delta}-e^{-\alpha T}}{1-e^{-\alpha\Delta}}\right)$$

(16.52)

*with* $\|F\| := \sup_{(s,x)\in[0,T]\times E}|F(s,x)|$ *and* $G_\Delta := \sup_{t\leq T-\Delta, u\in[0,\Delta], x\in E}|e^{-\alpha u}P_ug(t+u,x)-g(t,x)|$.

*Proof.* Since by (16.45)

$$\hat{w}_\Delta^\beta(T-\Delta,x) = \sup_{b\in[0,\beta]}\left[\int_0^\Delta e^{-(\alpha+\beta)u}\left(P_uf(T-\Delta+u,x)+bP_ug(s+u,x)\right)du\right.$$

$$\left.+e^{-(\alpha+\beta)\Delta}P_\Delta h(T,x)\right]$$

and

$$\bar{w}_\Delta^\beta(T-\Delta,x) = \sup_{b\in[0,\beta]}\left[e^{-b\Delta}\int_0^\Delta e^{-\alpha u}P_uf(T-\Delta+u,x)du\right.$$

$$\left.+(1-e^{-b\Delta})g(T-\Delta,x)+e^{-(\alpha+b)\Delta}P_\Delta h(T,x)\right],$$

we have

$$|\hat{w}_\Delta^\beta(T-\Delta,x)-\bar{w}_\Delta^\beta(T-\Delta,x)| \leq (1-e^{-\beta\Delta})\left(\int_0^\Delta e^{-\alpha u}\|P_uf\|du\right.$$

$$\left.+\sup_{u\in[0,\Delta]}|e^{-\alpha u}P_ug(T-\Delta+u,x)-g(T-\Delta,x)|\right).$$

(16.53)

By (16.45) and (16.51), we have

$$|\hat{w}_\Delta^\beta(n\Delta,x)-\bar{w}_\Delta^\beta(n\Delta,x)| \leq (1-e^{-\beta\Delta})\left(\int_0^\Delta e^{-\alpha u}\|P_uf\|du\right.$$

$$\left.+\sup_{u\in[0,\Delta]}|e^{-\alpha u}P_ug(T-\Delta+u,x)-g(T-\Delta,x)|\right)$$

$$+e^{-\alpha\Delta}|P_\Delta\hat{w}_\Delta^\beta((n+1)\Delta,x)-P_\Delta\bar{w}_\Delta^\beta((n+1)\Delta,x)|,$$

and using inductively (16.53) to the last term, we obtain (16.52). □

Note that by (16.8), $\bar{w}_\Delta^\beta$ coincides with the solution to the penalty equation (16.4) with $b = 1-e^{-\beta\Delta}$. It is just immediate from Theorem 16.2.1 that $\bar{w}_\Delta^\beta$ converges uniformly to $w_\Delta$ letting $\beta \to \infty$.

**Theorem 16.5.4.** *We have that* $\bar{w}_\Delta^\beta(s,x)$ *converges to* $w(s,x)$ *as* $\beta \to \infty$ *and* $\Delta \to 0$ *and the convergence is uniform on compact subsets of* $[0,T] \times E$. *Furthermore, the stopping time, defined when s is a multiplicity of* $\Delta$,

$$\bar{\tau}_\Delta^\beta(s) = \inf\left\{n\Delta : \bar{w}_\Delta^\beta(s+n\Delta, x(n\Delta)) \le g(s+n\Delta, x(n\Delta))\right\} \wedge (T-s) \qquad (16.54)$$

*is* $|w_\Delta^\beta(s,x) - w(s,x)|$ *optimal for the process* $(x(t))$ *starting from x at time s.*

*Proof.* Clearly, $\bar{w}_{\Delta 2^{-n}}^\beta$ is increasing in $\beta$ and in $n$. Therefore, we are allowed to change the order of limits. Letting $\beta \to \infty$, we have that $\bar{w}_{\Delta 2^{-n}}^\beta$ converges to $w_{\Delta 2^{-n}}$ uniformly, and by Lemma 16.4.2 and Theorem 16.4.3, we obtain uniform convergence on compact set of $w_{\Delta 2^{-n}}$ to $w$. Consequently, we have a uniform on compact sets convergence of $\bar{w}_{\Delta 2^{-n}}^\beta$ to $w$ as $\beta \to \infty$ and $n \to \infty$. The case with different $\Delta$ gives the same limit, and by **Lemmas 16.4.2 and** 16.4.3 and Theorem 16.4.3 for $\Delta$ and $\Delta'$, the difference between $w_{\Delta 2^{-n}}$ and $w_{\Delta' 2^{-n}}$ diminishes to 0, uniformly on compact sets as $n \to \infty$. To show near optimality of $\bar{\tau}_\Delta^\beta(s)$, notice that iterating the penalty equation (16.4) we have obtain

$$\bar{w}_\Delta^\beta(s,x) = E_{sx}\left\{\int_0^{\bar{\tau}_\Delta^\beta(s)} e^{-\alpha u} f(s+u,x(u))du + e^{-\alpha\bar{\tau}_\Delta^\beta(s)}\bar{w}_\Delta^\beta(s+\bar{\tau}_\Delta^\beta(s), x(\bar{\tau}_\Delta^\beta(s)))\right\}$$

$$\le E_{sx}\left\{\int_0^{\bar{\tau}_\Delta^\beta(s)} e^{-\alpha u} f(s+u,x(u))du + \chi_{\bar{\tau}_\Delta^\beta(s)<T-s} e^{-\alpha\bar{\tau}_\Delta^\beta(s)}g(s+\bar{\tau}_\Delta^\beta(s), x(\bar{\tau}_\Delta^\beta(s)))\right.$$

$$\left. + \chi_{\bar{\tau}_\Delta^\beta(s)=T-s} e^{-\alpha(T-s)}h(T,x(T-s))\right\}. \qquad (16.55)$$

Since $w(s,x) \le \bar{w}_\Delta^\beta(s,x) + |w(s,x) - \bar{w}_\Delta^\beta(s,x)|$ by (16.55), we complete the proof of Theorem 16.5.4.                                                                                             □

**Acknowledgements**  Research supported by MNiSzW grant NN 201 371836.

# References

1. Bensoussan, A., Lions, J.L.: Applications of Variational Inequalities in Stochastic Control. Elsevier North Holland (1982)
2. Davis, M., Karatzas, I.: A deterministic approach to optimal stopping with applications to a prophet inequality. P. Whittle Festschrift, in Wiley Ser. Probab. Math. Statist. Probab. Math. Statist., Wiley, Chichester, 455–466 (1994)
3. Dynkin, E.B.: Markov processes. Berlin, Gottingen, Heidelberg, Springer (1965)
4. Fakeev, A.G.: On the question of the optimal stopping of a Markov process. (Russian) Teor. Verojatnost. i Primenen. **16**, 708-710 (1971)

5. Haugh, M.B., Kogan, L.: Pricing American Options: a Duality Approach. Operations Research **52**, 258–270 (2004)
6. Krylov, N.V.: The problem with two free boundaries for an elliptic equation and optimal stopping of Markov processes. Dokl. Akad. Nauk USSR **194**, 1263–1265 (1970)
7. Meyer, P.A.: Processes de Markov, LNM 26, Berlin, Heidelberg, New York, Springer (1967)
8. Palczewski, J., Stettner, L.: Finite Horizon Optimal Stopping of Time - Discontinuous Functionals with Applications to Impulse Control with Delay. SIAM J. Control Optim. **48**, 4874–4909 (2010)
9. Palczewski, J., Stettner, L.: Stopping of discontinuous functionals with the first exit time discontinuity. Stoch. Proc. Their Appl., **121**, 2361–2392 (2011)
10. Robin, M.: Controle impulsionnel des processus de Markov (Thesis), University of Paris IX (1978)
11. Rogers, L.C.G.: Monte Carlo Valuation of American Options. Mathematical Finance **12**, 271–286 (2002)
12. Rogers, L.C.G.: Dual Valuation and Hedging of Bermidian Options. SIAM J. Financial Math. **1**, 604–608 (2010)
13. Stettner, L.: Zero-sum markov Games with Stopping and Impulsive Strategies. Appl. Math. Optimiz. **9**, 1–24 (1982)
14. Stettner, L.: Penalty method for finite horizon stopping problems. SIAM J. Control Optim. **49**, 1078–1999 (2011)

# Chapter 17
# A Direct Approach to the Solution of Optimal Multiple-Stopping Problems*

Richard H. Stockbridge and Chao Zhu

## 17.1  Introduction

With the deregulation of the energy markets in the United States, options to purchase electricity for a preset price have become an important risk-management tool; many of these options allow the holder the opportunity to exercise it each day during the contract period. In the world of water usage, rather than negotiate permanent sales of water rights, owners negotiate contracts in which the other party may divert a certain amount of water for other usage (such as from agricultural to urban), and these contracts often allow more than one diversion. Some employee compensation packages include stock options with the possibility of a number of reloads before expiration. A common feature of these various contracts is the opportunity for a decision-maker to act a finite number of times and receive some reward for each action. Rather than tie our presentation to a particular application, we examine a general formulation.

This chapter considers a broad class of optimal multiple-stopping problems, a natural extension to optimal (single-) stopping problems. Though the extension seems natural, there are nevertheless significant challenges to determining the value and optimal stopping policies. Our objective is to demonstrate a tractable method of

*It is with great pleasure that we contribute a paper to this Festschrift in honor of Onésimo Hernández-Lerma's 65th birthday. He has made many contributions to the stochastic control of Markov processes literature; our interests have many intersections with his work. This contribution honors his career at this important milestone and is dedicated to him.

R.H. Stockbridge (✉) • C. Zhu
Department of Mathematical Sciences, University of Wisconsin at Milwaukee,
Milwaukee, WI 53201, USA
e-mail: stockbri@uwm.edu; zhu@uwm.edu

D. Hernández-Hernández and A. Minjárez-Sosa (eds.), *Optimization, Control, and Applications of Stochastic Systems*, Systems & Control: Foundations & Applications, DOI 10.1007/978-0-8176-8337-5_17, © Springer Science+Business Media, LLC 2012

solution for models in which the distribution of the process is known. To establish
the problem, we assume $X$ is a solution of the stochastic differential equation

$$dX(t) = \mu(X(t))\,dt + \sigma(X(t))\,dW(t), \qquad X(0) = x \qquad (17.1)$$

in which $W$ is a standard Brownian motion process and the drift and diffusion
coefficients are such that $X$ takes values in an interval $(x_\ell, x_r) \subset \mathbb{R}$. The decision-
maker may select up to $N$ times (with $N$ fixed) at which to receive a reward.
However, after each decision time, a lag of at least $\delta > 0$ units of time (the refraction
period) must pass before the next decision to receive a reward is made; this time lag
increases the complexity of the problem. We assume the time horizon is $T = \infty$;
that is, there is no imposed limit on the time by which decisions must be made. Let
$\{\tau_n : n = 1, \ldots, N\}$ denote the decision times. Throughout this chapter, the subscript
will denote the number of remaining decisions, so $\tau_1$ is the last decision and $\tau_N$ is
the first. Note that for each $i = 1, \ldots, N-1$, $\tau_{i+1} < \tau_{i+1} + \delta \leq \tau_i$ on the set where
$\tau_{i+1}$ is finite. For $i = 1, \ldots, N$, let $R_i : (x_\ell, x_r) \to \mathbb{R}$ denote the payoff function for the
$i$th last decision. Letting $\alpha > 0$ denote the discount rate, the objective is to maximize
the expected payoff

$$\sum_{i=1}^{N} \mathbb{E}\left[ e^{-\alpha \tau_i} I_{\{\tau_i < \infty\}} R_i(X(\tau_i)) \right] \qquad (17.2)$$

over all decision times $\tau_1, \ldots, \tau_N$ satisfying the refraction period condition.

As indicated in the first paragraph, recent interest in multiple-stopping problems
has developed due to deregulation and new types of options, though multiple-
stopping problems have previously been studied in sequential analysis (see, e.g.,
Haggstrom [7]). Villinski [14] discusses contracts involving multiple decisions for
water rights from an economic point of view and describes a dynamic programming
formulation for the valuation of these contracts. From a more mathematical point of
view, Thompson [13] examines a discrete-time binomial tree model for the evolution
of the process and concentrates on developing a Monte Carlo method to value a path-
dependent contingent claim. Zeghal and Mnif [15] consider the valuation of swing
options for Lévy models using Snell envelopes and illustrates this approach using
Monte Carlo techniques on a put option having a maturity time of 1. Carmona and
Touzi [4] analyze the valuation of a perpetual put swing option with infinitely many
exercises in a continuous-time Black-Scholes market. The paper independently
develops a theoretical foundation to the solution using Snell envelopes and obtains
exercise rules by discrete approximation. The paper by Carmona and Dayanik [3]
examines the same type of problem for a more general one-dimensional diffusion
model having a more general reward function and determines the solution using
a generalized convex function approach. Dai and Kwok [5] examine the pricing
of reload and shout options in which the refraction period models the time until
the employee is vested. The paper uses a Black-Scholes model having continuous
dividend rate, approaches the solution using a variational inequality which is then
approximately solved using a binomial tree model and dynamic programming.
Interestingly, the authors relate the reload option to a lookback feature of the

stock price process. Aleksandrov and Hambly [1] use a dual approach to analyze multiple exercise options under constraints, though the formulation allows multiple exercises at the same time (no refraction period). The authors solve the problem by considering the marginal value of one additional exercise time. Kobylanski and Quenez [12] discuss the general theory of multiple-stopping time problems using Snell envelopes.

This paper seeks a numerically tractable approach to the solution of multiple-stopping problems. It considers the same model and general reward as Carmona and Dayanik [3] though it approaches the analysis of the problem using a quite different method. As in several of the aforementioned papers, the multiple-stopping problem is reduced to an iterated sequence of $N$ single-stopping problems through a conditioning argument. This paper then utilizes the results in Helmes and Stockbridge [8] to characterize the value of each single-stopping problem in two ways. This characterization enables the value function for each single-stopping problem to be determined in closed form for many payoff functions. We also employ the argument in Helmes and Stockbridge [9] in which we first obtain an upper bound on the value and then identify a stopping rule which achieves the bound. The problem formulation in terms of stochastic processes is given in Sect. 17.2 along with the reduction to the sequence of single-stopping problems. Section 17.3 then summarizes the approaches to determining the value function of Helmes and Stockbridge [8, 9]. The tractability of this method is then illustrated in Sect. 17.4.

The current paper is similar to Helmes and Stockbridge [10] in that both papers consider a finite number of decision times at which a reward is earned and analyze a sequence of single-stopping problems by solving nonlinear optimization problems. The significant difference is the requirement in this paper that successive decisions to stop must wait at least the length of the refraction period. The time lag increases the complexity of the analysis in a nontrivial way.

## 17.2   Problem Formulation

We begin with a precise formulation of the class of multiple-stopping problems examined in this chapter. We assume the coefficients $\mu$ and $\sigma$ of (17.1) are continuous and are such that $X$ takes values in some interval $(x_\ell, x_r) \subseteq \mathbb{R}$. The process $X$ has generator $A$ given by $Af(x) = (1/2)\sigma^2(x)f''(x) + \mu(x)f'(x))$ operating on $f \in C^2(x_\ell, x_r)$ (see [2, II.9, p. 17] for sufficient conditions).Further assume $X$ is a weak solution of (17.1) while $X(t) \in (x_\ell, x_r)$ (see Ethier and Kurtz [6, Sect. 5.3, p. 291] for details) and that the solution to (17.1) is unique in distribution. This existence and uniqueness imply that the martingale problem for $A$ is well posed and hence that $X$ is a strong Markov process (see [6, Theorem 4.4.2, p. 184]). We denote the filtration for the weak solution by $\{\mathscr{F}_t\}$. Throughout this chapter we assume $x_\ell < x < x_r$. We emphasize that $x$ will always represent the initial position for the multiple-stopping problem in this chapter.

A key additional assumption on the coefficients is required, which we separate out for later reference.

**Condition 17.2.1.** *The eigenvalue problem* $Af(\cdot) = \alpha f(\cdot)$ *has both a positive, strictly decreasing solution* $\phi$ *and a nonnegative, strictly increasing solution* $\psi$.

The conditions assumed in this paper are sufficient to imply Condition 17.2.1 (see Borodin and Salminen [2, II.10, p. 18,19]). The functions $\phi$ and $\psi$ depend on the discount factor $\alpha$; since we assume the discount factor is fixed, we omit this dependence from the notation.

Before proceeding further, we briefly digress to consider the boundary points. We restrict the models to those for which $x_\ell$ is either an entrance-not-exit boundary point or a natural boundary point [2, II.10, p. 14–19]. The analysis also applies when $x_\ell$ is an exit boundary point, but the expressions are slightly more complicated, so we have chosen this restriction on the type of boundary point for clarity of presentation. When $x_\ell$ is either an entrance or natural boundary, $X$ will almost surely never reach $x_\ell$ in finite time. The distinction between entrance and natural boundaries is that the process will immediately enter the interval $(x_\ell, x_r)$ when $x = x_\ell$ is an entrance point (we assume $x > x_\ell$), after which it will never return to the boundary, and thus $x_\ell$ is in the state space of the process. This behavior does not happen with a natural boundary point so such an $x_\ell$ will not be in the state space of $X$. We place the same restrictions on the model for $x_r$. In the event either $x_\ell = -\infty$ or $x_r = \infty$, we require these to be natural boundary points with the implication that the process $X$ will not "explode to $\infty$ or $-\infty$" in finite time.

The importance of the type of boundary points for this chapter is the properties that $\psi(x_\ell) \geq 0$ and $\phi(x_\ell+) = \infty$ [2, pp. 14–19]. When $x_\ell = -\infty$, the natural boundary point assumption implies $\phi(-\infty) = \infty$ and $\psi(-\infty) = 0$. Symmetric properties hold for $x_r$ with the roles of $\phi$ and $\psi$ reversed.

The reward earned by the decision-maker is the sum of the expected discounted payoffs at each decision time given in (17.2). Denote the optimal value by $V^{(N)}(x)$, in which the superscript indicates the number of decisions. We assume that for each $i = 1, \dots, N$, the reward function $R_i : (x_\ell, x_r) \mapsto \mathbb{R}$ is upper-semicontinuous, is positive for some $y \in (x_\ell, x_r)$, and satisfies

$$\lim_{y \searrow x_\ell} \frac{R_i(y)}{\phi(y)} = 0, \quad \text{and} \quad \lim_{y \nearrow x_r} \frac{R_i(y)}{\psi(y)} = 0. \tag{17.3}$$

We further assume that $\tau_1, \dots, \tau_N$ are $\{\mathscr{F}_t\}$-stopping times satisfying $0 \leq \tau_N$, and for each $i = 1, \dots, N-1$, on the set $\{\tau_{i+1} < \infty\}$ the stopping times satisfy $\tau_{i+1} < \tau_{i+1} + \delta \leq \tau_i$. Let $\mathscr{A}_N$ denote the set of these $N$-tuples of stopping times. Since the multiple-stopping problem will be reduced to a sequence of single-stopping problems, it will be beneficial to denote the set of nonnegative (single-) stopping times by $\mathscr{A}_1$, in which the subscript denotes that the set consists of stopping times and not $N$-tuples of stopping times.

We now present the key conditioning argument which reduces (17.2) to a sequence of single-stopping problems. The argument uses the strong Markov property so it is helpful to designate the expectation relative to the initial position of the process $X$ using a subscript. It is necessary to develop some additional notation.

Set $\tilde{V}_0^{(1)} \equiv 0$ and define the "modified" payoff function $\tilde{R}_1 = R_1 = R_1 + \tilde{V}_0^{(1)}$ for the reward received upon making the final decision. Define the corresponding value function $V_1^{(1)}$ by

$$V_1^{(1)}(y) = \sup_{\tau \in \mathscr{A}_1} \mathbb{E}_y[e^{-\alpha \tau} I_{\{\tau < \infty\}} \tilde{R}_1(X(\tau))], \qquad y \in (x_\ell, x_r).$$

Proceeding recursively, for $i = 2, \ldots, N$ and $y \in (x_\ell, x_r)$, define $\tilde{V}_i^{(1)}(y) = \mathbb{E}_y [e^{-\alpha \delta} V_i^{(1)}(X(\delta))]$, the modified payoff function $\tilde{R}_i = R_i + \tilde{V}_{i-1}^{(1)}$ and

$$V_i^{(1)}(y) = \sup_{\tau \in \mathscr{A}_1} \mathbb{E}_y[e^{-\alpha \tau} I_{\{\tau < \infty\}} \tilde{R}_i(X(\tau))]. \tag{17.4}$$

**Theorem 17.2.1.** *The value of optimal multiple-stopping problem of maximizing (17.2) over decision times $(\tau_1, \ldots, \tau_N) \in \mathscr{A}_N$ at which to stop the process $X$ satisfying (17.1) is obtained through recursion by solving the $N$ single-stopping problems; that is, $V^{(N)}(x) = V_N^{(1)}(x)$.*

*Proof.* Consider a single generic term of the form

$$\mathbb{E}_x \left[ e^{-\alpha \tau_i} I_{\{\tau_i < \infty\}} g(X(\tau_i)) \right],$$

in which $g$ is some measurable function such that the integrand is integrable and $\tau_i$ is one of the stopping times in an $N$-tuple $(\tau_1, \ldots, \tau_N) \in \mathscr{A}_N$ in which $i \in \{1, \ldots, N-1\}$. On the set $\{\tau_{i+1} < \infty\}$, notice that $\tau_i \geq \tau_{i+1} + \delta$ so we can define $\tilde{\tau}_i = \tau_i - \tau_{i+1} - \delta$ and have $\tilde{\tau}_i \in \mathscr{A}_1$, where the stopping times are relative to the filtration $\{\mathscr{G}_t\} = \{\mathscr{F}_{\tau_{i+1}+t}\}$. Using the strong Markov property of $X$ in the third equality below yields

$$\mathbb{E}_x \left[ e^{-\alpha \tau_i} I_{\{\tau_i < \infty\}} g(X(\tau_i)) \right]$$

$$= \mathbb{E}_x \left[ \mathbb{E}_x \left[ e^{-\alpha \tau_i} I_{\{\tau_i < \infty\}} g(X(\tau_i)) \middle| \mathscr{F}_{\tau_{i+1}+\delta} \right] \right]$$

$$= \mathbb{E}_x \left[ e^{-\alpha(\tau_{i+1}+\delta)} I_{\{\tau_{i+1} < \infty\}} \mathbb{E}_x \left[ e^{-\alpha \tilde{\tau}_i} I_{\{\tilde{\tau}_i < \infty\}} g(X(\tau_{i+1} + \delta + \tilde{\tau}_i)) \middle| \mathscr{F}_{\tau_{i+1}+\delta} \right] \right]$$

$$= \mathbb{E}_x \left[ e^{-\alpha(\tau_{i+1}+\delta)} I_{\{\tau_{i+1} < \infty\}} \mathbb{E}_{X(\tau_{i+1}+\delta)} \left[ e^{-\alpha \tilde{\tau}_i} I_{\{\tilde{\tau}_i < \infty\}} g(X(\tilde{\tau}_i)) \right] \right].$$

The key to the tractability of the problem lies in a second conditioning argument. Observe that for any integrable random variable $Y$,

$$\mathbb{E}_x \left[ \mathbb{E}_{X(\tau_{i+1}+\delta)}[Y] \right] = \mathbb{E}_x \left[ \mathbb{E}_x[\mathbb{E}_{X(\tau_{i+1}+\delta)}[Y] | \mathscr{F}_{\tau_{i+1}}] \right] = \mathbb{E}_x \left[ \mathbb{E}_{X(\tau_{i+1})}[\mathbb{E}_{X(\delta)}[Y]] \right]$$

and thus

$$
\mathbb{E}_x \left[ e^{-\alpha \tau_i} I_{\{\tau_i < \infty\}} g(X(\tau_i)) \right]
$$
$$
= \mathbb{E}_x \left[ e^{-\alpha \tau_{i+1}} I_{\{\tau_{i+1} < \infty\}} \mathbb{E}_{X(\tau_{i+1})} \left[ e^{-\alpha \delta} \mathbb{E}_{X(\delta)} \left[ e^{-\alpha \tilde{\tau}_i} I_{\{\tilde{\tau}_i < \infty\}} g(X(\tilde{\tau}_i)) \right] \right] \right].
$$

Now specify $i = 1$ and $g = R_1$. For $i = 1, \ldots, N$, define the set $\mathscr{A}_{N,i} = \{ \tau_i : (\tau_1, \ldots, \tau_N) \in \mathscr{A}_N \}$. Taking the supremum over $\tau_1 \in \mathscr{A}_{N,1}$ of the left-hand side and then over $\tau_2 \in \mathscr{A}_{N,2}$ and $\tilde{\tau}_1 \in \mathscr{A}_1$ on the right-hand side produces one inequality, whereas taking the suprema in the opposite order yields the opposite inequality and hence

$$
\sup_{\tau_1 \in \mathscr{A}_{N,1}} \mathbb{E}_x \left[ e^{-\alpha \tau_1} I_{\{\tau_1 < \infty\}} R_1(X(\tau_1)) \right]
$$
$$
= \sup_{\tau_2 \in \mathscr{A}_{N,2}} \mathbb{E}_x \left[ e^{-\alpha \tau_2} I_{\{\tau_2 < \infty\}} \mathbb{E}_{X(\tau_2)} \left[ e^{-\alpha \delta} V_1^{(1)}(X(\delta)) \right] \right]
$$
$$
= \sup_{\tau_2 \in \mathscr{A}_{N,2}} \mathbb{E}_x \left[ e^{-\alpha \tau_2} I_{\{\tau_2 < \infty\}} \tilde{V}_1^{(1)}(X(\tau_2)) \right]. \tag{17.5}
$$

An important reduction occurs when we consider the two successive terms of (17.2) involving $\tau_1$ and $\tau_2$. Observe

$$
\sup_{\substack{\tau_1 \in \mathscr{A}_{N,1}, \\ \tau_2 \in \mathscr{A}_{N,2}}} \mathbb{E}_x \left[ e^{-\alpha \tau_2} I_{\{\tau_2 < \infty\}} R_2(X(\tau_2)) + e^{-\alpha \tau_1} I_{\{\tau_1 < \infty\}} R_1(X(\tau_1)) \right]
$$
$$
= \sup_{\tau_2 \in \mathscr{A}_{N,2}} \mathbb{E}_x \left[ e^{-\alpha \tau_2} I_{\{\tau_2 < \infty\}} \left( R_2(X(\tau_2)) + \tilde{V}_1(X(\tau_2)) \right) \right]
$$
$$
= \sup_{\tau_2 \in \mathscr{A}_{N,2}} \mathbb{E}_x \left[ e^{-\alpha \tau_2} I_{\{\tau_2 < \infty\}} \tilde{R}_2(X(\tau_2)) \right]
$$

in which we recall $\tilde{R}_2(y) = R_2(y) + \tilde{V}_1(y)$. Using induction, we obtain

$$
V^{(N)}(x) = \sup_{(\tau_1, \ldots, \tau_N) \in \mathscr{A}_N} \sum_{i=1}^{N} \mathbb{E}_x [ e^{-\alpha \tau_i} I_{\{\tau_i < \infty\}} R_i(X(\tau_i)) ]
$$
$$
= \sup_{\tau_N \in \mathscr{A}_{N,N}} \mathbb{E}_x [ e^{-\alpha \tau_N} I_{\{\tau_N < \infty\}} \tilde{R}_N(X(\tau_N)) ]
$$
$$
= V_N^{(1)}(x). \qquad \square
$$

The implications of Theorem 17.2.1 is that the $N$-stopping problem can be solved using an iteration of three steps. First, obtain the value $V_i$ for the successor stopping time as a function of the initial position $y$; that is, determine the successor value function. Next, find the expected discounted value (discounted by the refraction time $\delta$) of this function evaluated at the new position $X(\delta)$ of the process. Finally, add this function to the predecessor (more decisions to make) payoff function $R_{i+1}$ to form a new payoff function for the predecessor stopping problem, leading again to a single-stopping problem.

Thus, the main tasks to solve the multiple-stopping problem are to determine the sequence of single-step value functions and to utilize the distribution of $X(\delta)$, parametrized by an arbitrary initial position $y \in (x_\ell, x_r)$.

## 17.3   Solution Approaches for Single-Stopping Problems

The single-stopping problem seeks to maximize

$$J(\tau; x) := \mathbb{E}_x[e^{-\alpha\tau} I_{\{\tau < \infty\}} R(X(\tau))] \tag{17.6}$$

over the set of all $\{\mathscr{F}_t\}$-stopping times $\tau$ in which $X$ is a weak solution of (17.1). Let $\mathscr{A}$ denote this set of stopping times and define $V(x) = \sup_{\tau \in \mathscr{A}} J(\tau; x)$. This section briefly states the line of reasoning in Helmes and Stockbridge [9] and then recalls the results in Helmes and Stockbridge [8]. The first method of solution identifies an upper bound on the value with the goal of identifying a stopping time that achieves this value. The second approach involves maximizing the expected reward over all two-point stopping rules, whereas the final technique utilizes duality theory. We wish to emphasize that the optimal stopping problem is solved for a single initial value $x$, rather than seeking the value function, though the structure of the values is such that the value function can often be determined.

### 17.3.1   Linear Programming Imbedding

A common imbedding of the stochastic problem underlies these methods. We briefly describe the derivation of the linear program and then, in the next sections, utilize this in two related ways. Applying Itô's formula to $e^{-\alpha t} f(X(t))$ for $f \in C_c^2(x_\ell, x_r)$ yields

$$e^{-\alpha t} f(X(t)) = f(x) + \int_0^t e^{-\alpha s}[Af - \alpha f](X(s))\,ds + \int_0^t e^{-\alpha s} f'(X(s))\,dW(s).$$

For any $\tau \in \mathscr{A}$, the optional sampling theorem indicates that

$$e^{-\alpha(t \wedge \tau)} f(X(t \wedge \tau)) - f(x) - \int_0^{t \wedge \tau} e^{-\alpha s}[Af - \alpha f](X(s))\,ds$$

is a mean 0 martingale, so taking expectations then letting $t \to \infty$ establishes Dynkin's formula

$$\mathbb{E}[e^{-\alpha\tau}I_{\{\tau<\infty\}}f(X(\tau))] - \mathbb{E}\left[\int_0^\tau e^{-\alpha s}[Af - \alpha f](X(s))\,\mathrm{d}s\right] = f(x). \qquad (17.7)$$

Defining $v_\tau$ to be the discounted (stopping) distribution of $X(\tau)$ and $\mu_0$ to be the expected, discounted occupation measure of $X$ over the interval $[0, \tau]$, (17.7) can be written as $\int f\,\mathrm{d}v_\tau - \int[Af - \alpha f]\,\mathrm{d}\mu_0 = f(x)$ and the single-stopping objective function (17.6) becomes $\int R\,\mathrm{d}v_\tau$. The optimal stopping problem is therefore imbedded in the infinite-dimensional linear program

$$\begin{cases} \text{Maximize} \quad \displaystyle\int R\,\mathrm{d}v_\tau \\ \text{Subject to} \quad \displaystyle\int f\,\mathrm{d}v_\tau - \int[Af - \alpha f]\,\mathrm{d}\mu_0 = f(x), \quad \forall f \in C_c^2(x_\ell, x_r). \end{cases} \qquad (17.8)$$

We note that the variables in this linear program are the measures $v_\tau$ and $\mu_0$ and that $v_\tau$ arises from the stopping time $\tau$ so is the decision variable.

### 17.3.2 Achieving an Upper Bound

A first auxiliary linear program is obtained by limiting the constraints to a single test function. One implication is that the feasible set of measures may be larger and hence the value of the auxiliary problem gives an upper bound for (17.8). We may take $f = \psi$ in (17.8) (see [9] for details justifying the use of $\psi$ as a test function since $\psi \notin C_c^2(x_\ell, x_r)$). The benefit of this choice is that $A\psi - \alpha\psi \equiv 0$ so the integral with respect to the occupation measure $\mu_0$ drops from the constraints. Notice the constraint can be written as

$$\int \psi/\psi(x)\,\mathrm{d}v_\tau = 1,$$

so the integrand forms the density for a probability measure $\tilde{v}_\tau$ on $(x_\ell, x_r)$.

**Proposition 17.3.1.** *Assume $X$ is a weak solution of (17.1) and Condition 17.2.1 is satisfied. Let $R$ satisfy the conditions in Sect. 17.2. Then*

$$V(x) \leq \sup_{y \in (x_\ell, x_r)} \frac{R(y)}{\psi(y)} \cdot \psi(x).$$

*In addition, if $\lim_{y \searrow x_\ell} R(y)/\psi(y) = 0$, then there exists a maximizer $y^*$ and $\tau_{y^*}$ is an optimal stopping rule when $x \leq y^*$.*

*Proof.* Examining the objective function, we have

$$\int R(y)\,v_\tau(\mathrm{d}y) = \int [R(y)\psi(x)/\psi(y)]\,\tilde{v}_\tau(\mathrm{d}y) \leq \sup_{y \in (x_\ell, x_r)} (R(y)/\psi(y)) \cdot \psi(x).$$

The conditions on $R$ imply the existence of a maximizer $y^*$ of $R(y)/\psi(y)$. It is well known (see [2]) that $\mathbb{E}_x[e^{-\alpha \tau_{y^*}}] = \psi(x)/\psi(y^*)$ when $x \le y^*$, so this stopping rule achieves the upper bound.                                                                                      □

It will be helpful to note that when $R$ is differentiable, an interior optimizer for the function $R(y)/\psi(y)$ occurs where $\psi(y)R'(y) - \psi'(y)R(y) = 0$. This necessary optimality condition implies the elasticities of the function $\psi$, and the payoff function $R$ must be the same at an optimizing level.

### 17.3.3  Maximization Over Two-Point Hitting Rules

The previous approach is sufficient when the structure of the problem is such that stopping to the right of the initial position is optimal. A second auxiliary problem provides a general solution and is also obtained from (17.8), this time by limiting the test functions to the pair $\phi$ and $\psi$.

Consider points $a$ and $b$ such that $x_\ell < a \le x \le b < x_r$ but $a < b$. Define $\tau_a = \inf\{t \ge 0 : X(t) = a\}$ and $\tau_b$ similarly. Define $\tau_{a,b} = \tau_a \wedge \tau_b$. The payoff associated with the decision rule $\tau_{a,b}$ is

$$J(\tau_{a,b};x) = R(a) \cdot \frac{\phi(x)\psi(b) - \phi(b)\psi(x)}{\phi(a)\psi(b) - \phi(b)\psi(a)} + R(b) \cdot \frac{\phi(a)\psi(x) - \phi(x)\psi(a)}{\phi(a)\psi(b) - \phi(b)\psi(a)}$$

$$= \frac{R(a)\psi(b) - R(b)\psi(a)}{\phi(a)\psi(b) - \phi(b)\psi(a)} \cdot \phi(x) + \frac{R(b)\phi(a) - R(a)\phi(b)}{\phi(a)\psi(b) - \phi(b)\psi(a)} \cdot \psi(x). \qquad (17.9)$$

Several observations are helpful. First, the fractional terms in the first expressions of (17.9) are the masses of $\nu_{\tau_{a,b}}$. Next, when $x = a$, the expression for $J(\tau_{a,b};x) = R(a)$ and similarly for $x = b$. This agrees with one's intuition that stopping occurs immediately resulting in a non-discounted payoff. Also when holding $b > x$ fixed and letting $a \to x_\ell$, the fractional terms in the first expression converge to 0 and $\psi(x)/\psi(b) = E[e^{-\alpha \tau_b}]$, respectively, and hence $J(\tau_{a,b};x) \to J(\tau_b;x)$. Similarly, when $b \to x_r$ with $a$ fixed, $J(\tau_{a,b};x) \to J(\tau_a;x)$. Finally, by examining the second expression of (17.9), one observes that, as a function of $x$, the value of $J(\tau_{a,b};x)$ is continuous on $(a,b)$.

**Proposition 17.3.2.** *Assume $X$ is a weak solution of* (17.1) *and Condition 17.2.1 is satisfied. Let $R$ satisfy the conditions in Sect. 17.2. Then*

$$V(x) = \sup_{a \le x \le b} J(\tau_{a,b};x).$$

*Moreover, there exist $a^*, b^* \in [x_\ell, x_r]$ such that $J(\tau_{a^*,b^*};x) = V(x)$; that is, $\tau_{a^*,b^*}$ is an optimal stopping rule.*

When $a* = x_\ell$, the two-point hitting rule is actually a one-point hitting rule at $b^*$ and hence $\tau_{b^*}$ is an optimal stopping time. Similar comments apply when

$b^* = x_r$. We observe that it will never occur that both $a^* = x_\ell$ and $b^* = x_r$ since, by assumption on the model, the process will never hit either $x_\ell$ or $x_r$ so the "stopping" time $\tau_{x_\ell, x_r} = \infty$ a.s. and the value is 0, but a positive value can be obtained by choosing to stop at a point where $R$ is strictly positive.

### 17.3.4  Minimization of α-Harmonic Functions

As indicated previously, to establish the optimality of a two-point hitting rule in [8], the stochastic problem is imbedded in an infinite-dimensional linear program, and an upper bound is obtained by restricting the constraints (and increasing the feasible set). A dual linear program to this auxiliary linear program is also derived for which it is easy to prove a weak duality result between the values of the linear programs, with more involved arguments establishing strong duality [8]. As a result, the optimal value can be obtained by solving the following two-dimensional linear program:

$$\begin{cases} \text{Minimize} & c_1\phi(x) + c_2\psi(x) \\ & c_1\phi(y) + c_2\psi(y) \ge R(y), \ \forall y \in (x_l, x_r), \\ \text{Subject to} & c_1, c_2 \text{ unrestricted.} \end{cases} \tag{17.10}$$

We note that this problem involves minimizing a linear combination of the functions $\phi$ and $\psi$ of Condition 17.2.1 evaluated at the initial position of the process. To be feasible, this linear combination is required to majorize the payoff function $R$.

A further observation will be helpful. As in Proposition 17.3.2, take $a^*$ and $b^*$ to be maximizers. Section 4.3 of [8] proves that when the payoff function $R$ is continuously differentiable in a neighborhood of $a^*$ and in a neighborhood of $b^*$, then these are points which satisfy the principle of smooth pasting; namely,

$$\begin{cases} c_1\phi(a) + c_2\psi(a) = R(a) \\ c_1\phi'(a) + c_2\psi'(a) = R'(a) \end{cases} \text{and} \quad \begin{cases} c_1\phi(b) + c_2\psi(b) = R(b) \\ c_1\phi'(b) + c_2\psi'(b) = R'(b). \end{cases} \tag{17.11}$$

To obtain this result, one analyzes the maximization over two-point stopping rules and shows how to optimally select $c_1$ and $c_2$. Notice there are four equations in the four variables $a$, $b$, $c_1$, and $c_2$.

It will be helpful to consider a particular case of smooth pasting more extensively. Consider the situation in which it is optimal to stop immediately at the initial time; this means that the smooth pasting conditions must be satisfied when $a = x$. In this case, the coefficients $c_1$ and $c_2$ are easily determined to be

$$c_1 = \frac{\psi'(x)R(x) - \psi(x)R'(x)}{\phi(x)\psi'(x) - \phi'(x)\psi(x)} \text{ and } c_2 = \frac{\phi(x)R'(x) - \phi'(x)R(x)}{\phi(x)\psi'(x) - \phi'(x)\psi(x)}. \tag{17.12}$$

Notice, in particular, that $c_2$ is always positive when $R$ is positive and increasing and similarly that $c_1$ is always positive when $R$ is positive and decreasing. Considering further the case that $R$ is positive and increasing, observe that the denominator of $c_1$ in (17.12) is always positive, so $c_1$ is positive when $\psi'(x)R(x) - \psi(x)R'(x) > 0$, and equals 0 when the same elasticity condition as in Sect. 17.3.2 is satisfied. Moreover, comparing the numerator of $c_1$ in (17.12) with the numerator of $(R/\psi)'$, we see that $c_1$ will be positive whenever $R/\psi$ is strictly decreasing. A similar comment holds for $c_2$ when $R$ is positive and decreasing by analyzing $(R/\phi)'$.

Finally, recall that the optimal stopping problem is solved for a single initial value $x$, rather than seeking the value function. But the structure of this approach typically determines the value for initial positions in regions, and hence the value function can be typically obtained through a limited number of optimizations. In fact, to determine the value function, it is often easiest to use different methods for $x$ in different regions.

## 17.4   Drifted Brownian Motion

The process $X$ satisfies $dX(t) = \mu \, dt + \sigma \, dW(t)$; that is, $X(t) = x + \mu t + \sigma W(t)$, in which $\mu \in \mathbb{R}$ and $\sigma \in \mathbb{R}_+$ and the process takes values in $\mathbb{R}$. It is easily verified that $\phi(y) = e^{\gamma_1 y}$ and $\psi(y) = e^{\gamma_2 y}$, where $\gamma_1 = -\frac{\mu}{\sigma^2} - \sqrt{\frac{\mu^2}{\sigma^4} + \frac{2\alpha}{\sigma^2}}$ and $\gamma_2 = -\frac{\mu}{\sigma^2} + \sqrt{\frac{\mu^2}{\sigma^4} + \frac{2\alpha}{\sigma^2}}$. We note that $\gamma_1 < 0 < \gamma_2$ and that these values are in fact the roots of the quadratic equation $(\sigma^2/2)y^2 + \mu y - \alpha = 0$.

We consider a triple-stopping problem so assume $N = 3$ and we take $R_i(y) = y^+$ for $i = 1, 2, 3$; recall throughout the paper, the subscript denotes the number of decisions that remain to be made. Proceeding in a recursive manner with the final stopping decision, we must first determine the value function $V_1^{(1)}(x)$.

Consider first the minimization approach to determining the value of this last stopping problem. To be feasible, the $\alpha$-harmonic function $c_1\phi + c_2\psi$ must lie above the payoff function $R_1(y) = y^+$. For $y \neq 0$, $R_1$ is differentiable, and hence we can apply the smooth pasting argument. Since both $\phi$ and $\psi$ are strictly positive functions, $R_1(x) = 0$ for $x < 0$, and it is not possible to find a linear combination $c_1\phi + c_2\psi$ which has $c_1\phi(x) + c_2\psi(x) = 0$ and majorizes $R_1$. Thus, the optimal value is not 0, a fact that also follows directly from the observation that using a stopping rule of $\tau_{y_0}$, where $R_1(y_0) > 0$, yields a strictly positive value.

We next investigate whether it is possible to have a feasible $\alpha$-harmonic function that equals the payoff function at $x$ when $x > 0$. The coefficients $c_1$ and $c_2$ must satisfy

$$\begin{cases} e^{\gamma_1 x}c_1 + e^{\gamma_2 x}c_2 = x, \\ \gamma_1 e^{\gamma_1 x}c_1 + \gamma_2 e^{\gamma_2 x}c_2 = 1. \end{cases}$$

The solution to this linear system is

$$c_1 = \frac{\gamma_2 x - 1}{(\gamma_2 - \gamma_1)e^{\gamma_1 x}} \quad \text{and} \quad c_2 = \frac{1 - \gamma_1 x}{(\gamma_2 - \gamma_1)e^{\gamma_2 x}}. \tag{17.13}$$

Since $R_1 \geq 0$ and $\phi(y) = e^{\gamma_1 y}$ and $\psi(y) = e^{\gamma_2 y}$ are both positive functions, to be feasible the coefficients $c_1$ and $c_2$ must both be nonnegative. The coefficient $c_2$ is always positive since $\gamma_1 < 0$ and $x > 0$. The coefficient $c_1$, however, is only non-negative when the initial value satisfies $x \geq 1/\gamma_2$. Thus, for $x$ in this range, the value of the single-optimal stopping problem is $V_1^{(1)}(x) = x$, and an optimal stopping rule is to stop immediately, $\tilde{\tau}_1^* = \tau_x$.

Now consider an initial position $x$ with $x < 1/\gamma_2$. First we note that the function $c_1^* \phi(y) + c_2^* \psi(y) := (\gamma_2 e)^{-1} e^{\gamma_2 y}$, which is obtained using the coefficients (17.13) with $x = 1/\gamma_2$, is feasible for the minimization problem. Now consider $c_1, c_2 > 0$ such that $c_1 \phi(x) + c_2 \psi(x) < c_2^* \psi(x)$. Simple algebra demonstrates that $c_1 \phi(x)/\psi(x) < c_2^* - c_2$ and hence $c_2^* - c_2 > 0$. Moreover, the inequality can be rearranged to show $c_1 < (c_2^* - c_2)\psi(x)/\phi(x)$, and thus evaluating the new $\alpha$-harmonic function at $1/\gamma_2$, we have

$$c_1 \phi(1/\gamma_2) + c_2 \psi(1/\gamma_2) < (c_2^* - c_2)\frac{\psi(x)\phi(1/\gamma_2)}{\phi(x)} + c_2 \psi(1/\gamma_2)$$

$$= (c_2^* - c_2)\left[\frac{\phi(1/\gamma_2)}{\phi(x)}\psi(x) - \psi(1/\gamma_2)\right] + c_2^* \psi(1/\gamma_2)$$

$$< R_1(1/\gamma_2);$$

the final inequality follows from the facts that $\phi$ is strictly decreasing and $\psi$ is strictly increasing along with the observation that $c_2^* \psi(1/\gamma_2) = R_1(1/\gamma_2)$. From this, we see that no linear combination with positive coefficients and $c_1 \phi(x) + c_2 \psi(x) < c_2^* \psi(x)$ is feasible for the minimization problem.

The above argument utilizes the minimization approach in both regions. Consider now the upper bound method of Sect. 17.3.2. Maximizing $h(y) := y^+/\psi(y)$ immediately results in a unique maximizer at $y_1^* = 1/\gamma_2$ and a corresponding upper bound of $(\gamma_2 e)^{-1} e^{\gamma_2 x}$. As noted in Proposition 17.3.1, this upper bound is achieved by the stopping rule $\tau_{(1/\gamma_2)}$ when $x \leq 1/\gamma_2$.

The value function is therefore

$$V_1^{(1)}(x) = \begin{cases} (1/\gamma_2)e^{\gamma_2 x - 1}, & \text{for } x \leq 1/\gamma_2, \\ x, & \text{for } x \geq 1/\gamma_2. \end{cases}$$

This value function is displayed in Fig. 17.1 along with the payoff function $\tilde{R}_1(y) = y^+$ (dotted). We also display the ratio $\tilde{R}_1/\psi$; notice one is able to observe the ratio achieves its maximum at the location of the maximizer $1/\gamma_2 \approx 20.5$ and the function is strictly decreasing above this maximizer, confirming graphically that the coefficient $c_1$ of (17.12) will be positive and that it is optimal to stop immediately.

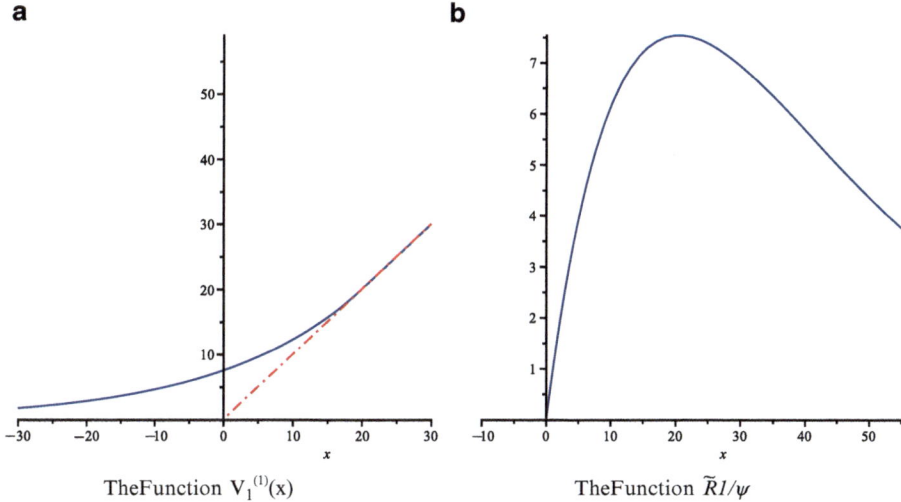

**Fig. 17.1** The value function and ratio for final stopping problem; $\mu = \sigma = \delta = 1$, $\alpha = 0.05$

The next stopping decision is the point at which one must take into account the refraction period. Since we have $V_1^{(1)}$ in explicit form, we can determine the function $\tilde{V}_1^{(1)}$. Notice that, with initial position $y$, $X(\delta)$ is $N(y + \mu\delta, \sigma^2\delta)$-distributed. Let $\Phi$ denote the standard normal distribution function and set $\overline{\Phi} = 1 - \Phi$. Now, recalling that $(\sigma^2/2)\gamma_2^2 + \mu\gamma_2 - \alpha = 0$,

$$\tilde{V}_1^{(1)}(y) = \mathbb{E}_y[e^{-\alpha\delta}V_3^{(1)}(X(\delta))]$$

$$= \int_{-\infty}^{1/\gamma_2}(1/\gamma_2)e^{\gamma_2 z - 1 - \alpha\delta} \cdot (2\pi\sigma^2\delta)^{-1/2}e^{-(z-y-\mu\delta)^2/(2\sigma^2\delta)}\,dz$$

$$+ \int_{1/\gamma_2}^{\infty} z e^{-\alpha\delta} \cdot (2\pi\sigma^2\delta)^{-1/2}e^{-(z-y-\mu\delta)^2/(2\sigma^2\delta)}\,dz$$

$$= (1/\gamma_2)e^{\gamma_2 y - 1}\,\Phi\left(\frac{-y - \mu\delta - \gamma_2\sigma^2\delta + \gamma_2^{-1}}{\sigma\sqrt{\delta}}\right)$$

$$+ e^{-\alpha\delta}(\sigma\sqrt{\delta/(2\pi)})\,e^{-(y+\mu\delta-\gamma_2^{-1})^2/(2\sigma^2\delta)}$$

$$+ e^{-\alpha\delta}(y + \mu\delta)\,\overline{\Phi}\left(\frac{-y - \mu\delta + \gamma_2^{-1}}{\sigma\sqrt{\delta}}\right).$$

It is easy to show that $\tilde{V}_1^{(1)}(y) \to 0$ as $y \to -\infty$ and that $\tilde{V}_1^{(1)}$ is asymptotic to the line $z = e^{-\alpha\delta}(y + \mu\delta)$ as $y$ goes to $\infty$.

Now recall $\tilde{R}_2(y) = R_2(y) + \tilde{V}_1^{(1)}(y)$, and the second last decision time is chosen to satisfy

$$J_2(\tau_2^*; y) = \sup_{\tau \in \mathscr{A}_1} \mathbb{E}_y \left[ e^{-\alpha \tau} I_{\{\tau < \infty\}} \tilde{R}_2(X(\tau)) \right].$$

The value of the modified payoff function $\tilde{R}_2(y)$ is asymptotic to $(1/(\gamma_2 e))e^{\gamma_2 y}$ as $y \to -\infty$ (and hence converges to 0) and is asymptotically linear as $y \to \infty$ with asymptote $z = (1 + e^{-\alpha \delta})y + \mu \delta e^{-\alpha \delta}$. Examining the value $\tilde{R}_2(1/\gamma_2)$, we have

$$\frac{\tilde{R}_2(1/\gamma_2)}{\psi(1/\gamma_2)} = (1/(\gamma_2 e)) + e^{-1} \tilde{V}_1^{(1)}(1/\gamma_2) > 1/(\gamma_2 e),$$

which implies the existence of some $y_2^* \in (x_\ell, x_r)$ at which $\tilde{R}_2(y)/\psi(y)$ achieves its maximum. Observe that since $y_2^*$ is an interior maximizer, $\psi(y_2^*)\tilde{R}_2'(y_2^*) - \psi'(y_2^*)\tilde{R}_2(y_2^*) = 0$. Using the upper bound approach of Sect. 17.3.2 therefore implies that for $x \le y_2^*$, $V_2^{(1)}(x) = e^{\gamma_2(x - y_2^*)} \tilde{R}_2(y_2^*)$ and an optimal stopping rule is given by $\tau_{y_2^*}$.

We believe that when $x > y_2^*$, an optimal value is obtained by stopping immediately. One way to verify this claim would be to show the existence of feasible $c_1$ and $c_2$ such that the smooth pasting conditions (17.11) are satisfied with $x = a$. Recalling the values $c_1$ and $c_2$ in (17.12), feasibility requires that $\psi'(x)\tilde{R}_2(x) - \psi(x)\tilde{R}_2'(x) > 0$ for $x > y_2^*$, and since $\psi'(x) = \gamma_2 \psi(x)$, we must examine the function

$$\gamma_2 \tilde{R}_2(y) - \tilde{R}_2'(y) = -e^{-\alpha \delta} \left[ (1 - \gamma_2 y) e^{\alpha \delta} + (1 - \gamma_2 y - \gamma_2 \alpha \delta) \Phi \left( \frac{y + \mu \delta - \gamma_2^{-1}}{\sigma \sqrt{\delta}} \right) \right. $$

$$\left. - \gamma_2 \sigma \sqrt{\delta/(2\pi)} \, e^{-(y + \mu \delta - \gamma_2^{-1})^2/(2\sigma^2 \delta)} \right]. \quad (17.14)$$

Note that $\gamma_2 \tilde{R}_2(y_2^*) - \tilde{R}_2'(y_2^*) = 0$ since $y_2^*$ is an interior maximizer. At this point, the dependence of the expression (17.14) on $y$ is such that a general proof is not clear, so numerical tractability becomes advantageous. Figure 17.2 displays the function $\gamma_2 \tilde{R}_2 - \tilde{R}_2'$ for a particular choice of parameters. Notice, in particular, for $y > y_2^*$ the values are positive and hence the value of $c_1$ is also positive resulting in a feasible solution to the minimization problem which has value $\tilde{R}_2(x)$. The value function $V_2^{(1)}$ is displayed in Fig. 17.2 as well for this choice of parameters.

Summarizing, the optimal value for the second last single-stopping problem is

$$V_2^{(1)}(x) = \begin{cases} \tilde{R}_2(y_2^*)e^{\gamma_2(x - y_2^*)}, & x \le y_2^*, \\ \tilde{R}_2(x), & x > y_2^*. \end{cases} \quad (17.15)$$

At this point, it is clear that determining closed-form expressions for the maximizer and the value function is not possible. However, some progress can be made theoretically, and one may also continue to employ numerical and graphical techniques for particular parameters. The analysis of the third single-stopping

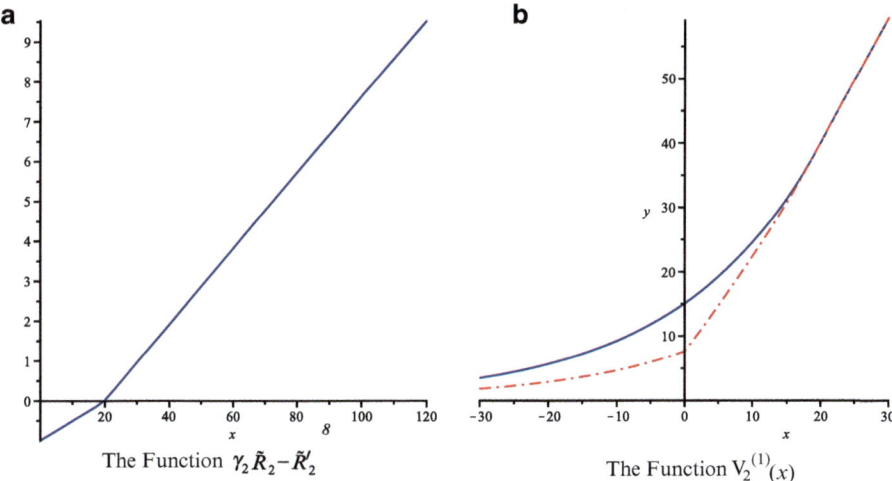

**Fig. 17.2** $\gamma_2 \tilde{R}_2 - \tilde{R}'_2$ and $V_1^{(1)}$ for the second stopping problem; $\mu = \sigma = \delta = 1$, $\alpha = 0.05$

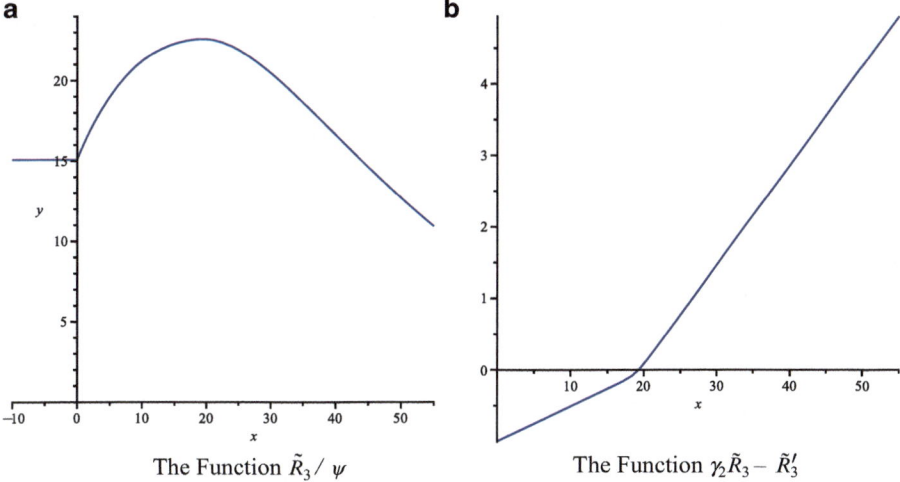

**Fig. 17.3** Checking for optimality; $\mu = 1$, $\sigma = 1$, $\alpha = 0.05$

problem follows along the same line as for the second. In particular, one may show that $\tilde{R}_3$ is asymptotic to $\tilde{R}_2(y_2^*)e^{\gamma_2(y-y_2^*)}$ as $y \to -\infty$ and has a linear asymptote as $y \to \infty$. Moreover, $\tilde{R}_3(y_2^*)/\psi(y_2^*) > \tilde{R}_2(y_2^*)e^{-\gamma_2 y_2^*}$ which implies the existence of some finite $y_3^*$ at which $\tilde{R}_3/\psi$ achieves its maximum. Therefore, the upper bound approach establishes that the value function is $\tilde{R}_3(y_3^*)e^{\gamma_2(y-y_3^*)}$ for $y \le y_3^*$. Figure 17.3 displays graphs of the ratio $\tilde{R}_3/\psi$ and the function $\gamma_2\tilde{R}_3 - \tilde{R}'_3$ to graphically verify that the form of the value function is the same as (17.15).

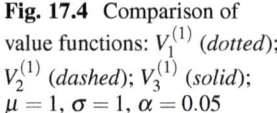

**Fig. 17.4** Comparison of
value functions: $V_1^{(1)}$ (*dotted*);
$V_2^{(1)}$ (*dashed*); $V_3^{(1)}$ (*solid*);
$\mu = 1, \sigma = 1, \alpha = 0.05$

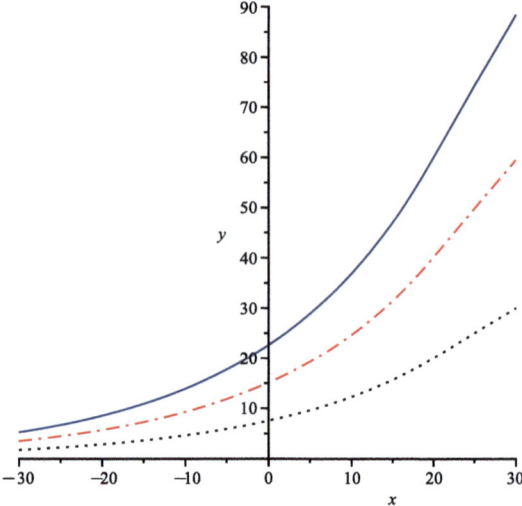

All three value functions are displayed in Fig. 17.4 for comparison purposes. In particular, one can notice the increase in the value functions as the number of available decisions increases. Finally, we identify optimal decision times for the original triple-stopping problem:

$$\tau_3^* = \inf\{t \geq 0 : X(t) \in [y_3^*, \infty)\},$$
$$\tau_2^* = \inf\{t \geq \tau_3^* + \delta : X(t) \in [y_2^*, \infty)\},$$
$$\tau_1^* = \inf\{t \geq \tau_2^* + \delta : X(t) \in [y_1^*, \infty)\},$$

where the critical values of the stopping locations are $y_3^* = 19.346$, $y_3^* = 19.888$, and $y_3^* = 20.488$, when $\mu = \sigma = \delta = 1$ and $\alpha = 0.05$.

## 17.5 Concluding Remarks

This chapter demonstrates that multiple-stopping problems of one-dimensional diffusions in the presence of refraction periods reduce to a sequence of single-stopping problems in which the reward for an earlier action must include the optimal payoff for the subsequent action. The presence of the refraction period introduces the need to evaluate the expectation of the value function for a later action according to the distribution of the process at a time dependent on the length of the refraction period. This becomes numerically tractable when this distribution is known. Three solution approaches to the single-stopping problems are briefly discussed based on an imbedding of the original stochastic problem in an infinite-dimensional linear program; a similar linear programming approach to stochastic control of discrete-time processes has been studied by O. Hernańdez-Lerma (e.g., [11]). Tractability of these type of problems is illustrated in detail for a particular example.

**Acknowledgements** The research of Richard H. Stockbridge was supported in part by the U.S. National Security Agency under Grant Agreement Number H98230-09-1-0002. The United States Government is authorized to reproduce and distribute reprints notwithstanding any copyright notation herein. The research of Chao Zhu was supported in part by a grant from the UWM Research Growth Initiative and under NSF grant DMS-1108782.

# References

1. N. Aleksandrov and B.M. Hambly, A dual approach to multiple exercise of option problems under constraints. *Math. Meth. Oper. Res.,* 71 (2010), 503–533.
2. A.N. Borodin and P. Salminen, *Handbook of Brownian Motion - Facts and Formulae,* 2nd. ed., Birkhäuser, Basel, (2002).
3. R. Carmona and S. Dayanik, Optimal multiple-stopping of linear diffusions and swing options, *Math. Oper. Res.,* 32 (2008), 446–460.
4. R. Carmona and N. Touzi, Optimal multiple stopping and valuation of swing options, *Math. Finance,* 18 (2008), 239–268.
5. M. Dai, and Y.K. Kwok, Optimal multiple stopping models of reload options and shout options, *J. Econom. Dyn. Control,* 32 (2008), 2269–2290.
6. S.N. Ethier and T.G. Kurtz, *Markov Processes: Characterization and Convergence,* Wiley, New York, 1986.
7. G. Haggstrom, Optimal sequential procedures when more than one stop is required, *Ann. Math. Stat.,* 38 (1967), 1618–1626.
8. K.L. Helmes and R.H. Stockbridge, Construction of the value function and stopping rules for optimal stopping of one-dimensional diffusions, *Adv. Appl. Prob.,* 42 (2010), 158–182.
9. K.L. Helmes and R.H. Stockbridge, Thinning and rotation of stochastic forest models, *J. Econ. Dyn. Control,* 35 (2011), 25–39.
10. K.L. Helmes, R.H. Stockbridge and H. Volkmer, Analysis of Production Decisions Under Budget Limitations, *Stochastics,* 83 (2011), 583–609.
11. O. Hernández-Lerma and J.B. Lasserre, Linear programming approximations for Markov control processes in metric spaces, *Acta Appl. Math.,* 51 (1998), 123–139.
12. M. Kobylanski, M-C Quenez and E. Rouy-Mironescu, Optimal multiple stopping time problems, *Ann. Appl. Probab.,* 21 (2011), 1365–1399.
13. A.C. Thompson, Valuation of path-dependent contingent claims with multiple exercise decisions over time: The case of take-or-pay. *J. Financial Quant. Anal.,* 30 (1995), 271–293.
14. M. Villinski, A methodology for valuing multipe-exercise option contracts for water Center for International Food and Agricultural Policy Working Paper WP-03-4.
15. A.B. Zeghal and M. Mnif, Optimal multiple stopping and valuation of swing options, *Int. J. Theor. Appl. Finance,* 9 (2006), 1267–1297.

# Chapter 18
# On the Regularity Property of Semi-Markov Processes with Borel State Spaces

Óscar Vega-Amaya

## 18.1 Introduction

A semi-Markov process (SMP) combines the probabilistic structure of a Markov chain and a renewal process as follows: it makes transitions according to a Markov chain, but the times spent between successive transitions are random variables whose distribution functions depend on the "present" state of the system. Observe that a continuous-time Markov chain is a SMP with exponentially distributed transition times. Thus, it is raised the question of whether the SMP experiences finite or infinitely many transitions in bounded time periods. If the former property holds, the SMP is said to be *regular* (or nonexplosive), and *irregular* (or explosive) otherwise.

A natural way to obtain the regularity property is to impose conditions that guarantee that transitions do not take place too quickly, and the most popular condition to do this is that used by Ross [7, Proposition 5.1, p. 88] and Çinlar [2, Chap. 10, Proposition 3.19, p. 327]. Roughly speaking, this condition requires the transition times to be greater than some $\gamma > 0$ with a probability of at least $\varepsilon > 0$, independently of the present state of the system [see (18.6) below]. Under this condition, both authors obtain the regularity of the SMP for the *countable state space* case only, but using a key remark of Bhattacharya and Majumdar [1] (see Remark 18.3.1, below), this result can also be proved for *Borel spaces* (see Theorem 18.3.2). It is worth mentioning that Çinlar's proof [2, Chap. 10, Proposition 3.19, p. 327] also extends directly to the general case of Borel spaces.

Moreover, for the countable state space case, Ross [7, Proposition 5.1, p. 88] and Çinlar [2, Chap. 10, Corollary 3.17, p. 327] prove that the regularity property holds whenever the "embedded" Markov chain reaches a recurrent state with probability

Ó. Vega-Amaya (✉)
Departamento de Matemáticas, Universidad de Sonora, Hermosillo, Sonora 83000, México
e-mail: ovega@gauss.mat.uson.mx

D. Hernández-Hernández and A. Minjárez-Sosa (eds.), *Optimization, Control, and Applications of Stochastic Systems*, Systems & Control: Foundations & Applications, DOI 10.1007/978-0-8176-8337-5_18, © Springer Science+Business Media, LLC 2012

one for every initial state. Thus, in particular, the regularity property holds if the embedded Markov chain is *recurrent*. However, their proofs cannot be extended, or at least not directly, to the case of Borel state space because they rely on the renewal process formed by the successive times at which a recurrent state is visited, which typically involves events of probability zero if the state space is uncountable. In fact, to the best of our knowledge, there is no counterpart of these results for Borel spaces.

The aim of this note is to fill this gap by extending the latter results to SMP with Borel state space. More precisely, imposing a fairly weak condition on the *sojourn* or *holding time distribution*, we show that the regularity property holds under each one of the following conditions: (a) the embedded Markov chain is *Harris recurrent*; (b) the embedded Markov chain is *recurrent* and the "recurrent part" of the state space is reached with probability one for each initial state; (c) the embedded Markov chain has a *unique invariant probability measure*. Under the latter condition, the regularity property is only ensured for almost all initial state with respect to the invariant probability measure.

## 18.2  Preliminary Concepts

This section briefly introduces the SMPs. The readers are referred to Limnios and Oprişan [5] for a rigorous and detailed description. Next, we have some notation which is used through the note. Let $(\mathbb{X}, \mathscr{B})$ be a measurable space where $\mathbb{X}$ is a Borel space and $\mathscr{B}$ is its Borel $\sigma$-algebra. We denote by $\mathbb{R}_+$ and $\mathbb{N}_0$ the sets of nonnegative real numbers and nonnegative integers, respectively, while $\mathbb{N}$ stands for the positive integers. Set $\Omega := (\mathbb{X} \times \mathbb{R}_+)^\infty$ and denote by $\mathscr{F}$ the corresponding product $\sigma$-algebra.

Consider a fixed stochastic kernel $Q(\cdot, \cdot | \cdot)$ on $\mathbb{X} \times \mathbb{R}_+$ given $\mathbb{X}$. Then, for each "initial" state $x \in \mathbb{X}$, there exists a probability measure $\mathbb{P}_x$ and a Markov chain $\{(X_n, \delta_{n+1}) : n \in \mathbb{N}_0\}$ defined on the *canonical* measurable space $(\Omega, \mathscr{F})$ such that

$$\mathbb{P}_x[X_0 = x] = 1, \tag{18.1}$$

$$\mathbb{P}_x[X_{n+1} \in B, \delta_{n+1} \leq t | X_n = y] = Q(B, [0,t] | y) \tag{18.2}$$

for all $B \in \mathscr{B}, t \in \mathbb{R}_+, y \in \mathbb{X}$.

The process $\{(X_n, \delta_{n+1}) : n \in \mathbb{N}_0\}$ is called *Markov renewal process* and usually thought of as a model of a stochastic system evolving as follows: it is observed at time $t = 0$ in some initial state $X_0 = x \in \mathbb{X}$ in which it remains up to a (nonnegative) random time $\delta_1$. The distribution function of $\delta_1$ is given by

$$F(t|x) := Q(\mathbb{X}, [0,t] | x) \quad \forall t \in \mathbb{R}_+, x \in \mathbb{X},$$

which is called the *sojourn* or *holding time distribution* in the state $x$. Thus, the *mean sojourn* or *holding time function* is defined as

$$\tau(x) := \int_{\mathbb{R}_+} t F(dt|x) \geq 0, \quad x \in \mathbb{X}.$$

Next, at time $\delta_1$, the system jumps to a new state, say $X_1 = y \in \mathbb{X}$, according to the probability measure

$$P(B|x) := Q(B, \mathbb{R}_+|x), \quad B \in \mathscr{B}, x \in \mathbb{X}.$$

Once the transition occurs, the system remains in the new state $X_1 = y$ up to a (nonnegative) random time $\delta_2$, and so on.

The state of the systems is tracked in continuous time by the process

$$Z_t := X_n \quad \text{if} \quad T_n \leq t < T_{n+1}$$

where

$$T_{n+1} := T_n + \delta_{n+1}, \quad n \in \mathbb{N}_0, \quad \text{and} \quad T_0 := 0.$$

The continuous-time process $\{Z_t : t \in \mathbb{R}_+\}$ is called *semi-Markov process* (SMP) with (semi-Markov) kernel $Q(\cdot, \cdot|\cdot)$.

Note, by (18.2), that the process $\{X_n : x \in \mathbb{N}_0\}$ is a *Markov chain* on $\mathbb{X}$ with one-step transition probability $P(\cdot|\cdot)$. Thus, it is called the *embedded Markov chain* in the SMP $\{Z_t : t \in \mathbb{R}_+\}$.

Now observe that the kernel $Q(\cdot, \cdot|\cdot)$ can be "disintegrated" as

$$Q(B, [0,t]|x) = \int_B G(t|x,y) P(dy|x) \quad \forall B \in \mathscr{B}, t \in \mathbb{R}_+, x \in \mathbb{X},$$

where $G(\cdot|x,y)$ is a distribution function on $\mathbb{R}_+$ for all $x, y \in \mathbb{X}$, while $G(t|\cdot, \cdot)$ is a measurable function on $\mathbb{X} \times \mathbb{X}$ for each $t \in \mathbb{R}_+$. Thus,

$$G(t|z,y) = \mathbb{P}_x[\delta_{n+1} \leq t | X_n = z, X_{n+1} = y] \quad \forall x, y, z \in \mathbb{X}, t \in \mathbb{R}_+. \tag{18.3}$$

Then, using the Markov property of the Markov renewal process and (18.3), it is easy to prove that the random variables $\{\delta_n : n \in \mathbb{N}\}$ are (conditionally) independent given the state process $\{X_n : n \in \mathbb{N}_0\}$ and also that

$$\mathbb{P}_x[\delta_1 \leq t_1, \ldots, \delta_n \leq t_n | X_0, X_1, \ldots, X_n] = \prod_{k=1}^{n} G(t_k | X_{k-1}, X_k). \tag{18.4}$$

## 18.3 The Regularity Property, Recurrence and Invariant Measures

Let $\{(X_n, \delta_{n+1}) : n \in \mathbb{N}_0\}$ be a Markov renewal process with stochastic kernel $Q(\cdot, \cdot|\cdot)$ on $\mathbb{X} \times \mathbb{R}_+$ given $\mathbb{X}$.

**Definition 18.3.1.** A state $x \in \mathbb{X}$ is said to be regular if

$$\lim_{n \to \infty} T_n = \infty \quad \mathbb{P}_x\text{-}a.s.$$

The SMP is said to be *regular* if every state $x \in \mathbb{X}$ is regular.

Define

$$\Delta(x) := \int_{\mathbb{R}_+} \exp(-t) F(dt|x), \quad x \in \mathbb{X}.$$

and observe that $0 < \Delta(\cdot) \leq 1$. Also note that

$$\Delta(x) = 1 \Leftrightarrow F(0|x) = 1 \Leftrightarrow \tau(x) = 0. \tag{18.5}$$

Clearly, to guarantee the regularity property holds, it is required to exclude this degenerate case occurs for all or "almost all" states. The most popular way to do this is by means of the following assumption: there exist positive constants $\gamma$ and $\varepsilon < 1$ such that

$$1 - F(\gamma|x) > \varepsilon \quad \forall x \in \mathbb{X}. \tag{18.6}$$

Ross [7, Proposition 5.1, p. 88] and Çinlar [2, Chap. 10, Proposition 3.19, p. 327] prove that the SMP is regular assuming condition (18.6) holds. Here, for the sake of completeness, we provide other proof based in the following remark due to Bhattacharya and Majumdar [1].

*Remark 18.3.1.* It follows from the conditional independence of the random variables $\{\delta_n : n \in \mathbb{N}\}$ and (18.4) that

$$\mathbb{E}_x[\exp(-T_{n+1})|X_0, X_1, \cdots, X_n] = \Delta(X_0) \cdots \Delta(X_n) \quad \forall n \in \mathbb{N}_0. \tag{18.7}$$

Hence,

$$T_n \to \infty \Leftrightarrow [\Delta(X_0) \cdots \Delta(X_n)] \to 0. \tag{18.8}$$

This follows directly from (18.7) after noting that $Z_n := \exp(-T_n)$ and $W_n := \Delta(X_0) \cdots \Delta(X_n), n \in \mathbb{N}$, are bounded and nonincreasing sequences.

**Theorem 18.3.1.** *If condition (18.6) holds, then the SMP is regular.*

*Proof of Theorem 18.3.1.* This follows directly from (18.8) after noting that condition (18.6) implies that

$$\sup_{x \in \mathbb{X}} \Delta(x) \leq (1 - \varepsilon) + \varepsilon \exp(-\gamma) < 1. \qquad \qquad \square$$

The regularity can also be guaranteed asking condition (18.6) holds only for states in a proper subset $C \subset \mathbb{X}$ provided it is accompanied by an appropriate "recurrence" property [see Remark 18.3.6(b) below].

Next, we prove the regularity of the SMP holds under some "recurrence" conditions which seems to be the weakest possible ones. To state these assumptions, we need several concepts and results from Markov chain theory which are collected from Hernández-Lerma and Lasserre[3] and Meyn and Tweedie [6].

A Markov chain $\{Y_n : n \in \mathbb{N}_0\}$ with state space $\mathbb{X}$ is said to be *irreducible* if there exists a nontrivial $\sigma$-finite measure $\nu(\cdot)$ on $(\mathbb{X}, \mathscr{B})$ such that

$$T(x,B) := \mathbb{E}_x \sum_{n=1}^{\infty} \mathbb{I}_B(Y_n) > 0 \quad \forall x \in \mathbb{X},$$

whenever $\nu(B) > 0$, $B \in \mathscr{B}$; in this case, $\nu(\cdot)$ is called an *irreducibility measure*. If the Markov chain $\{Y_n : n \in \mathbb{N}_0\}$ is irreducible, there exists a *maximal irreducibility measure* $\psi(\cdot)$, which means that $\psi(\cdot)$ is an irreducibility measure and that any other irreducibility measure $\nu(\cdot)$ is *absolutely continuous* with respect to $\psi(\cdot)$. Moreover, if $\psi(B) = 0$, then

$$\psi(\{y \in \mathbb{X} : T(y,B) > 0\}) = 0, \qquad \qquad (18.9)$$

which means that the set of initial states for which the Markov chain enters to a $\psi$-null set is also a $\psi$-null set [6, Proposition 4.2.2, p. 88].

Let $\{Y_n : n \in \mathbb{N}_0\}$ be an irreducible Markov chain and $\psi(\cdot)$ a maximal irreducibility measure. The Markov chain $\{Y_n : n \in \mathbb{N}_0\}$ is said to be *recurrent* if

$$\mathbb{E}_x \sum_{n=0}^{\infty} \mathbb{I}_A(Y_n) = \infty \quad \forall x \in \mathbb{X}, A \in \mathscr{B}^+, \qquad \qquad (18.10)$$

where $\mathscr{B}^+ := \{B \in \mathscr{B} : \psi(B) > 0\}$. Note that $\mathscr{B}^+$ is well defined because all maximal irreducibility measures are equivalent. If instead of condition (18.10) we have

$$\sum_{n=0}^{\infty} \mathbb{I}_A(Y_n) = \infty \quad \mathbb{P}_x\text{-a.s. } \forall x \in A, A \in \mathscr{B}^+,$$

then the Markov chain is said to be *Harris recurrent*. It is proved in Meyn and Tweedie [6, Theorem 9.1.4, p. 204] that a Harris recurrent Markov chain satisfies the (apparently) stronger condition

$$\sum_{n=0}^{\infty} \mathbb{I}_A(Y_n) = \infty \quad \mathbb{P}_x\text{-a.s. } \forall x \in \mathbb{X}, A \in \mathscr{B}^+.$$

We now come back to the discussion of the regularity property with the following remark.

*Remark 18.3.2.* Suppose the embedded Markov chain $\{X_n : n \in \mathbb{N}_0\}$ is irreducible. If the SMP is regular, due to property (18.8), the Markov chain $\{X_n : n \in \mathbb{N}_0\}$ visits the set

$$L := \{x \in \mathbb{X} : \Delta(x) < 1\}$$

infinitely often $\mathbb{P}_x$-a.s for every initial state $x \in \mathbb{X}$. Moreover, the set $L$ belongs to $\mathscr{B}^+$; otherwise, by (18.9),

$$\psi(\mathbb{X}) = \psi(\{y \in \mathbb{X} : T(y,L) > 0\}) = 0,$$

which obviously is a contradiction.

*Remark 18.3.3.* Suppose the embedded Markov chain $\{X_n : n \in \mathbb{N}_0\}$ is irreducible. Then, $L \in \mathscr{B}^+$ if and only if $B_\alpha := \{x \in \mathbb{X} : \Delta(x) \leq \alpha\} \in \mathscr{B}^+$ for some $\alpha \in (0,1)$. This claim follows noting that $L = \cup_{n=1}^\infty B_n$ where $B_n := \{x \in \mathbb{X} : \Delta(x) \leq \alpha_n\}$ and $\alpha_n \uparrow 1$.

We now state the first result of this note.

**Theorem 18.3.2.** *Suppose the embedded Markov chain is Harris recurrent. Then, the SMP is regular if and only if $L \in \mathscr{B}^+$.*

*Proof of Theorem 18.3.2.* Note that the "only if" part is proved in Remark 18.3.2. To prove the other part, take $B_\alpha$ as in Remark 18.3.3 and for each $n \in \mathbb{N}$ define

$$\sigma(1) := \inf\{k > 0 : X_k \in B_\alpha\}, \ \sigma(n+1) := \inf\{k > \sigma(n) : X_k \in B_\alpha\}$$

and

$$S_n := \sum_{k=1}^{n} \mathbb{I}_{B_\alpha}(X_k).$$

Now observe that

$$\Delta(X_0) \cdots \Delta(X_n) \leq \Delta(X_{\sigma(1)}) \Delta(X_{\sigma(2)}) \cdots \Delta(X_{\sigma(S_n)}) \leq \alpha^{S_n}$$

on the set $[S_n \neq 0]$. Thus, since the embedded Markov chain $\{X_n : n \in \mathbb{N}_0\}$ is Harris recurrent and $\psi(B_\alpha) > 0$, $S_n \to \infty$ $\mathbb{P}_x$-a.s. for all $x \in \mathbb{X}$; hence,

$$\Delta(X_0) \cdots \Delta(X_n) \to 0 \quad \mathbb{P}_x\text{-a.s. for all } x \in \mathbb{X},$$

which, by (18.8), proves that the process is regular. $\qquad\qquad\qquad\qquad\square$

The regularity property of the SMP can also be obtained assuming that the embedded Markov chain $\{X_n : n \in \mathbb{N}_0\}$ is recurrent. However, as in Ross [7, Proposition 5.1, p. 88] and Çinlar [2, Chap. 10, Corollary 3.17, p. 327], we need to assume additionally that the "recurrent part" of the state space is reached with probability one for every initial state. To state this condition precisely, we require the following important result (see, e.g., Hernández-Lerma and Lasserre [3, Proposition 4.2.12, p. 50] or Meyn and Tweedie [6, Theorem 9.0.1, p. 201]).

*Remark 18.3.4.* If the embedded Markov chain$\{X_n : n \in \mathbb{N}_0\}$ is recurrent, then

$$\mathbb{X} = H \cup N,$$

where the measurable set $H$ is *full* and *absorbing* (i.e., $\psi(N) = 0$ and $P(H|x) = 1$ for all $x \in H$, respectively). Moreover, the Markov chain restricted to $H$ is Harris recurrent, that is,

$$\sum_{n=0}^{\infty} \mathbb{I}_A(X_n) = \infty \quad \mathbb{P}_x\text{-a.s. } \forall x \in H, A \subset H, A \in \mathscr{B}^+.$$

**Theorem 18.3.3.** *If the embedded Markov chain is recurrent, $L \in \mathscr{B}^+$ and*

$$\sigma := \inf\{n \in \mathbb{N}_0 : X_n \in H\} < \infty \quad \mathbb{P}_x\text{-a.s. } \forall x \in \mathbb{X},$$

*then the SMP is regular.*

*Proof of Theorem 18.3.3.* The proof follows the same arguments given in the proof of Theorem 18.3.2 but considering $\overline{B}_\alpha := B_\alpha \cap H$ instead of the set $B_\alpha$.  □

Note that Theorems 18.3.2 and 18.3.3 state that the regularity property holds for all initial state $x \in \mathbb{X}$ under a recurrence condition independently of whether the embedded Markov chain admits an *invariant probability measure* $\mu(\cdot)$, that is, a probability measure satisfying the condition

$$\mu(B) = \int_{\mathbb{X}} P(B|x)\mu(\mathrm{d}x) \quad \forall B \in \mathscr{B}.$$

Recurrence (and then Harris recurrence) may be dispensed if one supposes the existence of a *unique* invariant probability measure with the cost that the regularity property will be ensured only for almost all initial states (see Theorem 18.3.4 below). The proof uses a *pathwise ergodic theorem* which is borrowed from Hernández-Lerma and Lasserre [3, Corollary 2.5.2]. To state this result, we need the following notation: for a measurable function $v(\cdot)$ and measure $\lambda(\cdot)$ on $(\mathbb{X}, \mathscr{B})$, let

$$\lambda(v) := \int_{\mathbb{X}} v(y)\lambda(\mathrm{d}y),$$

whenever the integral is well defined. Moreover, denote by $L_1(\lambda)$ the class of measurable functions $v(\cdot)$ on $\mathbb{X}$ such that $\lambda(|v|) < \infty$.

*Remark 18.3.5.* (a) Suppose that $\{X_n : n \in \mathbb{N}_0\}$ has a unique invariant probability measure $\mu(\cdot)$. Then, for each function $v \in L_1(\mu)$, there exists a set $B_v \in \mathcal{B}$, with $\mu(B_v) = 1$, such that

$$\frac{1}{n} \sum_{k=0}^{n-1} v(X_n) \to \mu(v) \quad \mathbb{P}_x\text{-a.s. } \forall x \in B_v. \tag{18.11}$$

(b) If in addition the Markov chain is Harris recurrent, then (18.11) holds for all $x \in \mathbb{X}$ (see Hernández-Lerma and Lasserre [3, Theorem 4.2.13, p.51]).

**Theorem 18.3.4.** *Suppose the following conditions hold: (a) the embedded Markov chain has a unique invariant probability measure $\mu(\cdot)$; (b) $\mu(\Delta) = \int_\mathbb{X} \Delta(x)\mu(dx) < 1$. Then, the SMP is regular for $\mu$-almost all $x \in \mathbb{X}$. If in addition the embedded Markov chain is Harris recurrent, then the regularity property holds for all $x \in \mathbb{X}$.*

*Proof of Theorem 18.3.4.* Observe that

$$[\Delta(X_0) \cdots \Delta(X_n)]^{1/(n+1)} \le \frac{1}{n+1} \sum_{k=0}^{n} \Delta(X_k) \quad \forall n \in \mathbb{N}_0.$$

Thus, by condition (a) and Remark 18.3.5(a), there exists a set $B_\Delta \in \mathcal{B}$ such that

$$\frac{1}{n+1} \sum_{k=0}^{n} \Delta(X_k) \to \mu(\Delta) < 1 \quad \mathbb{P}_x\text{-a.s. } \forall x \in B_\Delta,$$

with $\mu(B_\Delta) = 1$. Therefore,

$$\Delta(X_0) \cdots \Delta(X_n) \to 0 \quad \mathbb{P}_x\text{-a.s. for } \mu\text{-almost all } x \in \mathbb{X}.$$

The second statement of the theorem follows from Theorem 18.3.2 because the property $\mu(\Delta) < 1$ implies that $L \in \mathcal{B}^+$. □

*Remark 18.3.6.* (a) Let $\mu$ be a probability measure on $(\mathbb{X}, \mathcal{B})$. Observe that (18.5) implies that $\{x \in \mathbb{X} : \tau(x) > 0\} = \{x \in \mathbb{X} : \Delta(x) < 1\}$. Then

$$\mu(\tau) > 0 \Leftrightarrow \mu(\Delta) < 1.$$

Thus, the conclusions in Theorem 18.3.4 remain valid if condition (b) is replaced by the condition $\mu(\tau) > 0$.

(b) Schäl [8] and Jaśkiewicz [4] considered the following weakened version of condition (18.6): there exist positive constants $\gamma$, $\varepsilon < 1$, and a subset $C \in \mathcal{B}$ such that

$$1 - F(\gamma|x) > \varepsilon \quad \forall x \in C.$$

This condition by itself does not imply the regularity of the SMP (see the example in Ross [7, p. 87]); however, it does provided that $C \in \mathscr{B}^+$, and a suitable recurrence condition holds, e.g., the embedded Markov chain is Harris recurrent. To see this is true, note that

$$\sup_{x \in C} \Delta(x) \leq (1 - \varepsilon) + \varepsilon \exp(-\gamma) < 1,$$

which implies that $L = \{x \in \mathbb{X} : \Delta(x) < 1\} \in \mathscr{B}^+$. Hence, from Theorem 18.3.2, the SMP is regular.

**Acknowledgements**   The author takes this opportunity to thank to Professor Onésimo Hernández-Lerma for his constant encouragement and support during the author's academic career and, in particular, for his valuable comments on an early version of present work. Thanks are also due to the referees for their useful observations which improve the writing and organization of this note.

# References

1. Bhattacharya, R.N., Majumdar, M. (1989), *Controlled semi-Markov models–The discounted case*, J. Statist. Plann. Inf. **21**, 365–381.
2. Çinlar, E. (1975), *Introduction to Stochastic Processes*, Prentice-Hall, Englewood Cliffs, New Jersey.
3. Hernández-Lerma, O., Lasserre, J.B. (2003), *Markov Chains and Invariant Probabilities*, Birkhäuser, Basel.
4. Jaśkiewicz, A. (2004), *On the equivalence of two expected average cost criteria for semi-Markov control processes*, Math. Oper. Research 29, 326–338.
5. Limnios, N., Oprişan, G. (2001), *Semi-Markov Processes and Reliability*, Birkhaüser, Boston.
6. Meyn, S.P., Tweedie, R.L. (1993), *Markov Chains and Stochastic Stability*, Springer-Verlag, London.
7. Ross, S.M. (1970), *Applied Probability Models with Optimization Applications*, Holden-Day, San Francisco.
8. Schäl, M. (1992), *On the second optimality equation for semi-Markov decision models*, Math. Oper. Research 17, 470–486.

GPSR Compliance

*The European Union's (EU) General Product Safety Regulation (GPSR) is a set of rules that requires consumer products to be safe and our obligations to ensure this.*

*If you have any concerns about our products, you can contact us on ProductSafety@springernature.com*

In case Publisher is established outside the EU, the EU authorized representative is:

Springer Nature Customer Service Center GmbH
Europaplatz 3
69115 Heidelberg, Germany

**Batch number: 09484624**

Printed by Printforce, the Netherlands